Two Manuals Combined into One Book

Military Intelligence

And

Military Role in Civil Disturbance Operations

Department of Defense

Pentagon
Publishing

Change Number 1

Headquarters
Department of the Army
Washington, DC, 17 May 2004

Intelligence

Contents

Distribution Restriction: Approved for public release; distribution is unlimited.

*This publication supersedes FM 34-1, 27 September 1994.

Preface

FM 2-0 is the Army's keystone manual for military intelligence (MI) doctrine. It describes—

- The fundamentals of intelligence operations.
- The operational environment (OE).
- Intelligence in unified action.
- The Intelligence Battlefield Operating System (BOS).
- Intelligence considerations in strategic readiness.
- The intelligence process.
- MI roles and functions within the context of Army operations.

This manual conforms to the overarching doctrinal precepts presented in FM 3-0.

This manual provides doctrinal guidance for the Intelligence BOS actions in support of commanders and staffs. It also serves as a reference for personnel who are developing doctrine; tactics, techniques, and procedures (TTP); materiel and force structure; and institutional and unit training for intelligence operations.

This manual provides MI guidance for all commanders, staffs, trainers, and MI personnel at all echelons. It forms the foundation for MI and the Intelligence BOS doctrine development, and applies equally to the Active Component (AC), United States Army Reserve (USAR), and Army National Guard (ARNG). It is also intended for commanders and staffs of joint and combined commands, US Naval and Marine Forces, units of the US Air Force, and the military forces of multinational partners.

Headquarters, US Army Training and Doctrine Command is the proponent for this publication. The preparing agency is the US Army Intelligence Center and School. Send written comments and recommendations on DA Form 2028 (Recommended Changes to Publications and Blank Forms) directly to: Commander, ATZS-FDT-D (FM 2-0), 550 Cibeque Street, Fort Huachuca, AZ 85613-7017. Send comments and recommendations by e-mail to *ATZS-FDC-D@hua.army.mil*. Follow the DA Form 2028 format or submit an electronic DA Form 2028.

Unless otherwise stated, masculine nouns and pronouns do not refer exclusively to men.

This manual contains Army tactical task (ART) description taken verbatim from the *Army Universal Task List* (AUTL). ART task descriptions are followed by a reference to FM 7-15 in parentheses and the ART number; for example (See FM 7-15, ART 5.1.1.).

PART ONE

Intelligence in the Operational Environment

Part One discusses MI's role in peace, conflict, and war. Supporting the warfighter with effective intelligence is the primary focus of Military Intelligence. Intelligence provides commanders and decisionmakers with the requisite information facilitating their situational understanding so that they may successfully accomplish their missions in full spectrum operations.

Chapter 1 describes the operational environment and the roles of MI within the operational environment. It introduces the Intelligence warfighting function, the intelligence tasks, and the intelligence process, which are the mechanisms through which MI supports the warfighter. This chapter also introduces the intelligence disciplines, which are explained in detail in Part Three of this manual.

Chapter 2 describes the interaction of MI within the nation's intelligence community structure, providing an overview of the intelligence community at the national level and the unified action level—joint, multinational, and interagency aspects of full spectrum operations. This chapter also discusses the concepts and components of intelligence reach.

Chapter 1

Intelligence and the Operational Environment

ROLE OF INTELLIGENCE

1-1. The commander requires intelligence about the enemy and the area of operations (AO) prior to engaging in operations in order to effectively execute battles, engagements, and other missions across the full spectrum of operations. Intelligence assists the commander in visualizing the AO, organizing the forces, and controlling operations to achieve the desired tactical objectives or end-state. Intelligence supports force protection (FP) by alerting the commander to emerging threats and assisting in security operations.

1-2. The unit may need to deal with multiple threats. The commander must understand how current and potential enemies organize, equip, train, employ, and control their forces. Intelligence provides an understanding of the enemy, which assists in planning, preparing, and executing military operations. The commander must also understand the operational environment and its effects on both friendly and enemy operations. The commander receives mission-oriented intelligence on enemy forces and the

AO from the G2/S2. The G2/S2 depends upon the intelligence, surveillance, and reconnaissance (ISR) effort to collect and provide information on the enemy and AO.

1-3. One of the most significant contributions that intelligence personnel can accomplish is to accurately predict future enemy events. Although this is an extremely difficult task, predictive intelligence enables the commander and staff to anticipate key enemy events or reactions and develop corresponding plans or counteractions. Intelligence analysis conclusions, assessments, or products that attempt to define, describe, present, or portray the future situation or conditions of the enemy, terrain, weather, and civil considerations.

1-4. The most important purpose of intelligence is to influence decisionmaking. Commanders must receive the intelligence, understand it (because it is tailored to the commander's requirements), believe it, and act on it. Through this doctrinal concept, intelligence drives operations.

THE INTELLIGENCE WARFIGHTING FUNCTION

1-5. *The intelligence warfighting function is the related tasks and systems that facilitate understanding of the operational environment.* (FM 3-0) It includes tasks associated with intelligence, surveillance, and reconnaissance (ISR) operations, and is driven by the commander. Intelligence is more than just collection. It is a continuous process that involves analyzing information from all sources and conducting operations to develop the situation. The intelligence warfighting function includes the following tasks:

- Support to force generation.
- Support to situational understanding.
- Perform intelligence, surveillance, and reconnaissance.
- Support to targeting and information operations.

1-6. The intelligence warfighting function is one of six warfighting functions—movement and maneuver, intelligence, fires, sustainment, command and control, and protection. *A warfighting function is a group of tasks and systems (people, organizations, information, and processes) united by a common purpose that commanders use to accomplish missions and training objectives* (FM 3-0). (See FM 3-0, chapter 4, for a detailed discussion of the warfighting functions.) The intelligence warfighting function is a flexible force of personnel, organizations, and equipment that, individually or collectively, provide commanders with the timely, relevant, and accurate intelligence required to visualize the battlefield, assess the situation, and direct military actions. Additionally, the intelligence warfighting function is—

- A complex system that operates worldwide, from "mud-to-space," in support of an operation, to include the ability to leverage theater and national capabilities.
- Requires cooperation and division of labor internally, higher, lower, adjacent, and across components and the coalition.

1-7. The intelligence warfighting function not only includes assets within the MI branch but also includes the assets of all branches or warfighting functions that conduct intelligence warfighting function tasks. Every Soldier, as a part of a small unit, is a potential information collector and an essential component to help reach situational understanding. Each Soldier develops a special level of awareness simply due to exposure to events occurring in the AO and has the opportunity to collect and report information by observation and interaction with the population (see paragraph 3-3).

1-8. Planning and executing military operations will require intelligence regarding the threat (traditional, irregular, catastrophic, and disruptive) and the AO. The intelligence warfighting function generates intelligence and intelligence products that portray the enemy and aspects of the environment. These intelligence products enable the commander to identify potential courses of action (COAs), plan operations, employ forces effectively, employ effective tactics and techniques, and implement protection.

1-9. The intelligence warfighting function is always engaged in supporting the commander in offensive, defensive, stability, and, when directed, civil support operations. Intelligence provides products that are timely, relevant, accurate, and predictive. Hard training, thorough planning, meticulous preparation, and aggressive execution posture the Army for success. In the current environment we must maintain intelligence readiness to support operations on "no notice." This support is comprehensive and reaches across full spectrum operations and levels of war to produce the intelligence required to successfully accomplish the mission. A combination of space, aerial, seaborne, and ground-based systems provide the most comprehensive intelligence possible. During force projection operations, the intelligence warfighting function supports the commander with accurate and responsive intelligence from predeployment through redeployment.

1-10. The intelligence warfighting function architecture provides specific intelligence and communications structures at each echelon from the national level through the tactical level. These structures include intelligence organizations, systems, and procedures for collecting, processing, analyzing, and delivering intelligence and other critical information in a useable form to those who need it, when they need it. Effective communications connectivity and automation are essential components of this architecture.

INTELLIGENCE TASKS (METL)

1-11. The personnel and organizations within the Intelligence warfighting function conduct four primary intelligence tasks that facilitate the commander's visualization and understanding of the threat and the AO. These tasks are interactive and often take place simultaneously. (Refer to FM 7-15 for the complete subordinate task listing.) Figure 1-1 shows these tasks tailored to the commander's needs.

INTELLIGENCE TASKS	COMMANDER'S FOCUS	COMMANDER'S DECISIONS
Support to Situational Understanding - Perform IPB - Perform Situation Development - Provide Intelligence Support to Force Protection - Conduct Police Intelligence Operations	**Plan a mission.** **Execute the operation.** **Secure the force.**	**Which COA will I implement?** **Which enemy actions are expected?**
Support to Strategic Responsiveness - Perform I&W - Ensure Intelligence Readiness - Conduct Area Studies of Foreign Countries - Support Sensitive Site Exploitation	**Orient on contingencies.**	**Should I increase the unit's level of readiness?** **Should I implement the OPLAN?**
Conduct Intelligence, Surveillance, and Reconnaissance - Perform Intelligence Synchronization - Perform ISR Integration - Conduct Tactical Reconnaissance - Conduct Surveillance	**Plan the mission.** **Prepare.** **Execute.** **Assess.**	**Which DPs, HPTs, etc., are linked to the enemy actions?** **Are the assets available and in position to collect on the DPs, HPTs, etc.?** **Have the assets been repositioned for contingency mission?**
Provide Intelligence Support to Effects - Provide Intelligence Support to Targeting - Provide Intelligence Support to IO - Provide Intelligence Support to Combat Assessment	**Destroy/suppress/ neutralize targets.** **Reposition intelligence or attack assets.**	**Is my fire (lethal or non-lethal) and maneuver effective?** **Should I refire the same targets?**

Figure 1-1. Intelligence Tailored to the Commander's Needs.

SUPPORT TO SITUATIONAL UNDERSTANDING

1-12. This task centers on providing information and intelligence to the commander, which facilitates the commander's understanding of the enemy and the environment. It supports the command's ability to make sound decisions. Support to situational understanding comprises four subtasks: perform intelligence preparation of the battlefield (IPB), perform situation development, provide intelligence support to FP, and conduct police intelligence operations.

Perform Intelligence Preparation of the Battlefield

1-13. The G2/S2 is the staff proponent for IPB. IPB is the staff planning activity undertaken by the entire staff to define and understand the battlefield and the options it presents to friendly and threat forces. IPB includes input from the whole staff. There is only one IPB in each headquarters with inputs from all affected staff cells; they are not separate warfighting function or staff section IPBs throughout the headquarters. It is a systematic process of analyzing and visualizing the operational environment in a specific geographic area for a specific mission or in anticipation of a specific mission. By applying IPB, the commander and staff gain the information necessary to selectively apply and maximize combat power at critical points in time and space. IPB is most effective when it integrates each staff element's expertise into the final products. To conduct effective IPB, the G2/S2 must—

- Produce IPB products that support the staff's preparation of estimates and the military decision-making process (MDMP).

- Identify characteristics of the AO, including the information environment, that will influence friendly and threat operations.
- Establish the area of interest (AOI) in accordance with the commander's guidance.
- Identify gaps in current intelligence holdings.
- Determine multiple enemy COAs (ECOAs) by employing predictive analysis techniques to anticipate future enemy actions, capabilities, or situations.
- Establish a database that encompasses all relevant data sets within and related to the operational environment.
- Determine the enemy order of battle (OB), doctrine, and TTP. Identify any patterns in enemy behavior or activities.
- Accurately identify and report hazards within the AO, including the medical threat and toxic industrial material (TIM).
- Accurately identify threat capabilities, high-value targets (HVTs), and threat models.
- Integrate IPB information into COA analysis and the MDMP.
- Update IPB products as information becomes available.

Perform Situation Development

1-14. Situation development is a process for analyzing information and producing current intelligence about the enemy and environment during operations. The process helps the intelligence officer recognize and interpret the indicators of enemy intentions, objectives, combat effectiveness, and potential ECOAs. Situation development—

- Confirms or denies threat COAs.
- Provides threat locations.
- Explains what the threat is doing in relation to the friendly force operations.
- Provides an estimate of threat combat effectiveness.

1-15. Through situation development, the intelligence officer is able to quickly identify information gaps, and explain enemy activities in relation to the unit's own operations, thereby assisting the commander in gaining situational understanding. Situation development helps the commander make decisions and execute branches and sequels. This reduces risk and uncertainty in the execution of the plan. The intelligence officer maintains, presents, and disseminates the results of situation development through intelligence input to the common operational picture (COP) and other intelligence products.

Provide Intelligence Support to Force Protection

1-16. Provide intelligence in support of protecting the tactical force's fighting potential so that it can be applied at the appropriate time and place. This task includes the measures the force takes to remain viable and functional by protecting itself from the effects of or recovery from enemy activities. FP consists of those actions taken to prevent or mitigate hostile actions against Department of Defense (DOD) personnel (to include Department of Army

[DA] civilians, contractors, uniformed personnel, and family members), resources, facilities, and critical information. These actions—

- Conserve the force's fighting potential for application at the decisive time and place.
- Incorporate coordinated and synchronized offensive and defensive measures.
- Facilitate the effective employment of the joint force while degrading the capabilities of and opportunities for the threat.

1-17. Intelligence to FP consists of monitoring and reporting the activities, intentions, and capabilities of adversarial groups and determining their possible COAs. Detecting the adversary's methods in today's OE requires a higher level of situational understanding, informed by current and precise intelligence. This type of threat drives the need for predictive intelligence based on analysis of focused information from intelligence, law enforcement, and security activities.

1-18. Intelligence analysis in support of FP employs analytical methodologies and tools to provide situational understanding and to predict the adversary's actions. Modified or standard time-event charts, association matrixes, activity matrixes, link diagrams, and overlays are beneficial in monitoring the actions of the adversary. Overlays may include (but are not limited to) threat training camps, organizations, finances, personalities, industrial sites, information systems, decisionmaking infrastructures, specific activities, and locations of previous attacks.

Conduct Police Intelligence Operations

1-19. Police intelligence operations (PIO) are a military police (MP) function that supports, enhances, and contributes to the commander's force protection program, COP, and situational understanding. The PIO function ensures that information collected during the conduct of other MP functions—maneuver and mobility support, area security, law and order, and internment and resettlement—is provided as input to the intelligence collection effort and turned into action or reports. PIO has three components: (See FM 7-15, ART 1.1.4)

- Collect police information.
- Conduct Police Information Assessment Program (PIAP).
- Develop police intelligence products.

Notes:

US Code, Executive Orders, DOD Directives, and Army Regulations contain specific guidance regarding prohibition on the collection of intelligence information on US citizens, US corporations, and resident aliens. These laws and regulations include criminal penalties for their violation. Any PIO directed against US citizens should undergo competent legal review prior to their initiation.

The inclusion of the PIO task branch in the Intelligence warfighting function does not change the intelligence process described in this manual.

SUPPORT TO STRATEGIC RESPONSIVENESS

1-20. Intelligence support to strategic responsiveness supports staff planning and preparation by defining the full spectrum of threats, forecasting future threats, and forewarning the commander of enemy actions and intentions. Support to strategic responsiveness consists of four subtasks: Perform I&W, ensure intelligence readiness, conduct area studies of foreign countries, and support sensitive site exploitation. (See FM 7-15, ART 1.2)

Perform Indications and Warnings

1-21. This activity provides the commander with forewarning of enemy actions or intentions; the imminence of threat actions. The intelligence officer develops I&W in order to rapidly alert the commander of events or activities that would change the basic nature of the operations. It enables the commander to quickly reorient the force to unexpected contingencies and shape the battlefield. (See FM 7-15, ART 1.2.1)

1-22. The G2/S2 at the operational and strategic levels develops I&W in order to rapidly alert the commander of events or activities that would change the basic nature of the operations so the commander can initiate the appropriate action in a timely manner. I&W reduce the risk of enemy actions that are counter to planning assumptions. I&W enable the commander to quickly reorient the force to unexpected events and to shape the battlefield by manipulating enemy activities.

Ensure Intelligence Readiness

1-23. Intelligence readiness operations support contingency planning and preparation by developing baseline knowledge of multiple potential threats and operational environments. These operations and related intelligence training activities engage the Intelligence warfighting function to respond effectively to the commander's contingency planning intelligence requirements. (See FM 7-15, ART 1.2.2) While still in garrison, intelligence defines the full spectrum of threats and forecasts future threats and dangers. Intelligence readiness operations accomplish the following:

- Provide intelligence to support contingency-based training and staff planning.
- Identify, consider, and evaluate all potential threats to the entire unit.
- Provide a broad understanding of the operational environment of the contingency area, which is developed through continuous exchange of information and intelligence with higher echelon and joint intelligence organizations.

Conduct Area Studies of Foreign Countries

1-24. Study and understand the cultural, social, political, religious, and moral beliefs and attitudes of allied, host nation (HN), or indigenous forces to assist in accomplishing goals and objectives. (See FM 7-15, ART 1.2.3)

Note: The inclusion of this task does not change the support to strategic responsiveness provided by MI organizations described in this manual.

Support Sensitive Site Exploitation

1-25. Sensitive site exploitation consists of a related series of activities inside a sensitive site captured from an adversary. A sensitive site is a designated, geographically limited area with special military, diplomatic, economic, or information sensitivity for the United States. This includes factories with technical data on enemy weapon systems, war crimes sites, critical hostile government facilities, areas suspected of containing persons of high rank in a hostile government or organization, terrorist money laundering, and document storage areas for secret police forces. These activities exploit personnel, documents, electronic data, and material captured at the site, while neutralizing any threat posed by the site or its contents. While the physical process of exploiting the sensitive site begins at the site itself, full exploitation may involve teams of experts located around the world. (See FM 7-15, ART 1.2.4)

Note: The inclusion of this task does not change the support to strategic responsiveness provided by MI organizations described in this manual.

CONDUCT INTELLIGENCE, SURVEILLANCE, AND RECONNAISSANCE

1-26. With staff participation, the intelligence officer synchronizes intelligence support to the ISR effort by focusing the collection, processing, analysis, and intelligence products on the critical needs of the commander. The operations officer, in coordination with the intelligence officer, tasks and directs the available ISR assets to answer the commander's critical information requirements (CCIRs). Through various detection methods and systematic observation, reconnaissance and surveillance obtains the required information. A continuous process, this task has four subtasks: perform intelligence synchronization, perform ISR integration, conduct tactical reconnaissance, and conduct surveillance. (See FM 7-15, ART 1.3, and FM 3-90).

Perform Intelligence Synchronization

1-27. The intelligence officer, with staff participation, synchronizes the entire collection effort to include all assets the commander controls, assets of lateral units and higher echelon units and organizations, and intelligence reach to answer the commander's priority intelligence requirements (PIRs) and information requirements (IRs). (See FM 7-15, ART 1.3.1)

1-28. The intelligence officer, with staff participation, supports the G3/S3 in orchestrating the entire ISR effort to include all assets the commander controls, assets of lateral units and higher echelon units and organizations, and intelligence reach to answer the CCIRs (PIRs and friendly force information requirements [FFIRs]) and other intelligence requirements. Intelligence synchronization activities include the following:

- Conducting requirements management (RM): anticipate, develop, analyze, validate, and prioritize intelligence requirements. Recommend PIRs to the commander. Manage the commander's intelligence requirements, requests for information (RFIs) from subordinate and lateral organizations, and tasks from higher headquarters. Eliminate satisfied requirements and add new requirements as necessary.

- Developing indicators.
- Developing specific IRs (specific information requirements [SIRs]).
- Converting the SIRs into intelligence tasks or ISR tasks. (See Figure 1-2 for the ISR task development process.) The S2/G2 assigns intelligence production and reach tasks to subordinate intelligence elements or personnel, submits RFIs to higher and lateral echelons, and coordinates with (or assists) the G3/S3 to develop and assign ISR tasks.
- Comparing the ISR tasks to the capabilities and limitations of the available ISR assets (in coordination with the operations officer).
- Forwarding SIRs that cannot be answered by available assets to higher or lateral organizations as RFIs.
- Assessing collection asset reporting and intelligence production to evaluate the effectiveness of the ISR effort.
- Maintaining situational understanding to identify gaps in coverage and to identify the need to cue or redirect ISR assets.
- Updating the intelligence synchronization plan. The G2/S2 manages and updates the intelligence synchronization plan as PIRs are answered and new requirements arise.

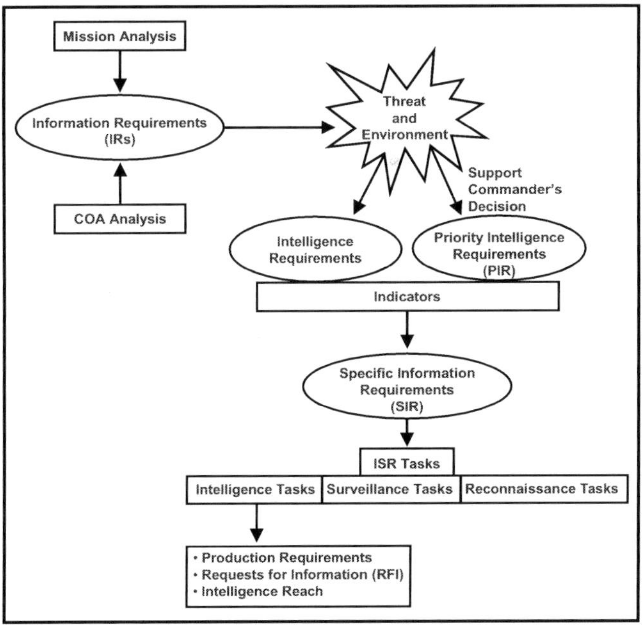

Figure 1-2. ISR Task Development Process.

1-29. **Intelligence Synchronization Considerations.** The G2/S2 generally follows six considerations in planning intelligence synchronization and ISR activities: anticipate, integrate, prioritize, balance, control, and reach. Refer to FM 2-01 for more information regarding intelligence synchronization.

- **Anticipate.** The intelligence staff must recognize when and where to shift collection or identify new intelligence requirements. The overall intent of this principle is to identify a new or adjust an existing requirement and present it to the commander for approval before waiting for the commander or staff to identify it.
- **Integrate.** The intelligence staff must be fully and continuously integrated into the unit's orders production and planning activities to ensure early identification of intelligence requirements. Early and continuous consideration of collection factors enhances the unit's ability to direct collection assets in a timely manner, ensures thorough planning, and increases flexibility in selecting assets.
- **Prioritize.** Prioritize each intelligence requirement based on its importance in supporting the commander's intent and decisions. Prioritization, based on the commander's guidance and the current situation, ensures that limited ISR assets and resources are directed against the most critical requirements.
- **Balance.** ISR capabilities complement each other. The intelligence staff should resist favoring or becoming too reliant on a particular unit, discipline, or system. Balance is simply planning redundancy, when required, eliminating redundancy when not desired, and ensuring an appropriate mix of ISR assets or types. The intelligence synchronization matrix (ISM) is useful in determining or evaluating balance.
- **Control.** To ensure timely and effective responses to intelligence requirements, a unit should first use ISR assets it controls. These assets usually are more responsive to their respective commander and also serve to lessen the burden on the ISR assets of other units, agencies, and organizations.
- **Reach.** Although usually not as responsive as a unit's own assets, intelligence reach may be the only way to satisfy an intelligence requirement. If at all possible, one should not depend solely on intelligence reach to answer a PIR.

1-30. **Develop IRs.** The intelligence staff develops a prioritized list of what information needs to be collected and produced into intelligence. Additionally, the intelligence staff dynamically updates and adjusts those requirements in response to mission adjustments and changes. This list is placed against a latest time intelligence is of value to ensure intelligence and information are reported to meet operational requirements. (See FM 7-15, ART 1.3.1) Figure 1-3 shows a comparison of IR, PIR, and intelligence requirements.

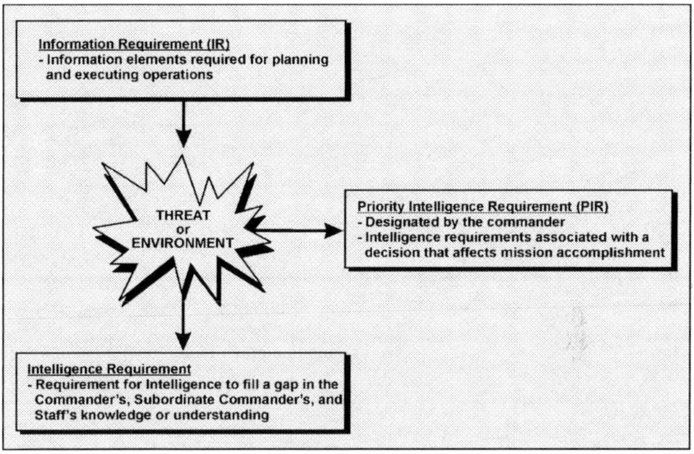

Figure 1-3. IR, PIR, and Intelligence Requirements Comparison.

1-31. An effective discussion of ISR has to include an understanding of the CCIRs. The CCIRs are elements of information required by commanders that directly affect decisionmaking and dictate the successful execution of military operations. The commander decides what information is critical based on the commander's experience, the mission, the higher commander's intent, and the staff's input (initial IPB, information, intelligence, and recommendations). Refer to FM 3-0 for more information regarding CCIRs.

1-32. Based on the CCIRs, two types of supporting IRs are generated: PIRs and FFIRs. However, commanders may determine that they need to know whether one or more essential elements of friendly information (EEFI) have been compromised or that the enemy is collecting against a designated EEFI. In those cases, commanders may designate that question as one of their CCIRs. Figure 1-4 shows the CCIR composition.

1-33. IRs are all of the information elements required by the commander and staff for the successful planning and execution of operations; that is, all elements necessary to address the factors of mission, enemy, terrain and weather, troops and support available, time available, civil considerations (METT-TC). Vetting by the commander or a designated representative turns an IR into either a PIR or an intelligence requirement. IRs are developed during COA analysis based on the factors of METT-TC.

1-34. PIRs are those intelligence requirements for which a commander has an anticipated and stated priority during the task of planning and decisionmaking. PIRs are associated with a decision based on an enemy action or inaction or the AO that will affect the overall success of the commander's mission. **The commander designates intelligence requirements tied directly to decisions as CCIR (PIR and FFIR).**

Answers to the PIRs help produce intelligence essential to the commander's situational understanding and decisionmaking. For information on PIR development, see FM 2-01.

1-35. The G2/S2 recommends to the commander those IRs produced during the MDMP that meet the criteria for PIR. They do not become CCIR (PIR and FFIR) until approved by the commander. Additionally, The commander may unilaterally designate PIRs. The IRs that are not designated by the commander as PIRs become intelligence requirements. The intelligence requirement is a gap in the command's knowledge or understanding of the AO or threat that the Intelligence warfighting function must fill.

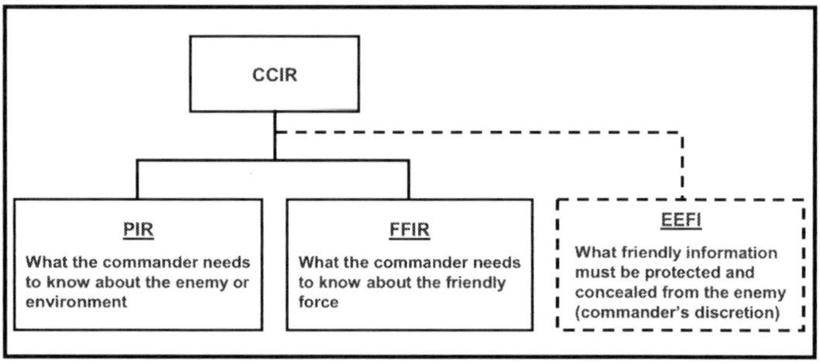

Figure 1-4. Commander's Critical Information Requirements Composition.

1-36. The G3/S3 then tasks the unit's assets to answer both the PIR and intelligence requirements through the ISR plan. PIR should—

- Ask only one question.
- Support a decision.
- Identify a specific fact, event, activity (or absence thereof) which can be collected.
- If linked to an ECOA, indicate an ECOA prior to, or as early as possible in, its implementation.
- Indicate the latest time the information is of value (LTIOV). The LTIOV is the absolute latest time the information can be used by the commander in making the decision the PIR supports. The LTIOV can be linked to time, an event, or a point in the battle or operation.

1-37. **Friendly Force IRs.** The staff also develops FFIRs which, when answered, provide friendly force information that the commander and staff need to achieve situational understanding and to make decisions.

1-38. **Essential Elements of Friendly Information**. EEFI establish information to protect, not information to obtain. However, commanders may determine that they need to know whether one or more EEFI have been compromised or that the enemy is collecting against a designated EEFI. In those cases, commanders may designate that question as one of their CCIRs, which generates PIRs and/or FFIRs. For example, a commander may determine that if the enemy discovers the location and movement of the friendly reserve, the operation is at risk. In this case, the location and movement of the friendly reserve are EEFI. The commander designates determining whether the enemy has discovered the location and movement of the friendly reserve as one of the CCIR. That CCIR, in turn, generates PIR and FFIR to support staff actions in determining whether the EEFI has been compromised.

1-39. **Develop the Intelligence Synchronization Plan.** The entire unit staff develops their IRs and determines how best to satisfy them. The staff uses reconnaissance and surveillance assets to collect information. The intelligence synchronization plan includes all assets that the operations officer can task or request and coordination mechanisms to ensure adequate coverage of the AOIs. (See FM 7-15, ART 1.3.1.2)

1-40. The intelligence synchronization plan, often presented in a matrix format as an ISM, aids in synchronizing the entire ISR effort with the overall operation and the commander's decisions and/or decision points (DPs). The intelligence synchronization plan is often produced in conjunction with the ISR plan. However, before performing intelligence synchronization and finalizing the intelligence synchronization plan, the G2/S2 must have—

- The CCIR (PIR and FFIR).
- A prioritized list of the remaining intelligence requirements.
- Evaluated ISR assets and resources.
- All of the assigned ISR tasks.

Perform ISR Integration

1-41. The operations officer, in coordination with the intelligence officer and other staff members, orchestrates the tasking and directing of available ISR assets to answer the CCIR. The operations officer, with input from the intelligence officer, develops tasks from the SIRs which coincide with the capabilities and limitations of the available ISR assets and the latest time information is of value (LTIOV). Intelligence requirements are identified, prioritized, and validated and an ISR plan is developed and synchronized with the scheme of maneuver. (See FM 7-15, ART 1.3.2, and FM 3-90)

1-42. The G3/S3, in coordination with the G2/S2 and other staff members, orchestrates the tasking and directing of available ISR assets to answer the CCIRs (PIRs and FFIRs) and IRs. The result of this process is the forming of the ISR plan. The ISR plan provides a list of all the ISR tasks to be accomplished. The G2/S2 and the G3/S3 develop tasks from the SIRs. These tasks are then assigned based on the capabilities and limitations of the available ISR assets and the LTIOV.

1-43. **Develop the ISR Plan.** The operations officer is responsible for developing the ISR plan. The entire unit staff analyzes each requirement to determine how best to satisfy it. The staff will receive orders and RFIs from both subordinate and adjacent units and higher headquarters. The ISR plan includes all assets that the operations officer can task or request and coordination mechanisms to ensure adequate coverage of the area of interest. (See FM 7-15, ART 1.3.2.1, and FM 3-90)

1-44. **The ISR Plan**. The ISR plan is usually produced as the ISR Annex to an OPORD (Annex L, Intelligence Surveillance, and Reconnaissance). Refer to FM 5-0 for specific and authoritative information on the ISR Annex. ISR is a continuous combined arms effort led by the operations and intelligence staffs in coordination with the entire staff that sets reconnaissance and surveillance in motion. The PIRs and other intelligence requirements drive the ISR effort. The commander takes every opportunity to improve situational understanding and the fidelity of the COP about the enemy and terrain through the deployment of ISR assets. Commanders integrate reconnaissance and surveillance to form an integrated ISR plan that capitalizes on their different capabilities. The ISR plan is often the most important part of providing information and intelligence that contributes to answering the CCIRs. For the G2/S2, an effective ISR plan is critical in answering the PIR. Also see FM 3-55 for more information on the ISR plan.

1-45. The ISR plan is not an MI-specific product—the G3/S3 is the staff proponent of the ISR plan— it is an integrated staff product executed by the unit at the direction of the commander. The G2/S2, however, must maintain situational understanding in order to recommend to the commander and G3/S3 changes or further development of the ISR plan. Based on the initial IPB and CCIRs, the staff—primarily the G2/S2—identifies gaps in the intelligence effort and develops an initial ISR plan based on available ISR assets. The G3/S3 turns this into an initial ISR Annex that tasks ISR assets as soon as possible to begin the collection effort.

1-46. The G3/S3, assisted by the G2/S2, uses the ISR plan to task and direct the available ISR assets to answer the CCIRs (PIRs and FFIRs) and intelligence requirements. Conversely, the staff revises the plan as other intelligence gaps are identified if the information is required to fulfill the CCIRs or in anticipation of future intelligence requirements. With staff participation, the G2/S2 intelligence officer synchronizes the ISR effort through a complementary product to the ISR plan—the intelligence synchronization plan.

1-47. **Execute and Update the ISR Plan**. The operations officer updates the ISR plan based on information received from the intelligence officer. The operations officer is the integrator and manager of the ISR effort through an integrated staff process and procedures. As PIRs are answered and new information requirements arise, the intelligence officer updates intelligence synchronization requirements and provides the new input to the operations officer who updates the ISR plan. He works closely with all staff elements to ensure the unit's organic collectors receive appropriate taskings. This ISR reflects an integrated collection strategy and employment, production and dissemination scheme that will effectively answer the commander's PIR. (See FM 7-15, ART 1.3.2.2, and FM 3-90)

Conduct Tactical Reconnaissance

1-48. To obtain, by visual observation or other detection methods, such as signals, imagery, measurement of signature or other technical characteristics, human interaction and other detection methods about the activities and resources of an enemy or potential enemy, or to secure data concerning the meteorological, hydrographic, or geographic characteristics and the indigenous population of a particular area. This task includes the conduct of NBC reconnaissance and the tactical aspects of SOF special reconnaissance. The five subtasks are—

- Conduct a Zone Reconnaissance.
- Conduct an Area Reconnaissance.
- Conduct a Reconnaissance in Force.
- Conduct a Route Reconnaissance.
- Conduct a Reconnaissance Patrol.

Note: This task branch includes techniques by which ART 1.1.4.1 (Collect Police Information) may be performed. (FM 7-15)

1-49. Reconnaissance is a mission undertaken to obtain by visual observation or other detection methods, information about the activities and resources of an enemy or potential enemy, and about the meteorological, hydrographic, or geographic characteristics of an AO. MI personnel and organizations can conduct reconnaissance through obtaining information derived from signals, imagery, measurement of signatures, technical characteristics, human interaction, and other detection methods. When performing reconnaissance, it is important to—

- Orient the reconnaissance asset on the named area of interest (NAI) and/or reconnaissance objective in a timely manner.
- Report all information rapidly and accurately.
- Complete the reconnaissance mission not later than (NLT) the time specified in the order.
- Answer the requirement that prompted the reconnaissance task.

Conduct Surveillance

1-50. To systematically observe the airspace, surface, or subsurface areas, places, persons, or things in the AO by visual, aural (audio), electronic, photographic, or other means. Other means may include but are not limited to space-based systems, and using special NBC, artillery, engineer, SOF, and air defense equipment. (See FM 7-15, ART 1.3.4, and FM 3-90)

Note: This task is a technique by which ART 1.1.4.1 (Collect Police Information) may be performed.

1-51. Conducting surveillance is systematically observing the airspace, surface, or subsurface areas, places, persons, or things in the AO by visual, aural, electronic, photographic, or other means. Surveillance activities include—

- Orienting the surveillance asset on the NAI and/or the surveillance objective in a timely manner.
- Reporting all information rapidly and accurately.
- Completing the surveillance mission NLT the time specified in the order.
- Answering the requirement that prompted the surveillance task.

PROVIDE INTELLIGENCE SUPPORT TO EFFECTS

1-52. The task of providing the commander information and intelligence support for targeting of the threat's forces, threat organizations, units and systems through lethal and non-lethal fires to include electronic attack and information operations. This task includes three subtasks: provide intelligence support to targeting, provide intelligence support to information operations, and provide intelligence support to combat assessment. (See FM 7-15, ART 1.4)

Provide Intelligence Support to Targeting

1-53. The intelligence officer, supported by the entire staff, provides the commander information and intelligence support for targeting the threat's forces and systems through employment of direct, indirect lethal and non-lethal fires. It includes identification of threat capabilities and limitations. The Intelligence warfighting function plays a crucial role by providing intelligence throughout the steps of the targeting process: decide, detect, deliver, assess. Support to targeting includes identifying threat capabilities and limitations. The G2/S2 supports the unit's lethal and non-lethal targeting effort through target development, target acquisition (TA), electronic warfare (EW), and combat assessment. This task has two subtasks:

- *Provide Intelligence Support to Target Development* is the systematic analysis of the enemy forces and operations to determine HVTs, systems, and system components for potential attack through maneuver, fires, or information.
- *Provide Intelligence Support to Target Detection* establishes procedures for dissemination of targeting information. The targeting team develops the sensor/attack system matrix to determine the sensor required to detect and locate targets. The intelligence officer places the following requirements into the integrated ISR plan.
  - Requires reconnaissance and surveillance operations to identify, locate, and track high-payoff targets (HPTs) for delivery of lethal or non-lethal effects.
  - Includes employing fires, offensive IO, and other attack capabilities against enemy C2 systems as part of the unit's FS plan and IO objectives.

Provide Intelligence Support to Information Operations

1-54. IO are actions taken to affect adversary information, influence other's decisionmaking processes and information systems while protecting one's own information and information systems. Overall operational continuity and mission success requires close, mutual coordination and synchronization of intelligence plans and operations with IO elements and related activities. Refer to Appendix A, this manual, for more information. This task has three subordinate tasks: (See FM 7-15, ART 1.4.2)

- Provide Intelligence Support to Offensive IO.
- Provide Intelligence Support to Defensive IO.
- Provide Intelligence Support to Activities Related to IO.

Provide Intelligence Support to Combat Assessment

1-55. Combat assessment is the determination of the overall effectiveness of force employment during military operations. The objective of combat assessment is to identify recommendations for the course of military operations. It answers the question, "Were the objectives met by force employment?" Although the assessment is primarily an intelligence responsibility, it requires input from and coordination with operations and fire support staffs. Combat assessment consists of conducting physical damage, functional damage, and target system assessments. (See FM 7-15, ART 1.4.3)

1-56. The staff determines how combat assessment relates to a specific target by conducting physical damage, functional damage, and target system assessments.

- **Conduct Physical Damage Assessment (PDA).** A PDA is an estimate of the extent of physical damage to a target based upon observed or interpreted damage. This post-attack target analysis is a coordinated effort among all units.
- **Conduct Functional Damage Assessment (FDA).** The FDA estimates the remaining functional or operational capability of a targeted facility or object. The staff bases FDA on the assessed physical damage and estimates of the threat's ability to recuperate (to include the time required to resume normal operations). Multiple echelons typically conduct this all-source analysis. The targeting or combat assessment cell integrates the initial target analyses with other sources, including intelligence, and then compares the original objective with the current status of the target to determine if the objective has been met.
- **Conduct Target System Assessment (TSA).** The TSA is an estimate of the overall impact of force employment against an adversary's target system. The unit, supported by higher echelon assets, normally conducts this assessment. The analyst combines all combat assessment reporting on functional damage to targets within a target system and assesses the overall impact on that system's capabilities. This process lays the groundwork for future recommendations for military operations in support of operational objectives.

1-57. **Munitions Effects Assessment (MEA).** MEA takes place concurrently and interactively with combat assessment since the same signatures used to determine the level of physical damage also give clues to munitions effectiveness. MEA is primarily the responsibility of operations and FS personnel, with input from the G2/S2. After the same weapon is used to attack several targets of a specific type, MEA should be accomplished to evaluate weapon performance. MEA analysts seek to identify through systematic trend analysis any deficiencies in weapon system and munitions performance or combat tactics by answering the question, "Did the weapons employed perform as expected?" Using a variety of input (targeting analysts, imagery analysts, structural engineers, and mission planners) analysts prepare a report assessing munitions performance. If combat troops capture attacked targets, it is then possible to collect detailed information on the target. Reports should detail weapon performance against specified target types. This information could have a crucial impact on future operations.

1-58. **Re-attack Recommendation**. Re-attack recommendations follow directly from both battle damage assessment (BDA) and MEA efforts. Basically re-attack recommendations answer the question, "Have we achieved the desired effects against our targeted objectives?" Evolving objectives, target selection, timing, tactics, weapons, vulnerabilities, and munitions are all factors in the new recommendations, combining both operations and intelligence functions.

1-59. **Targeting Meeting**. The role of the G2/S2 in targeting meetings is critical to ensuring intelligence supports the targeting process. The G2/S2 must come to the targeting meeting prepared. Additionally, the G2/S2 must understand the steps of the targeting process and tailor the intelligence products and G2/S2 participation according to the targeting process steps of: decide, detect, deliver, and assess.

1-60. There are two major areas in which the G2/S2 must prepare in order to support the targeting meeting.

- **Analysis:** Determine what the enemy is doing (current situation) and anticipated threat models; ECOA situation templates, ECOA sketches and statements, HVTs, and other associated products.
- **IPB Products:** Updated from the initial IPB effort, including analytical results (see above bullet), and tailored to the targeting requirements.

1-61. **Decide**. At the targeting meeting, the G2/S2 should be prepared to do the following:

- Brief the current and projected future enemy situation with an event template.
- Brief any combat assessments.
- Brief HVTs for potential selection of HPTs.
- Brief the current PIR portion of CCIRs, the PIRs of the CCIRs for the next phase or in accordance with the same timeframe of the targeting meeting.
- Brief the current ISM or intelligence synchronization plan.
- Participate in wargaming.
- Refine the initial ISR plan (in conjunction with the FS and G3/S3 representatives).

1-62. **Detect**. Based on information from the decide portion of the targeting meeting, the G2/S2 should be prepared to do the following:

- Identify specific ISR assets.
- Develop IRs and propose pertinent IRs to the commander for designation as PIRs.
- Complete the initial ISR plan; ensure it addresses the HPTs.

1-63. **Electronic Warfare**. EW provides targeting information to the commander based on transmitter location data. There are three major subdivisions within EW: electronic support (ES), electronic protection (EP), and electronic attack (EA).

- **ES** – Also referred to as electronic warfare support (EWS) in JP 1-02, involves actions to search for, intercept, identify, and locate or localize sources of intentional and unintentional radiated electromagnetic energy. In addition to supporting the overall ISR effort, ES is needed to produce intelligence required to support EA missions.
- **EP** – Involves passive and active means taken to protect personnel, facilities, and equipment from any effects of friendly or enemy employment of EW that degrades, neutralizes, or destroys friendly combat capabilities.
- **EA** – Involves the use of electromagnetic energy, directed energy, or anti-radiation weapons to attack personnel, facilities, or equipment with the intent of degrading, neutralizing, or destroying enemy combat capability and is considered a form of fires. EA includes actions taken to prevent or reduce an enemy's effective use of the electromagnetic spectrum (EMS), such as jamming and electromagnetic deception.

1-64. **Deliver**. The G2/S2 should be prepared to do the following:

- Brief how the Intelligence warfighting function will work to detect and track the target through the entire targeting process.
- Brief the implementation of the intelligence synchronization and ISR plans, including the LTIOV and communications structure.

1-65. **Assess**. Depending on the factors of METT-TC and the PIRs, the same ISR assets that detect and track the targets may be required to support combat assessments in order to determine if re-attack is required. Conversely, if re-attack is a decision that the commander will make based upon the effects on the enemy, combat assessment should be reflected in the PIR, and supported in the corresponding ISR plan.

THE OPERATIONAL ENVIRONMENT

1-66. The operational environment can be defined as a composite of the conditions, circumstances, and influences that affect the employment of capabilities and bear on the decision of the commander. The operational environment encompasses physical areas and factors of the air, land, maritime, and space domains. It also includes the information environment and enemy, adversary, friendly, and neutral systems.

OPERATIONAL VARIABLES (PMESII-PT)

1-67. Analysis of the operational environment—in terms of political, military, economic, social, infrastructure, and information with the addition of physical environment and time (PMESII-PT) variables—provides relevant information that senior commanders use to understand, visualize, and describe the operational environment. As a set, these operational variables are often abbreviated as PMESII-PT. A description of each of the PMESII-PT variables follows:

- **Political**. Describes the distribution of responsibility and power at all levels of governance or cooperation.
- **Military**. Explores the military capabilities of all relevant actors in a given operational environment.
- **Economic**. Encompasses individual behaviors and aggregate phenomena related to the production, distribution, and consumption of resources.
- **Social**. Describes the cultural, religious, and ethnic makeup within an operational environment.
- **Information**. Describes the nature, scope, characteristics, and effects of individuals, organizations, and systems that collect, process, disseminate, or act on information.
- **Infrastructure**. Is composed of the basic facilities, services, and installations needed for the functioning of a community or society.
- **Physical Environment**. Defines the physical circumstances and conditions that influence the execution of operations throughout the domains of air, land, sea, and space.
- **Time**. Influences military operations within an operational environment in terms of the decision cycles, operational tempo, and planning horizons. (FM 3-0)

1-68. For the intelligence portion, when identifying threats based on the systems approach to PMESII-PT, there are four primary components of the operational environment. These are the threat, terrain, weather, and civil considerations.

1-69. While the operational variables are directly relevant to campaign planning, they are too broad to be applied directly to tactical planning. That does not mean they are not of value at all levels; they are fundamental to developing the understanding of the operational environment necessary to conduct planning at any level, in any situation. The degree to which each operational variable provides useful information depends on the situation and echelon. For example, social and economic variables often receive close analysis as part of enemy and civil considerations at brigade and higher levels. They may affect the training and preparation of small units. However, they are not relevant to a small-unit leader's mission analysis. That leader may only be concerned with such questions as "Who is the tribal leader for this village?" "Is the electrical generator working?" "Does the enemy have antitank missiles?"

MISSION VARIABLES

1-70. Upon receipt of a warning order (WARNO) or mission, Army leaders narrow their focus to six mission variables. Mission variables are those aspects of the operational environment that directly affect a mission. They outline the situation as it applies to a specific Army unit. The mission variables are mission, enemy, terrain and weather, troops and support available, time available, and civil considerations (METT-TC). These are the categories of relevant information used for mission analysis. Army leaders use the mission variables to synthesize operational variables and tactical-level information with local knowledge about conditions relevant to their mission. The METT-TC categories are listed below:

- **Mission**. The mission is the task, together with the purpose, that clearly indicates the action to be taken and the reason.
- **Enemy**. Relevant information regarding the enemy may include the following:
  - Threat characteristics (previously order of battle [OB] factors).
    - Doctrine (or known execution patterns).
    - Personal habits and idiosyncrasies.
    - Equipment, capabilities, and vulnerabilities.
  - Threat COAs.
- **Terrain and Weather**. Terrain and weather are natural conditions that profoundly influence operations. Terrain and weather are neutral; they favor neither side unless one is more familiar with—or better prepared to operate in—the environment.
- **Troops and Support Available**. The number, type, capabilities, and condition of available friendly troops and support available. These include the resources available from joint, interagency, intergovernmental, multinational, host nation (HN), commercial (via contracting), and private organizations.
- **Time Available**. Time is critical to all operations. Controlling and exploiting time is central to initiative, tempo, and momentum. By exploiting time, commanders can exert constant pressure, control the relative speed of decisions and actions, and force exhaustion on enemy forces.
- **Civil Considerations**. Civil considerations comprise six characteristics expressed in the memory aid ASCOPE:
  - Areas.
  - Structures.
  - Capabilities.
  - Organizations.
  - People.
  - Events.

Note. For additional information on ASCOPE and the IPB process, see FM 2-01.3. Understanding the operational environment requires understanding the civil aspects of the AO. Civil considerations reflect how the manmade infrastructure, civilian institutions, and the attitudes and activities of the civilian leaders, populations, and organizations within an AO influence the conduct of military operations.

1-71. METT-TC enables leaders to synthesize operational level information with local knowledge relevant to their missions and tasks in a specified AO. Tactical and operational leaders can then anticipate the consequences of their operations before and during execution. See FM 3-0 for a detailed discussion of PMESII-PT and METT-TC.

THE INTELLIGENCE PROCESS

1-72. Intelligence operations generally include the five functions that constitute the intelligence process: **plan, prepare, collect, process,** and

produce. Additionally, there are three common tasks that occur across the five functions of the intelligence process: **analyze, disseminate,** and **assess**. The three common tasks are discussed after the last function, produce. The intelligence process functions are not necessarily sequential; this is what differentiates the Army's intelligence process from the Joint intelligence cycle. The intelligence process provides a common model with which to guide one's thinking, discussing, planning, and assessing about the threat or AOI environment. The intelligence process generates information about the threat, the AOI, and the situation, which allows the commander and staff to develop a plan, seize and retain the initiative, build and maintain momentum, and exploit success.

PLAN

1-73. The plan step consists of the activities that identify pertinent IRs and develop the means for satisfying those requirements. The CCIRs (PIRs and FFIRs) drive the planning of the ISR effort. The intelligence officer supports the G3/S3 in arranging the ISR effort, based on staff planning, to achieve the desired collection effects. Planning activities include, but are not limited to—

- Conducting IPB.
- Managing requirements.
- Submitting RFIs and using intelligence reach to fill information gaps.
- Evaluating reported information.
- Establishing the intelligence communications and dissemination architecture.
- Developing, managing, and revising the intelligence synchronization plan and the ISR plan as mission requirements change.

PREPARE

1-74. The failure of MI units or personnel to accomplish their tasks or missions can often be attributed to their failure to prepare. The prepare step includes those staff and leader activities which take place upon receiving the operation order (OPORD), operation plan (OPLAN), warning order (WARNO), or commander's intent in order to improve the unit's ability to execute tasks or missions.

1-75. The most habitual and egregious preparation failures committed by leaders (as evidenced by performance at the combat training centers) is not the conduct of specific tasks, but the failure to adequately coordinate for more generic combat requirements such as—

- Friendly forward unit liaison.
- Departing or reentering friendly lines or AOs.
- Fire support: enroute, at mission location, and return.
- Casualty evacuation procedures.
- Passwords (running, forward of friendly lines), recognition signals, call signs, and frequencies.
- Resupply.
- Movement through friendly AOs.

COLLECT

1-76. The collect step involves collecting and reporting information in response to ISR tasking. ISR assets collect information and data about threat forces, activities, facilities, and resources as well as information concerning the environmental and geographical characteristics of a particular AO. A successful ISR effort results in the timely collection and reporting of relevant and accurate information. This collected information forms the foundation of intelligence databases, intelligence production, and the G2's/S2's situational understanding. The requirements manager evaluates the reported information for its responsiveness to the CCIRs (PIRs and FFIRs).

PROCESS

1-77. Processing involves converting collected data and information into a form that is suitable for analyzing and producing intelligence. Examples of processing include enhancing imagery, translating a document from a foreign language, converting electronic data into a standardized report that can be analyzed by a system operator, and correlating information.

1-78. Processing data and information is performed unilaterally and cooperatively by both humans and automated systems.

PRODUCE

1-79. The produce step involves evaluating, analyzing, interpreting, synthesizing, and combining information and intelligence from single or multiple sources into intelligence or intelligence products in support of known or anticipated requirements. Production also involves combining new information and intelligence with existing intelligence in order to produce intelligence in a form that the commander and staff can apply to the MDMP and that supports and helps facilitate situational understanding. During the produce step, the intelligence staff manipulates information by—

- Analyzing the information to isolate significant elements.
- Evaluating the information to determine accuracy, timeliness, usability, completeness, precision, and reliability.
- Combining the information with other relevant information and previously developed intelligence.
- Applying the information to estimate possible outcomes.
- Presenting the information in a format that will be most useful to its eventual user.

1-80. The intelligence staff deals with numerous and varied production requirements based on PIRs and intelligence requirements; diverse missions, environments, and situations; and presentation requirements. Through analysis, collaboration, and intelligence reach, the G2/S2 and the staff use the collective intelligence production capability of higher, lateral, and subordinate echelons to meet the production requirements. Proficiency in these techniques and procedures facilitates the intelligence staff's ability to answer command and staff requirements regardless of the factors of METT-TC.

1-81. The three common tasks, discussed below, can occur throughout the intelligence process.

Analyze

1-82. The intelligence staff analyzes intelligence and information about the enemy's capabilities, friendly vulnerabilities, and the battlefield environment as well as issues and problems that arise within the intelligence process itself to determine their nature, origin, and interrelationships. This analysis enables commanders, staffs, and leaders to determine the appropriate action or reaction and to focus or redirect assets and resources to fill information gaps or alleviate pitfalls. It is also within the analyze function that intelligence analysts sort through large amounts of collected information and intelligence to obtain only that information which pertains to the CCIRs (PIRs and FFIRs), maintenance of the COP, and facilitates the commander's situational understanding.

Disseminate

1-83. Disseminating is communicating relevant information of any kind from one person or place to another in a usable form by any means to improve understanding or to initiate or govern action. Disseminating intelligence entails using information management techniques and procedures to deliver timely, relevant, accurate, predictive, and usable intelligence to the commander. Determining the product format and selecting the means to deliver it are key aspects of dissemination. Information presentation may be in a verbal, written, interactive, or graphic format. The type of information, the time allocated, and the individual preference of the commander all influence the information format.

Assess

1-84. Assessment plays an integral role in all aspects of the intelligence process. Assessing includes evaluating the effectiveness of intelligence in supporting the operation. Assessing the situation and available information begins upon receipt of the mission and continues throughout the intelligence process. The continual assessment of intelligence operations and ISR efforts, available information and intelligence, and various aspects of the factors of METT-TC are critical to ensure—

- The G2/S2 answers all CCIRs (PIRs and FFIRs).
- The G2/S2 and G3/S3 redirect ISR assets to support changing requirements.
- Using information and intelligence properly.

INTELLIGENCE DISCIPLINES

1-85. Intelligence disciplines are categories of intelligence functions. The Army's intelligence disciplines are All-Source Intelligence, Human Intelligence (HUMINT), Geospatial Intelligence (GEOINT), Imagery Intelligence (IMINT), Signals Intelligence (SIGINT), Measurement and Signatures Intelligence (MASINT), Open-Source Intelligence (OSINT), Technical Intelligence (TECHINT), and Counterintelligence (CI). For more information regarding the intelligence disciplines, see Part Three of this manual, as well as the respective manuals, which covers each individual intelligence discipline.

ALL-SOURCE INTELLIGENCE

1-86. All-source intelligence is defined as the intelligence products, organizations, and activities that incorporate all sources of information and intelligence, including open-source information, in the production of intelligence. All-source intelligence is a separate intelligence discipline, as well as the name of the task used to produce intelligence from multiple intelligence or information sources.

HUMAN INTELLIGENCE

1-87. HUMINT is the collection of foreign information—by a trained HUMINT Collector—from people and multimedia to identify elements, intentions, composition, strength, dispositions, tactics, equipment, personnel, and capabilities. It uses human sources as a tool, and a variety of collection methods, both passively and actively, to collect information.

GEOSPATIAL INTELLIGENCE

1-88. Title 10 U.S. Code § 467 establishes GEOINT. The definition of geospatial intelligence is—

> *The exploitation and analysis of imagery and geospatial information to describe, assess, and visually depict physical features and geographically referenced activities on the Earth. GEOINT consists of imagery, imagery intelligence, and geospatial information.*

1-89. The Army implementation of GEOINT is a result of the Army's organization, manning, and training. There are multiple types of data and information that various Army units and organizations collect, provide, and analyze in order to support the GEOINT enterprise. The two primary GEOINT service providers in the Army are MI units and organizations and Engineer (topographic) units and organizations. MI units and organizations provide imagery and IMINT to the enterprise. Engineer (topographic) units and organizations provide geospatial data and information to the enterprise. Therefore, while some of the collection, analysis, and exploitation of imagery and geospatial information occur within the intelligence warfighting function; some of the collection, analysis, and exploitation of imagery and geospatial information occur outside intelligence.

IMAGERY INTELLIGENCE

1-90. IMINT is intelligence derived from the exploitation of imagery collected by visual photography, infrared, lasers, multi-spectral sensors, and radar. These sensors produce images of objects optically, electronically, or digitally on film, electronic display devices, or other media.

SIGNALS INTELLIGENCE

1-91. SIGINT is a category of intelligence comprising either individually or in combination all communications intelligence (COMINT), electronic intelligence (ELINT), and foreign instrumentation signals intelligence (FISINT), however transmitted. SIGINT is derived from communications, electronics, and foreign instrumentation signals.

MEASUREMENT AND SIGNATURES INTELLIGENCE

1-92. MASINT is technically derived intelligence that detects, locates, tracks, identifies, and/or describes the specific characteristics of fixed and dynamic target objects and sources. It also includes the additional advanced processing and exploitation of data derived from IMINT and SIGINT collection. MASINT collection systems include but are not limited to radar, spectroradiometric, electro-optical (E-O), acoustic, radio frequency (RF), nuclear detection, and seismic sensors, as well as techniques for gathering chemical, biological, radiological, and nuclear (CBRN), and other material samples.

OPEN-SOURCE INTELLIGENCE

1-93. The *National Defense Authorization Act for Fiscal Year 2006* states:

> *Open source intelligence is produced from publicly available information that is collected, exploited, and disseminated in a timely manner to an appropriate audience for the purpose of addressing a specific intelligence requirement.*

1-94. Expressed in terms of the Army intelligence process, OSINT is relevant information derived from the systematic collection, processing, and analysis of publicly available information in response to intelligence requirements.

1-95. The Army does not have a specific military occupational specialty (MOS), additional skill identifier (ASI), or special qualification identifier (SQI) for OSINT. With the exception of the Asian Studies Detachment, the Army does not have base tables of organization and equipment (TOE) for OSINT units or staff elements. OSINT missions and tasks are imbedded within existing missions and force structure or accomplished through task organization.

TECHNICAL INTELLIGENCE

1-96. TECHINT is intelligence derived from the collection and analysis of threat and foreign military equipment and associated materiel for the purposes of preventing technological surprise, assessing foreign scientific and technical (S&T) capabilities, and developing countermeasures designed to neutralize an adversary's technological advantages.

COUNTERINTELLIGENCE

1-97. CI counters or neutralizes intelligence collection efforts through collection, counterintelligence investigations, operations, analysis, and production, and functional and technical services. CI includes all actions taken to detect, identify, track, exploit, and neutralize the multidiscipline intelligence activities of friends, competitors, opponents, adversaries, and enemies; and is the key intelligence community contributor to protect US interests and equities. CI assists in identifying EEFI, identifying vulnerabilities to threat collection, and actions taken to counter collection and operations against US forces.

UNIFIED ACTION

1-98. Unified action is the synchronization, coordination, and/or integration of the activities of governmental and nongovernmental entities with military operations to achieve unity of effort. (JP 1) Joint operations focus and maximize the complementary and reinforcing effects and capabilities of each service. Joint force commanders (JFCs) synchronize the complementary capabilities of the service components that comprise the joint force.

1-99. The employment of MI in campaigns and major operations must be viewed from a joint perspective, and the "mud-to-space" intelligence concept must establish a fully interoperable and integrated joint intelligence capability. ARFOR intelligence assets work with multinational and interagency partners to accomplish their missions. Ideally, multinational and interagency intelligence partners provide cultures, perspectives, and capabilities that reinforce and complement Army MI strengths and capabilities. Close intelligence coordination is the foundation of successful unified action. (See chapter 2.)

1-100. A critical part of current operations is the execution of the joint doctrinal concept of persistent surveillance. Joint doctrine defines persistent surveillance as a collection strategy that emphasizes the ability of some collection systems to linger on demand in an area to detect, locate, characterize, identify, track, target, and possibly provide battle damage assessment and re-targeting in real or near-real time. Persistent surveillance facilitates the formulation and execution of preemptive activities to deter or forestall anticipated adversary courses of action (JP 2-0).

THE NATURE OF LAND OPERATIONS

1-101. Landpower is the ability—by threat, force, or occupation—to gain, sustain, and exploit control over land, resources, and people (FM 3-0). Army operations reflect expeditionary and campaign capabilities that constantly adapt to each campaign's unique circumstances. Expeditionary capabilities require forces organized to be modular, versatile, and rapidly deployable. Rapidly deployed expeditionary force packages provide immediate options for seizing or retaining the operational initiative and also allow the conduct of sustained operations for as long as necessary. Army forces are organized, trained, and equipped for endurance. The Army's preeminent challenge is to balance expeditionary agility and responsiveness with the endurance and adaptability needed to complete a campaign successfully.

1-102. The capability to prevail in close combat is indispensable and unique to land operations. The outcome of battles and engagements depends on Army forces' ability to prevail in close combat. Many factors inherent in land combat combine to complicate the situation. These factors include chaos, complexity, insufficient intelligence, errors in understanding or planning, difficult terrain, the civilian population, and an adaptive and lethal enemy. The axiom "intelligence drives operations" continues to be true especially for land operations; operations and intelligence are inextricably linked.

1-103. Four considerations are preeminent in generating expeditionary capabilities and Army force packages.

- **Scope.** Considers and strives to understand an enemy throughout the depth of an operational area. Commanders rely on intelligence in order to use maneuver, fires, and other elements of combat power to defeat or destroy enemy forces.
- **Duration.** Forces routinely conduct missions prior to, during, and after the commitment of land combat forces. Intelligence is always engaged.
- **Terrain.** Missions occur among a complex variety of natural and manmade features. Employing forces in the complexity of the ground environment requires thorough planning.
- **Permanence.** Forces are integrated with, or assigned to, land combat forces as they seize or secure ground.

1-104. Several attributes of the land environment affect the application of landpower. These attributes include—

- The requirement to deploy and employ Army forces rapidly.
- The requirement for Army forces to operate for protracted periods.
- The nature of close combat.
- Uncertainty, chance, friction, and complexity.

1-105. Reconnaissance, Surveillance, and Target Acquisition (RSTA)/ISR is the means the Army uses to implement the joint doctrinal concept of persistent surveillance in support of tactical operations. Dependable technology and responsive intelligence lessen the effects of uncertainty, chance, friction, and complexity. Complex and dynamic Army tactical operations require extensive ISR capabilities to satisfy the commander's information requirements to detect, locate, characterize, identify, track, and target HPTs, and to provide combat assessment in real time within a very fluid operational environment.

1-106. *Reconnaissance, Surveillance, and Target Acquisition/ Intelligence, Surveillance, and Reconnaissance* is a full spectrum combined arms mission that integrates ground and air capabilities to provide effective, dynamic, timely, accurate, and assured combat information and multidiscipline actionable intelligence for lethal and non-lethal effects and decisions in direct support of the ground tactical commander.

FORCE PROJECTION OPERATIONS

1-107. Force projection is the military component of power projection. It is a central element of the national military strategy. Army organizations and installations, linked with joint forces and industry, form a strategic platform to maintain, project, and sustain ARFOR, wherever they deploy. Force projection operations are inherently joint and require detailed planning and synchronization. Force projection encompasses a range of processes—mobilization, deployment, employment, sustainment, and redeployment—discussed below.

1-108. The Army must change its mindset from depending on an "intelligence buildup" to performing intelligence readiness on a daily basis in order to meet the requirements for strategic responsiveness. MI personnel, even in garrison at the lowest tactical echelons, must use their analytic and other systems and prepare for possible operations on a daily basis.

1-109. Built on a foundation of intelligence readiness, the Intelligence warfighting function provides the commander with the intelligence needed to plan, prepare, and execute force projection operations. Successful intelligence during force projection operations relies on continuous collection and intelligence production before and during the operation. In a force projection operation, higher echelons will provide intelligence to lower echelons until the tactical ground force completes entry and secures the lodgment area. The joint force J2 must exercise judgment when providing information to subordinate G2s/S2s so not to overwhelm them. Figure 1-5 is an example of force projection intelligence.

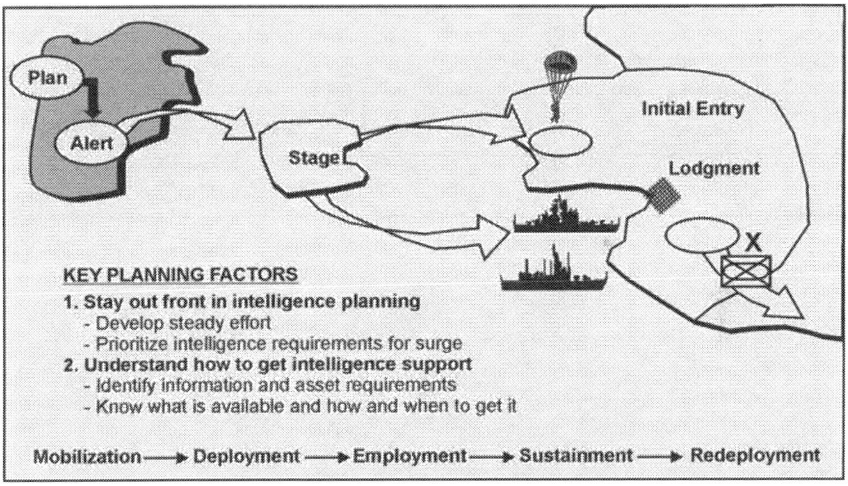

Figure 1-5. Force Projection Intelligence.

1-110. The G2/S2 must anticipate, identify, consider, and evaluate all threats to the entire unit throughout force projection operations. This is critical during the deployment and entry operations stages of force projection. During these stages, the unit is particularly vulnerable to enemy actions because of its limited combat power and knowledge of the AO. Intelligence personnel must, therefore, emphasize the delivery of combat information and intelligence products that indicate changes to the threat or AO developed during predeployment IPB.

1-111. The G2/S2 should—

- Review available databases on assigned contingency AOIs, conduct IPB on these AOIs, and develop appropriate IPB products.
- Comply with higher headquarters standing operating procedures (SOPs) and manuals for specific intelligence operations guidance.
- Coordinate for and rehearse electronic message transfers (for example, Internet Protocol addresses, routing indicators) using the same communications protocols with theater, higher headquarters, subordinate, and lateral units that the unit would use when deployed.
- Plan, train, and practice surging intelligence functions on likely or developing contingency crises.
- Prepare and practice coordination from predeployment through redeployment with other elements and organizations (for example, HUMINT, IMINT, SIGINT, MASINT, IO, staff weather officer [SWO], civil affairs [CA], psychological operations [PSYOP], and special operations forces [SOF] units, to include databases and connectivity).
- Include the following as a part of daily (sustainment) operations:
  - US Army Reserve and other augmentation.
  - A linguist plan with proficiency requirements. (Alert linguists through early entry phases of deployment.)
  - Training (individual and collective).
- Establish formal or informal intelligence links, relationships, and networks to meet developing contingencies.
- Forward all RFIs to higher headquarters in accordance with SOPs.
- Establish statements of intelligence interests (SIIs), other production, and I&W requirements.

1-112. To draw intelligence from higher echelons and focus intelligence downward, based on the commander's needs, the G2/S2 must—

- Understand the J2's multiple echelon and broadcast dissemination capability to ensure near-real time (NRT) reporting to all deployed, in transit, or preparing to deploy forces.
- Maintain or build intelligence databases on the environment and threats for each probable contingency.
- State and record the CCIR (as a minimum, list the PIRs and ISR tasks or requests).

1-113. Until the unit's collection assets become operational in the AO, the G2/S2 will depend upon intelligence from the ARFOR or JTF to answer the unit's intelligence needs. The following paragraphs describe the intelligence and ISR considerations during force projection.

MOBILIZATION

1-114. Mobilization is the process by which the armed forces or part of them are brought to a state of readiness for war or other national emergency. It assembles and organizes resources to support national objectives. Mobilization includes activating all or part of the US Army Reserve, and assembling and organizing personnel, supplies, and materiel. A unit may be brought to a state of readiness for a specific mission or other national emergency. This process, called mobilization, is where specific US Army

Active or Reserve units, capabilities, and personnel are identified and integrated into the unit. During mobilization, the G2/S2 must—

- Monitor intelligence reporting on threat activity and I&W data.
- Manage IRs and RFIs from their unit and subordinate units to include updating ISR planning.
- Establish habitual training relationships with their US Army Active and Reserve augmentation units and personnel as well as higher echelon intelligence organizations as identified in the existing OPLAN. Support the US Army Reserve units and augmentation personnel by preparing and conducting intelligence training and threat update briefings and by disseminating intelligence.
- Identify ISR force requirements for the different types of operations and contingency plans (CONPLANs).
- Identify individual military, civilian, and contractor augmentation requirements for intelligence operations. The Army, and the Intelligence warfighting function in particular, cannot perform its missions without the support of its Department of the Army Civilians (DACs) and contractors. The force increasingly relies on the experience, expertise, and performance of non-uniformed personnel and has fully integrated these non-uniformed personnel into the warfighting team.

1-115. During mobilization the G2/S2, in conjunction with the rest of the staff, must ensure the adequacy of training and equipping of US Army Active and Reserve MI organizations and individual augmentees to conduct intelligence operations.

1-116. The G2/S2 supports peacetime contingency planning with IPB products and databases on likely contingency areas. The G2/S2 establishes an intelligence synchronization plan that will activate upon alert notification. For smooth transition from predeployment to entry, the G2/S2 must coordinate intelligence synchronization and communications plans before the crisis occurs. The intelligence synchronization plan identifies the intelligence requirements supporting those plans, to include—

- ISR assets providing support throughout the AOI.
- Command and support relationships of ISR assets at each echelon.
- Report and request procedures not covered in unit SOPs.
- Sequence of deployment of ISR personnel and equipment. Early deployment of key ISR personnel and equipment is essential for force protection and combat readiness. Composition of initial and follow-on deploying assets is influenced by the factors of METT-TC, availability of communications, and availability of lift.
- Communications architecture supporting both intelligence staffs and ISR assets.
- Friendly vulnerabilities to hostile intelligence threats and plans for conducting FP measures. The staff must begin this type of planning as early as possible to ensure adequate support to FP of deploying and initial entry forces.
- Monitor time-phased force and deployment data (TPFDD) and recommend changes in priority of movement, unit, or capability to enable ISR operations.

1-117. The G2/S2 must continually monitor and update the OPLANs to reflect the evolving situation, especially during crisis situations. National intelligence activities monitor regional threats throughout the world and can answer some intelligence requirements supporting the development of OPLANs.

1-118. Upon alert notification, the G2/S2 updates estimates, databases, IPB products, and other intelligence products needed to support command decisions on force composition, deployment priorities and sequence, and the AOI. Units reassess their collection requirements immediately after alert notification. The G2/S2 begins verifying planning assumptions within the OPLANs. CI and ISR personnel provide FP support and antiterrorism measures.

1-119. Throughout mobilization, unit intelligence activities will provide the deploying forces with the most recent intelligence on the AO. The intelligence staff will also update databases and situation graphics. The G2/S2 must—

- Fully understand the unit, ARFOR, and joint force intelligence organizations.
- Revise intelligence and intelligence-related communications architecture and delete or integrate any new systems and software with the current architecture.
- Support 24-hour operations and provide continuous intelligence.
- Plan all required intelligence reach procedures.
- Determine transportation availability for deployment and availability when deployed.
- Determine all sustainability requirements.
- Determine intelligence release requirements and restrictions; releasability to multinational and HN sources.
- Review status of forces agreements (SOFAs), rules of engagement (ROE), international laws, and other agreements, emphasizing the effect that they have on intelligence collection operations. (Coordinate with the staff judge advocate [SJA] on these issues.)
- Ensure ISR force deployment priorities are reflected in the TPFDD to support ISR operations based upon the factors of METT-TC.
- Ensure intelligence links provide the early entry commander vital access to multi-source army and joint intelligence collection assets, processing systems, and databases.
- Review the supporting unit commanders' specified tasks, implied tasks, task organization, scheme of support, and coordination requirements with forward maneuver units. Address issues or shortfalls and direct or coordinate changes.
- Establish access to national HUMINT, IMINT, SIGINT, MASINT, and CI databases, as well as automated links to joint service, multinational, and HN sources.

DEPLOYMENT

1-120. Deployment is the movement of forces and materiel from their point of origin to the AO. This process has four supporting components:

predeployment activities, fort to port, port to port, and port to destination; these components are known collectively as reception, staging, onward movement, and integration (RSO&I) activities. Success in force projection operations hinges on timely deployment. The size and composition of forces requiring lift are based on the factors of METT-TC, availability of pre-positioned assets, the capabilities of HN support, and the forward presence of US forces. Force or tactical tailoring is the process used to determine the correct mix and sequence of deploying units.

1-121. During deployment, intelligence organizations at home station or in the rear area take advantage of modern SATCOM, broadcast technology, and automated data processing (ADP) systems to provide graphic and textual intelligence updates to the forces enroute. Enroute updates help eliminate information voids and, if appropriate, allow the commander to adjust the plan prior to arrival in theater in response to changes in the OE or enemy actions.

1-122. Intelligence units extend established networks to connect intelligence staffs and collection assets at various stages of the deployment flow. Where necessary, units establish new communications paths to meet unique demands of the mission. The theater and corps analysis and control elements (ACEs) play a critical role in making communications paths, networks, and intelligence databases available to deploying forces.

1-123. Space-based systems are key to supporting intelligence during the deployment and the subsequent stages of force projection operations by—

- Monitoring terrestrial AOIs through ISR assets to help reveal enemy location and disposition, attempting to identify the enemy's intent.
- Providing communications links between forces enroute and in the continental United States (CONUS).
- Permitting MI collection assets to accurately determine their position through the Global Positioning System (GPS).
- Providing timely and accurate data on meteorological, oceanographic, and space environmental factors that might affect operations.
- Providing warning of theater ballistic missile launches.
- Providing timely and accurate weather information to all commanders through the Integrated Meteorological System (IMETS).

1-124. Situation development dominates intelligence operations activities during initial entry operations. The G2/S2 attempts to identify all threats to arriving forces and assists the commander in developing FP measures. During entry operations, echelons above corps (EAC) organizations provide intelligence. This support includes providing access to departmental and joint intelligence and deploying scalable EAC intelligence assets. The entire effort focuses downwardly to provide tailored support to deploying and deployed echelons in response to their CCIRs (PIRs and FFIR).

1-125. Collection and processing capabilities are enhanced, as collection assets build up in the deployment area, with emphasis on the build-up of the in-theater capability required to conduct sustained ISR operations. As the build-up continues, the G2/S2 strives to reduce total dependence on extended split-based intelligence from outside the AO. As assigned collection assets arrive into the theater, the G2/S2 begins to rely on them for tactical intelligence although higher organizations remain a source of intelligence.

1-126. As the ARFOR enter the theater of operations, the joint force J2 implements and, where necessary, modifies the theater intelligence architecture. Deploying intelligence assets establishes liaison with staffs and units already present in the AO. Liaison personnel and basic communications should be in place prior to the scheduled arrival of parent commands. ISR units establish intelligence communications networks.

1-127. CONUS and other relatively secure intelligence bases outside the AO continue to support deployed units. Systems capable of rapid receipt and processing of intelligence from national systems and high capacity, long-haul communications systems are critical to the success of split-based support of a force projection operation. These systems provide a continuous flow of intelligence to satisfy many operational needs.

1-128. The G2/S2, in coordination with the G3/S3, participates in planning to create conditions for decisive operations. The G2/S2 also adjusts collection activities as combat strength builds. During entry operations the G2/S2—

- Monitors FP indicators.
- Monitors the ISR capability required to conduct sustained intelligence operations.
- Monitors intelligence reporting on threat activity and I&W data.
- Develops measurable criteria to evaluate the results of the intelligence synchronization plan.
- Assesses—
  - Push versus pull requirements of intelligence reach.
  - Effectiveness of the intelligence communications architecture.
  - Reporting procedures and timelines.
  - Intelligence to OPLANs and OPORDs, branches, and sequels (to include planning follow-on forces).

EMPLOYMENT

1-129. Employment is the conduct of operations to support a JFC commander. Employment encompasses an array of operations, including but not limited to—

- Entry operations (opposed or unopposed).
- Shaping operations (lethal and non-lethal).
- Decisive operations (combat or support).
- Postconflict operations (prepare for follow-on missions or redeployment).

Entry Operations

1-130. Enemies often possess the motives and means to interrupt the deployment flow of ARFOR. Threats to deploying forces may include advanced conventional weaponry (air defense, mines, etc.) and WMD. Sea and air ports of debarkation (PODs) should be regarded as enemy HPTs because they are the entry points for forces and equipment. PODs are vulnerable because they are fixed targets with significant machinery and equipment that is vulnerable to attack; in addition to military forces and materiel, HN support personnel, contractors, and civilians may all be working there. An enemy attack, or even the threat of an enemy attack, on a POD can have a major impact on force projection momentum. Commanders at all levels require predictive intelligence so that they may focus attention on security actions that reduce vulnerabilities. To avoid, neutralize, or counter threats to entry operations, the commanders rely on the ability of the G2/S2 to support future operations by accurately identifying enemy reactions to US actions, anticipating their response to our counteractions and predicting additional ECOAs.

1-131. Predictive intelligence also supports the decisions the commander and staff must make about the size, composition, structure, and deployment sequence of the force in order to create the conditions for success. Commanders rely on predictive intelligence to identify potential friendly decisions before the actual event. While thorough planning develops friendly COAs to meet possible situations, the nature of an operation can change significantly before execution. The G2/S2 must provide timely, accurate, and predictive intelligence to ensure the commander can retain the initiative to implement the plan or make decisions before losing the opportunity to do so.

Shaping Operations

1-132. Shaping operations create and preserve conditions for the success of the decisive operation. Shaping operations include lethal and non-lethal activities conducted throughout the AO. They support the decisive operation by affecting enemy capabilities and forces, or by influencing enemy decisions. G2/S2 intelligence analysis and ISR activities support the development and execution of shaping operations by identifying threat centers of gravity and decisive points on the battlefield. The G2/S2 also ensures the intelligence process focuses on the CCIRs (PIRs and FFIRs). It is critical that the G2/S2 provide timely, accurate, and predictive intelligence to the commander to facilitate the execution of shaping operations. Predictive intelligence should provide sufficient time for the commander to understand how the enemy will react to US COAs so that appropriate shaping operations can be implemented.

Decisive Operations

1-133. Decisive operations are those that directly accomplish the task assigned by the higher headquarters. Decisive operations conclusively determine the outcome of major operations, battles, and engagements. Continuously synchronized ISR activities, coupled with predictive intelligence results and products during all stages of force projection, combine to ensure the commander is prepared to employ the right forces with the right support to conduct decisive operations at the most appropriate place and time. In addition to coordinating with the G3/S3 to synchronize all ISR activities, the G2/S2 must answer the CCIRs (PIRs and FFIRs) and provide continuous assessments of the enemy's current situation and predict the enemy's subsequent COA, branches, and/or sequels.

Postconflict Operations

1-134. Upon cessation of hostilities or truce, deployed forces enter a new stage of force projection operations. Postconflict operations focus on restoring order, reestablishing HN infrastructure, preparing for redeployment of forces, and planning residual presence of US forces. While postconflict operations strive to transition from conflict to peace, there remains a possibility of resurgent hostilities by individuals and forces. The ISR effort, particularly predictive intelligence analysis, remains just as critical in postconflict operations as in the other employment operations. ISR operations support the postconflict emphasis on restoration operations exemplified by commanders redirecting their CCIRs (PIRs and FFIRs) and IRs to support units conducting these missions. These operations might include—

- Engineer units conducting mine clearing or infrastructure reconstruction operations.
- Medical and logistics units providing humanitarian relief.
- MP units providing law and order assistance.
- CA units reestablishing local control and preparing for the orderly transition of local governments and to support civil-military operations (CMO).

SUSTAINMENT

1-135. Sustainment involves providing and maintaining levels of personnel and materiel required to sustain the operation throughout its duration. It is essential to generating combat power. CSS may be split-based between locations within and outside CONUS. These operations include ensuring units have the MI assets required to accomplish the mission, such as personnel (including linguists), communications systems, ISR systems, and appropriate maintenance support.

REDEPLOYMENT

1-136. Redeployment is the process by which units and materiel reposture themselves in the same theater; transfer forces and materiel to support another JFC's operational requirements; or return personnel and materiel to the home or demobilization station upon completion of the mission. Redeployment operations encompass four phases:

- Recovery, reconstitution, and pre-deployment activities.
- Movement to and activities at the port of embarkation (POE).
- Movement to the POD.
- Movement to home station.

1-137. As combat power and resources decrease in the AO, FP and I&W become the focus of the commander's intelligence requirements. This in turn drives the selection of those assets that must remain deployed until the end of the operation and those that may redeploy earlier. The S2—

- Monitors intelligence reporting on threat activity and I&W data.
- Continues to conduct intelligence to FP.
- Requests ISR support (theater and national systems) and intelligence in support of redeployment.

1-138. After redeployment, MI personnel and units recover and return to predeployment activities. ISR units resume contingency-oriented peacetime intelligence operations. US Army Reserve ISR units demobilize and return to peacetime activities. G2/S2s must—

- Monitor intelligence reporting on threat activity and I&W data.
- Update or consolidate databases.
- Maintain intelligence readiness.
- Provide their input into the Force Design Update (FDU) process to refine modified table of organizations and equipment (MTOE) and evaluate the need for Individual Mobilization Augmentee (IMA) personnel.

Chapter 2

Intelligence and Unified Action

UNIFIED ACTION

2-1. As discussed in JP 1-02 and Chapter 1, unified action describes the wide scope of actions (including the synchronization of activities with governmental and NGO agencies) taking place within unified commands, subordinate unified commands, or JTFs under the overall direction of the commanders of those commands. Under unified action, commanders integrate joint, single-service, special, and supporting intelligence operations with interagency, nongovernmental, and multinational operations. The ARFOR often brings unique ISR capabilities to unified action operations.

2-2. In a unified action, the intelligence staff takes on additional responsibilities, relationships, and infrastructure to execute multi-service and multinational intelligence operations. This chapter discusses the synchronization of Army intelligence efforts with joint and other national and international partners to achieve unity of effort and to accomplish the commander's intent.

THE LEVELS OF WAR

2-3. The levels of war are doctrinal perspectives that clarify the links between strategic objectives and tactical actions. Although there are no finite limits or boundaries between them, the three levels of war are strategic, operational, and tactical (see Figure 2-1).

- The strategic level is that level at which a nation, often as one of a group of nations, determines national and multinational security objectives and guidance and develops and uses national resources to accomplish them.
- The operational level is the level at which campaigns and major operations are conducted and sustained to accomplish strategic objectives within theaters or AOs. It links the tactical employment of forces to strategic objectives.
- The tactical level is the employment of units in combat. It includes the ordered arrangement and maneuver of units in relation to each other, the terrain, and the enemy to translate potential combat power into victorious battles and engagements.

2-4. Understanding the interdependent relationship of all three levels of war helps commanders visualize a logical flow of operations, allocate resources, and assign tasks. Actions within the three levels are not associated with a particular command level, unit size, equipment type, or force or component type. The concept of strategic, operational, and tactical intelligence operations aids JFCs and their J2s in visualizing the flow of intelligence from one level to the next. The concept facilitates allocating required collection, analysis, production, and dissemination resources; and facilitates the assignment of appropriate intelligence tasks to national, theater, component, and supporting intelligence elements.

STRATEGIC

2-5. The President and the Secretary of Defense use strategic intelligence to develop national strategy and policy, monitor the international situation, prepare military plans, determine major weapon systems and force structure requirements, and conduct strategic operations.

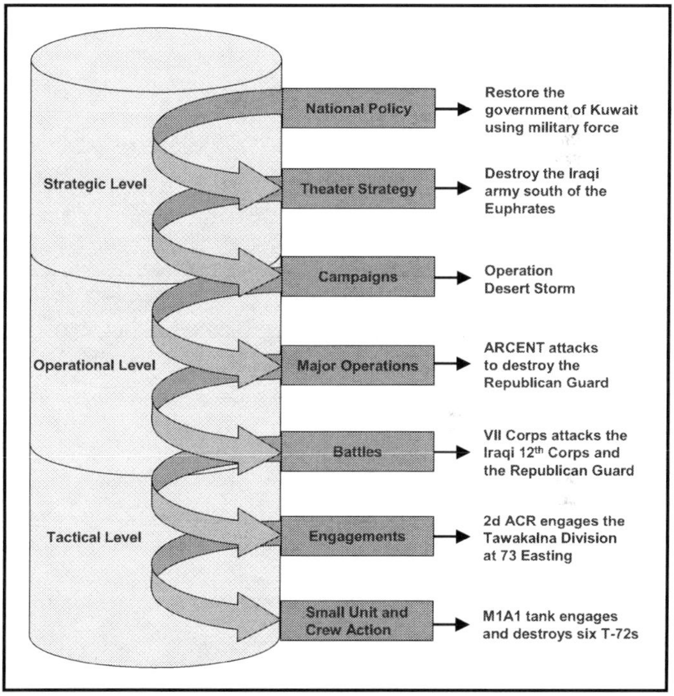

Figure 2-1. The Levels of War.

2-6. Intelligence supports joint operations across the full spectrum of military operations. It determines the current capabilities, and forecasts future developments, of adversaries or potential adversaries that could affect the national security and interests. A strategic operation also produces the intelligence required by combatant commanders to prepare strategic estimates, strategies, and plans to accomplish missions assigned by higher authorities. Theater intelligence includes determining when, where, and in what strength the adversary will stage and conduct theater level campaigns and strategic unified operations. The intelligence staff should also focus predictive analysis efforts on identifying strategic threat events and how these events will impact US actions at the strategic, operational, and tactical levels. Intelligence operations support IO as well. Intelligence operations support strategic planning by—

- Developing strategic intelligence policy.
- Preparing the strategic collection plan.
- Allocating national intelligence resources.

2-7. Intelligence requirements are those requirements generated from the staff's IRs regarding the enemy and environment that are not a part of the CCIRs (PIRs and FFIRs). Intelligence requirements require collection and can provide answers in order to identify indicators of enemy actions or intent, which reduce the uncertainties associated with an operation. Significant changes (that is, branches and sequels) with an operation usually lead to changes in intelligence requirements. Of particular importance is information relating to enemy or threat strategic vulnerabilities, strategic forces, strategic centers of gravity, and any capabilities relating to the development and employment of CBRNE.

2-8. Global and regional issues and threats are identified and reported to the President and the Secretary of Defense, as well as to the senior military leadership and the combatant commanders. Intelligence requirements include any foreign developments that could threaten the US, its citizens abroad, or multinational military, political, or economic interests. Intelligence also includes identifying hostile reactions to US reconnaissance activities and indications of impending terrorist attacks (I&W). For a complete listing of strategic tasks, refer to CJCSM 3500.04C.

OPERATIONAL

2-9. Combatant commanders and subordinate JFCs and their component commanders are the primary users of intelligence. At the operational echelons, intelligence—

- Focuses on the military capabilities and intentions of enemies and threats.
- Provides analysis of events within the AOI and helps commanders determine when, where, and in what strength the adversary might stage and conduct campaigns and major operations.
- Supports all phases of military operations, from mobilization all the way through redeployment of US forces, and continues during sustainment.
- Supports all aspects of the joint campaign.
- Identifies adversary centers of gravity and HVTs.
- Provides critical support to friendly IO.

2-10. The JFC and staff allocate intelligence resources and request support from national agencies, other theaters, and multinational partners. During stability operations and support operations, operational intelligence includes training and assisting multinational partners in conducting intelligence operations.

2-11. Combatant commanders use intelligence concerning the nature and characteristics of the battlespace to determine the type and scale of operations. Intelligence also aids in determining the impact of significant regional features and hazards on the conduct of both friendly and adversary operations. Significant regional factors include the natural environment, political, informational, economic, industrial, infrastructure, geographic, demographic, topographic, hydrographic, climatic, populace, cultural, medical, lingual, historical, and psychological aspects of the AOI. Intelligence analysis also assists in determining the ROE and other restrictions which will affect operations in the JFC's AO.

2-12. Intelligence relating to the adversary's military and nonmilitary capabilities assists in determining the adversary's ability to conduct military operations. Factors that operational intelligence addresses include mobilization potential, force structure (including alliance forces), force dispositions, equipment, military doctrine, C2 structure, and their MDMP. Intelligence includes the continuous refinement of the OBs for the entire array of the enemy's forces in the AOI. For a complete listing of operational tasks, refer to CJCSM 3500.04C.

US Army Intelligence and Security Command (INSCOM)

2-13. The U.S. Army has vested its intelligence at the operational level with INSCOM, the major Army command (MACOM) responsible for the Army's intelligence forces above Corps. INSCOM's mission is to conduct and support dominant intelligence, security, and information operations for military commanders and national decisionmakers. The INSCOM strategy is to provide superior information and information capabilities to Army commanders, while denying the same to adversaries. Headquarters (HQ), INSCOM, in coordination with its major subordinate commands (MSCs), provides a myriad of general intelligence support operations. INSCOM is providing a globally focused, rapidly deployable, knowledge based, adaptively force packaged capability, supporting commanders and leaders with actionable intelligence at the point of decision. INSCOM serves as the national to tactical Intelligence bridge.

Army Space Program Office (ASPO)

2-14. ASPO executes the Army's Tactical Exploitation of National Capabilities Program (TENCAP). The program focuses on exploiting current and future tactical potential of national systems and integrating the capabilities into the Army's tactical decisionmaking process. Army TENCAP systems enable the tactical commander maximum flexibility to satisfy intelligence needs under a wide range of operational scenarios. ASPO is the point of contact (POC) for all tactical activities between MACOMs or users and the National Reconnaissance Office (NRO).

TACTICAL

2-15. Tactical commanders use intelligence for planning and conducting battles and engagements. Relevant, accurate, predictive, and timely intelligence allows tactical units to achieve an advantage over their adversaries. Precise and predictive intelligence, on the threat and targets, is essential for mission success. Predictive intelligence also enables the staff to better identify or develop ECOAs. Tactical intelligence—

- Identifies and assesses the enemy's capabilities, COAs, and vulnerabilities, as well as describes the battlespace.
- Seeks to identify when, where, and in what strength the enemy will conduct tactical level operations.
- Provides the commander with information on imminent threats to the force including those from terrorists, saboteurs, insurgents, and foreign intelligence collection.
- Provides critical support to friendly IO.
- Develops and disseminates targeting information and intelligence.

2-16. Intelligence provides the tactical commander with the information and intelligence he requires to successfully employ combat forces against enemy forces. Thus, intelligence tasks support the execution of battles and engagements. These intelligence tasks are different from those at other levels due to their ability to immediately influence the outcome of the tactical commander's mission. They include information gathered from tactical sources, such as combat information, interrogations, debriefings, and eliciting information from captured or misplaced personnel. For a complete listing of tactical tasks, refer to FM 7-15.

CATEGORIES OF INTELLIGENCE

2-17. Directly or indirectly, the unit receives intelligence from throughout the US intelligence community. The intelligence operations and products of national, joint, and service organizations that make up the intelligence community fall into one of six categories:

- I&W.
- Current intelligence.
- GMI.
- Target intelligence.
- S&T intelligence.
- CI.

2-18. The categories of intelligence are distinguishable from each other primarily by the purpose of the intelligence product. The categories can overlap and some of the same intelligence is useful in more than one category. Depending upon the echelon, intelligence organizations use specialized procedures to develop each category of intelligence. The following information describes each category.

INDICATIONS AND WARNINGS

2-19. Indications and warnings (I&W) are those intelligence activities intended to detect and report time-sensitive intelligence information on foreign developments that could involve a threat to the US or allied and/or coalition military, political, or economic interests or to US citizens abroad. I&W includes forewarning of enemy actions or intentions; the imminence of hostilities; insurgency; nuclear or non-nuclear attack on the US, its overseas forces, or allied and/or coalition nations; hostile reactions to US reconnaissance activities; terrorist attacks; and other similar events. (JP 3-13). I&W comes from time-sensitive information and analysis of developments that could involve a threat to the US and multinational military forces, US political or economic interests, or to US citizens. While the G2/S2 is primarily responsible for producing I&W intelligence, each element, such as the MPs conducting PIO, within every unit contributes to I&W through awareness of the CCIRs and reporting related information.

CURRENT INTELLIGENCE

2-20. Current intelligence involves the integration of time-sensitive, all-source intelligence and information into concise, accurate, and objective reporting on the battlespace and current enemy situation. One of the most important forms of current intelligence is the enemy situation portion of the COP. The G2/S2 is responsible for producing current intelligence for the unit. Current intelligence supports ongoing operations across full spectrum operations. In addition to the current situation, current intelligence should provide projections of the enemy's anticipated situations (estimates) and their implications on the friendly operation.

GENERAL MILITARY INTELLIGENCE

2-21. GMI is intelligence concerning military capabilities of foreign countries or organizations or topics affecting potential US or multinational military operations relating to armed forces capabilities, including OB, organization, training, tactics, doctrine, strategy, and other factors bearing on military strength and effectiveness and area and terrain intelligence. This broad category of intelligence is normally associated with long-term planning at the national level. However, GMI is also an essential tool for the intelligence staff and should be in place long before the start of preparations for a particular military operation. An up-to-date, comprehensive intelligence database is critical to the unit's ability to plan and prepare rapidly for the range of operations and global environments in which it may operate. GMI supports the requirement to quickly respond to differing crisis situations with corresponding intelligence spanning the globe. One of the many places to get information for GMI is the medical intelligence database. The G2/S2 planner develops his initial IPB from GMI products.

2-22. The G2/S2 develops and maintains the unit's GMI database on potential threat forces and environments based on the commander's guidance. As an essential component of intelligence readiness, this database supports the unit's planning, preparation, execution, and assessment of operations. The G2/S2 applies and updates the database as it executes its intelligence production tasks.

TARGET INTELLIGENCE

2-23. Target intelligence is the analysis of enemy units, dispositions, facilities, and systems to identify and nominate specific assets or vulnerabilities for attack, re-attack, or exploitation (for intelligence). It consists of two mutually supporting production tasks: target development and combat assessment.

- **Target development** is the systematic evaluation and analysis of target systems, system components, and component elements to determine HVTs for potential attack through maneuver, fires, or non-lethal means.
- Once attacked, **combat assessment** provides a timely and accurate estimate of the affects of the application of military force (lethal or non-lethal) an IO on targets and target systems based on predetermined objectives.

SCIENTIFIC AND TECHNICAL INTELLIGENCE

2-24. Scientific and technical intelligence (S&TI) is the product resulting from the collection, evaluation, analysis, and interpretation of foreign S&T information which covers foreign developments in basic and applied research and in applied engineering techniques and S&T characteristics, capabilities, and limitations of all foreign military systems, weapons, weapon systems, and materiel, the research and development (R&D) related thereto, and the production methods employed for their manufacture.

2-25. S&T intelligence concerns foreign developments in basic and applied sciences and technologies with warfare potential. It includes characteristics, capabilities, vulnerabilities, and limitations of all weapon systems, subsystems, and associated materiel, as well as related R&D. S&T intelligence also addresses overall weapon systems and equipment effectiveness. Specialized organizations—such as the Defense Intelligence Agency (DIA) Missile and Space Intelligence Center (MSIC), INSCOM, Air Missile Defense (AMD), Area Air and Missile Defense Command (AAMDC), and National Ground Intelligence Center (NGIC)—produce this category of intelligence. The G2/S2 establishes instructions within SOPs, orders, and plans for handling and evacuating captured enemy materiel (CEM) for S&T intelligence (S&TI) exploitation.

COUNTERINTELLIGENCE

2-26. As previously defined at the end of Chapter 1, CI attempts to identify and recommend countermeasures to the threat posed by foreign intelligence security services (FISS) and the ISR activities of non-state entities such as organized crime, terrorist groups, and drug traffickers. CI analysis incorporates information from all sources as well as the results of CI collection, investigations, and operations to analyze the multidiscipline threat posed by foreign intelligence services and activities.

INTELLIGENCE COMMUNITY

2-27. There are many organizations in the intelligence community that support military operations by providing specific intelligence products and services. The J2/G2/S2 and his staff must be familiar with these organizations and the methods of obtaining information from them as necessary. Figure 2-2 shows organizations that compose the intelligence community.

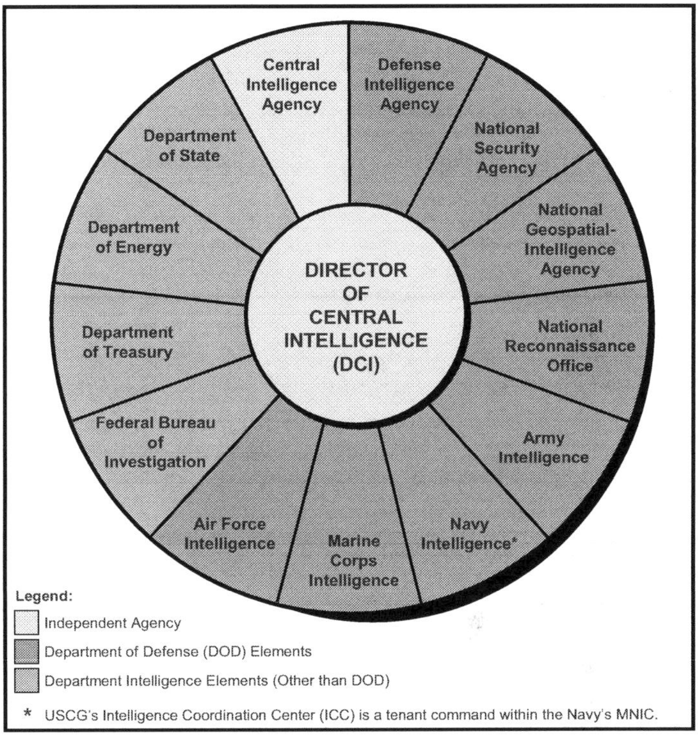

Figure 2-2. Intelligence Community Membership.

DOD AGENCIES

Defense Intelligence Agency (DIA)

2-28. DIA is a combat support agency and a major collector and producer in the defense intelligence community. The DIA supports the full spectrum of operations to include basic MI, counterterrorism, counterdrugs, medical intelligence, WMD and proliferation, United Nations (UN) peacekeeping and multinational support, missile and space intelligence, noncombatant evacuation operations (NEOs), targeting, combat assessment, and battle damage assessment (BDA).

National Security Agency (NSA)

2-29. NSA ensures cryptologic planning and support for joint operations. Working with the tactical cryptologic units of a command, the NSA provides SIGINT and information security (INFOSEC), encompassing communications security (COMSEC) and computer security, as well as telecommunications support and operations security (OPSEC). The people and equipment providing SIGINT, INFOSEC, and OPSEC constitute the United States Cryptologic System (USCS). The NSA, through the USCS, fulfills cryptologic command and/or management, readiness, and operational responsibilities in support of military operations according to the Secretary of Defense tasking, priorities, and standards of timeliness.

National Geospatial-Intelligence Agency (NGA)

2-30. The NGA (formerly National Imagery and Mapping Agency [(NIMA)] mission is to provide timely, relevant, and accurate intelligence and geospatial information in support of national security objectives of the United States. The Director of NGA advises the Secretary of Defense, Director Central Intelligence (DCI), Chairman of the Joint Chiefs of Staff (JCS), and the combatant commanders on imagery, IMINT, and geospatial information. The Operations Directorate, Customer Support Office, is the focal point for interface with external customers, including the JCS, combatant commands, services, and national and defense agencies.

National Reconnaissance Office (NRO)

2-31. The NRO develops and operates unique and innovative space reconnaissance systems and conducts intelligence-related activities essential for US security. The role of the NRO is to enhance US government and military information superiority across full spectrum operations. NRO responsibilities include supporting I&W, monitoring arms control agreements, and performing crisis support to the planning and conduct of military operations. The NRO accomplishes its mission by building and operating IMINT and SIGINT reconnaissance satellites and associated communications systems

US Navy (USN)

2-32. Naval intelligence products and services support the operating naval forces, the Department of the Navy, and the maritime intelligence requirements of national level agencies. Naval intelligence operates the National Maritime Intelligence Center (NMIC). Naval intelligence responsibilities include maritime intelligence on global merchant affairs, counter-narcotics, fishing issues, ocean dumping of radioactive waste, technology transfer, counter-proliferation, cryptologic related functions, CI, I&W support, management of Coast Guard collection, and development of new weapons systems and countermeasures.

US Marine Corps (USMC)

2-33. USMC intelligence provides pre-deployment training and force contingency planning for requirements that are not satisfied by theater, other service, or national capabilities. The Marine Corps Intelligence Agency (MCIA) handles the integration, development, and application of GMI, technical information, all-source production, and open-source information.

US Air Force (USAF)

2-34. USAF ISR assets fill a variety of roles to meet US national security requirements. The USAF operates worldwide ground sites and an array of airborne ISR platforms to meet national level intelligence requirements. To support day-to-day USAF operations and to meet specific USAF requirements, intelligence professionals at the wing and squadron levels use suites of interoperable analysis tools and dissemination systems to tailor information received from all levels and agencies in the intelligence community. USAF responsibilities include all-source information on aerospace systems and potential adversaries' capabilities and intentions, cryptologic operations, I&W, IO, and criminal investigative and CI services.

NON-DOD AGENCIES

2-35. Although the primary focus of non-DOD members of the intelligence community is strategic intelligence and support to the President and the Secretary of Defense, these agencies also produce intelligence and intelligence products that support operational and tactical ARFOR. This responsibility includes assessing potential issues and situations that could impact US national security interests and objectives. These agencies identify global and regional issues and threats. Some of the intelligence products and services these agencies provide are essential to accurate assessment of the threat and battlespace, particularly during stability operations and support operations.

Central Intelligence Agency (CIA)

2-36. The CIA's primary areas of expertise are in HUMINT collection, imagery, all-source analysis, and the production of political and economic intelligence. CIA and military personnel staff the CIA's Office of Military Affairs (OMA). As the CIA's single POC for military support, OMA negotiates, coordinates, manages, and monitors all aspects of agency support for military operations. This support is a continuous process, which the agency enhances or modifies to respond to a crisis or developing operation. Interaction between OMA and the DCI representatives to the Office of the Secretary of Defense (OSD), the Joint Staff, and the combatant commands facilitates providing national level intelligence in support of joint operations, contingency and operations planning, and exercises.

Department of State, Bureau of Intelligence and Research

2-37. The Bureau of Intelligence and Research coordinates programs for intelligence, analysis, and research; it produces intelligence studies and current intelligence analyses essential to foreign policy determination and execution. Its subordinate Bureau of International Narcotics Matters develops, coordinates, and implements international narcotics control assistance activities. It is the principal POC and provides policy advice on international narcotics control matters for the Office of Management and Budget (OMB), the NSC, and the White House Office of National Drug Control Policy (ONDCP). The Bureau also oversees and coordinates the international narcotics control policies, programs, and activities of US agencies.

Department of Energy, Office of Nonproliferation and National Security

2-38. The Department of Energy's Office of Nonproliferation and National Security assists in the development of the State Department's policy, plans, and procedures relating to arms control, nonproliferation, export controls, and safeguard activities. Additionally, this office is responsible for—

- Managing the department's R&D program.
- Verifying and monitoring arms implementation and compliance activities.
- Providing threat assessments and support to headquarters and field offices.

Department of the Treasury

2-39. The US Treasury Department's intelligence-related missions include producing and disseminating foreign intelligence relating to US economic policy and participating with the Department of State in the overt collection of general foreign economic information.

Federal Bureau of Investigation (FBI)

2-40. The FBI is the principal investigative arm of the Department of Justice (DOJ) and has primary responsibility for CI and counterterrorism operations conducted in the US. CI operations contemplated by any other organizations in the US must be coordinated with the FBI. Any overseas CI operation conducted by the FBI must be coordinated with the CIA.

Department of Homeland Security

2-41. The mission of the Department of Homeland Security is to develop and coordinate the implementation of a comprehensive national strategy to secure the US from terrorist threats or attacks. The organizational construct of homeland security mission is a framework of prepare, deter, preempt, defend, and respond. Component agencies will analyze threats and intelligence, guard our borders and airports, protect our critical infrastructure, and coordinate the response of our nation for future emergencies. The component agencies of the department include—

- US Coast Guard (USCG).
- US Customs Service.
- US Border Patrol.
- Federal Emergency Management Agency (FEMA).
- Immigration and Naturalization Service (INS).
- Transportation Security Administration.
- Federal Protective Service.
- Office of Domestic Preparedness.

United States Coast Guard (USCG)

2-42. The USCG, subordinate to the Department of Homeland Security, has unique missions and responsibilities as both an armed force and a law enforcement agency (LEA), which makes it a significant player in several national security issues. The USCG intelligence program supports counter-drug operations, mass seaborne migration operations, alien migration interdiction operations, living marine resource enforcement, maritime intercept operations, port status and/or safety, counterterrorism, coastal and harbor defense operations, and marine safety and/or environmental protection.

Other Agencies

2-43. There are a number of US Government agencies and organizations, not members of the intelligence community, that are responsible for collecting and maintaining information and statistics related to foreign governments and international affairs. Organizations such as the Library of Congress, the Departments of Agriculture and Commerce, the National Technical Information Center, US Information Agency, US Information Service, and the US Patent Office are potential sources of specialized information on political, economic, and military-related topics. The intelligence community may draw on these organizations to support and enhance research and analysis and to provide relevant information and intelligence for planners and decisionmakers. Many other US Government agencies may become directly involved in supporting DOD especially during stability operations and support operations. (See JP 2-02 for a description of agency support to joint operations and intelligence.) These organizations include—

- Department of Transportation.
- Disaster Assistance Response Team within the Office of Foreign Disaster.
- US Agency for International Development.
- NGA.

UNIFIED ACTION INTELLIGENCE OPERATIONS

2-44. In unified action, ARFOR synchronize their actions with those of other participants to achieve unity of effort and to accomplish the combatant commander's objectives. Unified action links subordinates to the combatant commander under combatant command (command authority) (COCOM). Multinational, interagency, and nonmilitary forces work with the combatant commander through cooperation and coordination. Combatant commanders form theater strategies and campaigns, organize joint forces, designate operational areas, and provide strategic guidance and operational focus to subordinates. The aim is to achieve unity of effort among many diverse agencies in a complex environment. Subordinate JFCs synchronize joint operations in time and space, direct the action of foreign military forces (multinational operations), and coordinate with governmental and NGOs (interagency coordination) to achieve the same goal.

2-45. The J2 staff provides intelligence promptly, in an appropriate form, and by any suitable means to those who need it. Intelligence personnel ensure that the consumers understand the intelligence and assist them as they apply the intelligence to their operations.

2-46. Dissemination requires establishing appropriate communications systems and procedures. The J2 and other intelligence personnel must fully participate in all operation planning and execution, and develop close working relationships with the JFC and other staff elements.

2-47. The commander and staff assess intelligence operations to determine their effectiveness and to make any necessary improvements. The intelligence process functions presented in Chapter 4 provide the criteria for evaluating intelligence operations.

JOINT INTELLIGENCE OPERATIONS

2-48. The JTF commanders and his intelligence staff must—

- Understand the intelligence requirements of superior, subordinate, and component commands.
- Identify organic intelligence capabilities and shortfalls.
- Access theater and national systems to ensure appropriate intelligence and CI products are available to the JTF.

2-49. The JTF's intelligence effort focuses on integrating multi-source information and multi-echelon intelligence into all-source intelligence products that provide clear, relevant, and timely knowledge of the enemy and battlespace. These products must be in formats that are readily understood and directly usable by the recipient in a timely manner. They must neither overload the user nor the communications architecture.

2-50. The J2 is the JTF commander's focal point for intelligence. The J2 directly supports the JFC's responsibilities for determining objectives, directing operations, and evaluating the effects of those operations. The J2 supports the execution of the plan with the intelligence needed to sustain the operation, attain joint force objectives, provide support to subordinate commands, and continually support FP efforts. The J2 analyzes the potential threat situation and provides assessments to support friendly opportunities. The J2 then supports the execution of the plan with the operational intelligence needed to sustain the operations, attain joint force objectives, and support FP. To maintain the initiative, the JFC will seek to get inside the adversary's decisionmaking cycle; that is, the JFC will seek to develop procedures and an organization in order to receive new and accurate intelligence and respond to the new situation faster than the adversary. The J2 must help in identifying the adversary's decisionmaking cycle and identifying vulnerabilities that may be exploited. The J2 also—

- Ensures the provision of the required ISR support to the JTF and its subordinate functional and service components.
- Assists the JTF commander in defining intelligence responsibilities and PIRs.
- Actively participates in joint staff planning and in planning, preparing, executing, and assessing ISR efforts. This includes leading the JTF's joint collection board and providing representation at the joint targeting board.

2-51. Figure 2-3 shows a typical JTF J2 organization. The overall organization of the JTF and operations will dictate actual composition of the J2. At a minimum, a core element of analytical, ISR management, and administrative capabilities is required.

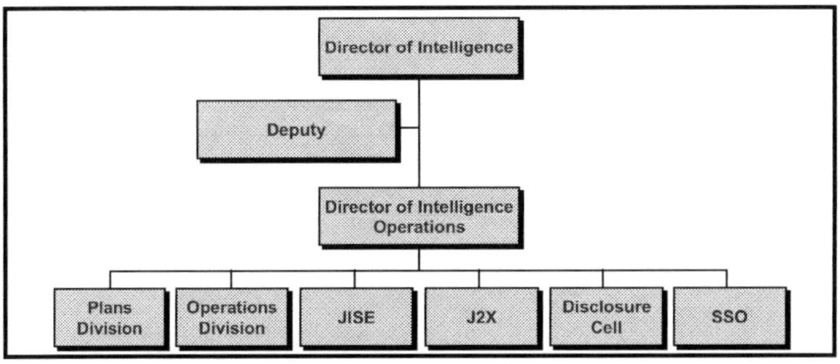

Figure 2-3. Typical Joint Task Force J2 Organization.

Considerations In Joint Intelligence Operations

2-52. The Army, Air Force, Navy, Marine Corps, and SOF must provide globally responsive assets to support the combatant commanders' theater strategies and the national security strategy. The capabilities of the other armed forces complement those of ARFOR. During joint operations, they provide support consistent with JFC-directed missions. When conducting joint intelligence operations, there are a number of unique problems that can arise due to the complexity of integrating the efforts of the different services and commands. Elements affecting joint intelligence operations among the different services include the following:

- Intelligence liaison is critical to the success of intelligence operations and requires early establishment, particularly between units that have not routinely trained together and possess differing capabilities. As a minimum, organizations exchange liaison teams with the higher echelon organization. Additional liaison may be necessary to facilitate joint force collection, production, or dissemination requirements. Liaison teams—
  - Support planning and C2 of intelligence operations.
  - Ensure timely two-way flow of intelligence between commands.
  - Manage intelligence and resource requirements of the subordinate command.
  - Advise the commander on service ISR capabilities, limitations, and employment.
- Commanders and staffs use IPB to understand the battlespace and develop or refine plans and orders. IPB products exchanged between echelons ensure a common picture of the battlespace and estimate of the situation.
- Communications considerations for joint operations include—
  - Planning for intelligence communications transition to facilitate execution of branches or sequels to the plan or to accommodate shifting of the main effort from one force to another.

    - Identifying the initial communications architecture to include establishing procedures and protocols for information exchanges (databases, text, imagery, voice, and video).
    - Balancing the availability of service-unique intelligence systems between echelons or services. This may require each service providing additional resources. The senior commander is responsible for allocating resources.
    - Disseminating intelligence between commands and services. Additional communications equipment, intelligence terminals, and personnel may be required to balance capabilities between services.
- Identifying the databases each service possesses or has access to; determining which databases will support the operation and, if necessary, merge them into a single database; and ensure access by the entire force.
- Providing a focal point for subordinate command access to national or joint intelligence is essential. The senior commander will request and allocate resources required to support this access.

2-53. The JFC's intelligence requirements, concept of operation, and intent drive the ISR effort. The different organizations and services participating in joint intelligence operations must continuously share information, intelligence, and products to satisfy requirements. See FM 2-01 for details on intelligence requirements and RM. Activities required to facilitate an effecttive joint collection effort include—

- Coordinating intelligence and ISR operations to optimize capabilities of collection assets and reduce duplication of effort.
- Integrating supporting national and theater intelligence collection assets into the intelligence synchronization plan.
- Establishing procedures for tracking and handing off HPTs between services and echelons.
- Establishing procedures for cueing Army and other service collection assets.
- Coordinating airspace for ISR assets.
- Maximizing available linguist capabilities. Shortages of military linguists trained in target languages may require cross-leveling Army and other service linguists.

2-54. Reporting and intelligence production considerations in joint operations include—

- Establishing production criteria and thresholds that produce timely and relevant intelligence keyed to the commander's intelligence and targeting requirements. The ISM, attack guidance matrix (AGM), and HPT list are examples of tools used to support joint intelligence and targeting efforts. The sharing of these products to all echelons and services is crucial.
- Establishing common methodology and criteria for producing the BDA and supporting the combat assessment function.
- Establishing set standards for the number and frequency of periodic reports such as intelligence summaries (INTSUMs).

Joint Intelligence Architecture

2-55. In addition to the J2 staffs at every joint level of command, the key organizations in the joint intelligence architecture are the National Military Joint Intelligence Center (NMJIC) (see Figure 2-4), the Joint Intelligence Centers (JICs) or Joint Analysis Centers (JACs) of the unified commands, and, when formed, the JTF's joint intelligence support element (JISE). Working together, these organizations play the primary role in managing and controlling joint intelligence operations. The formal relationships which link these organizations facilitate information management and optimize complementary intelligence functions by echelon without obstructing the timely flow of intelligence up, down, or laterally.

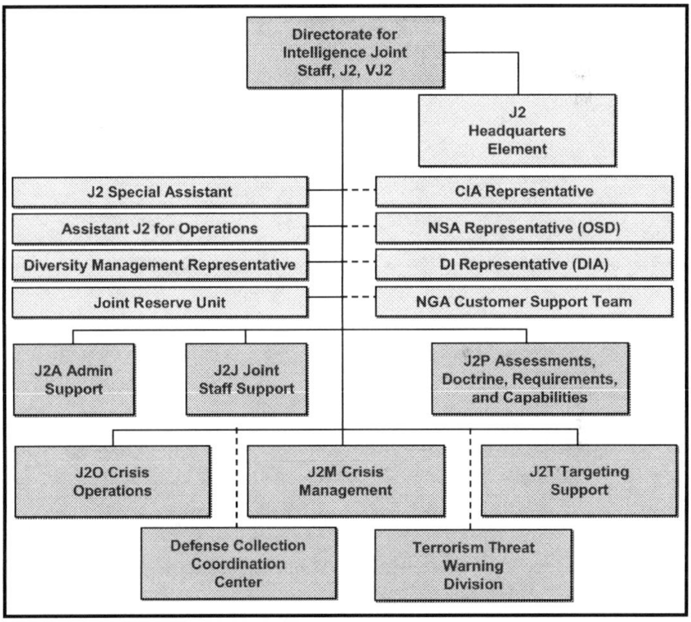

Figure 2-4. National Military Joint Intelligence Center.

2-56. The NMJIC is the focal point for intelligence activities in support of joint operations. The NMJIC is the CJCS J2 intelligence watch within the National Military Command Center (NMCC) in the Pentagon. When the assigned or attached assets cannot satisfy the combatant commander and the JTF commander's crisis-related and time-sensitive intelligence requirements, the NMJIC, as part of the J2's staff, sends tasks to appropriate agencies to fill the requirements. During crises, the NMJIC expands as necessary to establish a working group, intelligence task force, or an expanded intelligence task force. The NMJIC participates in targeting by developing national level target lists. When requested, the NMJIC supports the theater in performing BDA. Besides supporting the combatant commands and JTFs, the NMJIC supports any multinational partners and prescribed international organizations.

2-57. The JIC is organized in accordance with the combatant commander's prerogatives, but normally performs the general functions described in JP 2-0 and specific unified command intelligence TTP. The JIC is responsible for providing intelligence to JTF and its subordinates during military operations. If the JIC cannot meet the combatant commander's requirements, the JIC forwards requests to the NMJIC or to subordinate command levels through established channels, using standard command procedures. In some cases, the JIC may also ensure timely support by approving a direct communication path, in advance, between requesters such as the JTF JISE and outside producers. In this case, the parties must inform the JIC of all requests as they occur. This method is most appropriate when the parties require products and services, which the JIC does not routinely produce.

2-58. At the discretion of the JTF commander, the J2 can establish a JISE during the crisis or preparation stage for operations to augment the J2 staff. The JISE is a tailored subset of the JIC or the intelligence organization of the service component designated as the JTF headquarters.

2-59. Under the direction of the J2, the JISE normally manages the intelligence collection, production, and dissemination for the JTF. The JISE provides intelligence to JTF operational forces and performs common intelligence functions. By design, the JISE is scaleable and can expand to meet the needs of the JTF and the operating environment. It is composed of analytical experts, analysis teams, and ISR managers that provide services and products, which the JTF, JTF staff, and subordinate components require. These experts, mentioned above, focus on solving the operational intelligence problems of concern to the JTF commander. The JISE's capability to perform all-source analysis and ISR synchronization is key to producing operational intelligence that is timely, relevant, and complete. Figure 2-5 illustrates the features of a typical JISE.

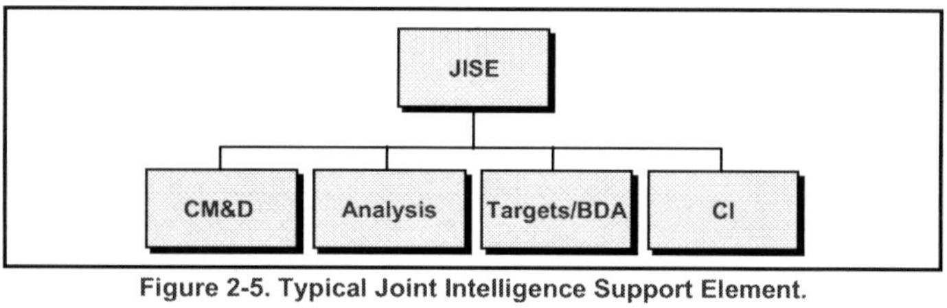

Figure 2-5. Typical Joint Intelligence Support Element.

Joint Task Force Intelligence Organizations

2-60. In addition to the JISE, the JTF commander and J2 may require other supporting JICs or teams based on projected operations. The JTF commander may make a request to the NMJIC for specific national intelligence agency capabilities. The NMJIC evaluates and coordinates the JTF commander's requirements with the J3, J5, and national intelligence agencies and tailors the composition of the deployment packages to meet those needs. The deployment packages, such as the National Intelligence Support Team (NIST), provide access to the entire range of capabilities resident in the national intelligence agencies and can focus those capabilities on the JTF commander's intelligence requirements. The J2X manages and coordinates the HUMINT and CI activities of national, theater, and service components operating within the JTF's JOA. The Joint Captured Materiel Exploitation Center (JCMEC) assists in management of recovery, exploitation, and disposal of captured enemy equipment (CEE). The JTF commander's requirements dictate the composition and tailoring of such deployment packages.

Augmentation Considerations

2-61. Depending on the scale of the operations, the intelligence organizations described above and those of the JTF's subordinate command may require personnel augmentation. Optimum use of available intelligence assets is essential to ensure quality support in meeting the JTF commander's requirements.

2-62. The demand for additional intelligence increases significantly during crisis and wartime operations. As the need arises for more intelligence personnel, the intelligence presence also increases at all command levels. Locating additional intelligence personnel and knowing how to integrate those personnel into the operation is vital. The JTF J2 should identify intelligence personnel augmentation requirements in accordance with the Chairman, Joint Chiefs of Staff (CJCS) Instruction 1301.01. The combatant commander's joint table of mobilization and distribution (JTMD) also contains the standing augmentation requirements. The JTMD should reflect the need for either IMA or Individual Ready Reserve (IRR) personnel. The combatant commander and the JTF refine personnel requirements and initiate requests when they anticipate or start an operation.

2-63. A consideration for the JTF when requesting support or augmentation is that these national level teams and individual augmentees are not totally self-contained elements; rather they require logistic, information, and other support from the supported command. Each deployment is unique based on mission, duration, team composition, and capabilities required. A full NIST, for example, requires a private, access-controlled area within a sensitive compartmented information facility (SCIF) work environment and dedicated secure communications.

2-64. For more information on intelligence operations as they apply to other armed services, see the individual service intelligence doctrine. See also JP 2-0 series, JP 3-0, JP 3-55, and JP 5-0 for more details on joint intelligence operations and considerations.

Multinational Intelligence

2-65. Multinational intelligence operations take place within the structure of an alliance or coalition. Some multinational military organizations, such as the North Atlantic Treaty Organization (NATO) and the UN Command in the Republic of Korea (ROK), are highly structured and enduring. Others, such as the coalition formed during the Gulf War, are less formal and temporary.

2-66. In multinational operations, the multinational force commander (MNFC) exercises command authority over a military force composed of elements from two or more nations. Therefore, in most multinational operations, the JTF must share intelligence, as necessary, for mission accomplishment with foreign military forces and coordinate exchange of intelligence with those forces.

2-67. In some circumstances, the JTF may need to seek authority to go outside the usual political-military channels to provide information to NGOs. The JTF must tailor intelligence policy and dissemination criteria to each multinational operation.

INTELLIGENCE REACH

2-68. Military forces use intelligence reach to rapidly access information, receive support, and conduct collaboration and information sharing with other units and organizations unconstrained by geographic proximity, echelon, or command.

2-69. Intelligence reach supports minimizing the deployed footprint of ISR assets. By providing enhanced information and tailored intelligence products, intelligence reach can greatly enhance the intelligence capabilities of the unit and play a significant role in improving the commander's ability to make decisions on the battlefield. It allows the commander to harness national, joint, foreign, and other military organization resources. Table 2-1 shows examples of partners and sources for intelligence reach.

2-70. Detailed planning and training are critical to the success of intelligence reach operations. The following are steps that the staff can take to ensure optimal use, operability, and effectiveness of intelligence reach:

- Establish data exchange methods and procedures.
- Establish electronic message transfer procedures.
- Establish homepages for identified forces.
- Establish POCs for I&W centers, production centers, Theater JICs, DIA, INSCOM, and their major subordinate commands such as NGIC and the theater intelligence brigades and groups.
- Ensure the intelligence staff has the necessary personnel, training, automated systems, bandwidth, and resources to conduct intelligence reach.
- Determine IRs through staff planning. Develop production requirements for identified intelligence gaps.
- Order geospatial products for the projected joint AOI.
- Establish and maintain a comprehensive directory of intelligence reach resources before deployment and throughout operations. The value of intelligence reach will greatly increase as the staff develops and maintains ready access to rich information resources. These resources are numerous and may include, for example, Army, Joint, DOD, non-DOD, national, commercial, foreign, and university research programs. Know what types of information the resources can provide. Continuously expand the resource directory through identification of new resources.

- Use intelligence reach first to fill intelligence gaps and requirements and answer RFIs. This technique can preclude unnecessary tasking or risk to limited ISR assets.
- Maintain continuous situational understanding and anticipate intelligence requirements. Use intelligence reach to fulfill these requirements and provide the results to the commander and staff for planning and decisionmaking.
- Exchange intelligence reach strategies with other units.
- Present the information retrieved through intelligence reach in a usable form. Share the information derived from intelligence reach with subordinate, lateral, and higher echelons. Ensure follow-on forces have all information as well.

Table 2-1. Examples of Intelligence and Reach Partners and Sources.

ARMY	SERVICES	JOINT	DOD
ACE ISE TSE (TIB/TIG) 902d NGIC ATCAE ARISCs	ONI NMIC AIA NAIC MCIA	USEUCOM JAC USSOUTHCOM JIC USSOCOM JIC USSPACECOM CIC USJFCOM AIC USCENTCOM JIC USTRANSCOM JIC USSTRATCOM USPACOM JIC USNORTHCOM JIC	DIA CMO MSIC AFMIC DHS NGA NSA RSOCs
NON-DOD	**NATIONAL**	**COMMERICAL**	**FOREIGN**
DOE FBI DOS DEA FEMA INS	DCI NIC CIA NRO INR DOT, Office of Intel Support	RAND Jane's Defence Weekly Economic Intelligence Unit CNN Reuters Associated Press United Press International	DIS NFHQ DIO

INTELLIGENCE REACH COMPONENTS

2-71. Intelligence reach requires the G2/S2 to develop a strategy on how best to support the unit's mission with intelligence reach capabilities. There are six basic components of the strategy:

- Push
- Database Access.
- Pull.
- Broadcast Services.
- Collaborative Tools.
- Requirements Management.

Push

2-72. Push occurs when the producers of intelligence or information are knowledgeable of the customer's requirements and are able to send the desired intelligence to the customer without further requests. Push is accomplished through the most appropriate or efficient dissemination means. Unless coordinated as an acceptable dissemination method, push is not accomplished solely by posting intelligence products on a web page. The entity that posts the document must ensure the intended recipient is aware of the product's location or has received the product.

2-73. Push begins with the statement of intelligence interest (SII). The SII establishes the unit's intelligence and IRs. The SII is prepared by intelligence staff organizations to register their interest in receiving recurring hardcopy and softcopy reports, studies, and publications covering a wide variety of intelligence subjects. The J2/G2/S2 works with the Department of the Army Production and Dissemination Management (DAPDM) to establish a profile for information transfer. Intelligence reports or information meeting the unit's requirements are then automatically sent directly to the unit's classified or unclassified networks or communication systems.

2-74. Organizations that push information and data down to subordinate units must be careful of the amount, detail, and focus of the information they are sending. Too much information will overwhelm the subordinate unit. Crucial information may be lost in the midst of an overabundant flow of information, much of which is of little use. Intelligence reach often runs more efficiently if the user pulls the information or at least focuses the intelligence producers to send the appropriate information with the correct level of detail at the appropriate time.

Database Access

2-75. Access to local, theater, DOD, non-DOD, and commercial databases allows analysts to leverage stored knowledge on topics ranging from basic demographics to OB information. A validated DIA Customer Number (acquired by the J2/G2/S2) in combination with Secret Internet Protocol Network (SIPRNET) and Joint Worldwide Intelligence Communications System (JWICS) connectivity establishes access to most of the databases online. The challenge for an analyst is to gain an understanding of the structure, contents, strengths, and weaknesses of the database regardless of database type. Additionally, the procedures are often difficult for extracting portions or downloading and transferring the database to unit level automated information systems.

2-76. Each intelligence discipline has unique databases established and maintained by a variety of agencies. Database access is accomplished through unit or agency homepages via SIPRNET (Intelink-S) and JWICS (Intelink). The DAPDM office, upon approving the unit's SII, validates the requirement for access to the majority of these databases. Units coordinate with the DA dissemination and program manager (DPM) and/or the agency for access to those databases requiring passwords and permissions beyond normal Intelink-S and Intelink access.

Pull

2-77. Pull occurs when the requestor is familiar enough with existing databases and products to be able to anticipate the location of the desired information. Knowledge of both the types and locations of intelligence databases can greatly increase the efficiency of the Intelligence BOS by saving time and effort on the part of analysts at every echelon.

2-78. The G2/S2 must also ensure that intelligence reach capability extends to multinational forces, augmentees, and other services or organizations working as part of the JTF. The G2/S2 must forward relevant intelligence to units which do not possess the necessary automation to conduct intelligence reach.

Broadcast Services

2-79. Broadcast services are an integrated, interactive dissemination system, focusing on tactical user's IRs using joint message formats. The theater broadcast's data prioritization and processing occur through information management. Selected tactical, theater, DOD, and national collectors broadcast their messages using prescribed message formats. The analyst decides which types of messages are required. Broadcast service data is responsive to the operational commander's information needs and allows tactical users to construct successively detailed pictures of the battlespace.

2-80. Tactical users require intelligence and information to be available at the lowest classification level possible (for example, collateral SECRET) and releasable, as necessary, to multinational partners. Broadcast service information management will adhere to national policy as it relates to the provision of data to NATO or multinational partners.

Collaborative Tools

2-81. Collaboration is the sharing of knowledge, expertise, and information, normally online, and may take numerous forms. Collaborative tools are computer-based tools (groupware) that help individuals work together and share information. They allow for virtual online meetings and data sharing. As an example of the use of collaborative tools, the President or the Secretary of Defense and DOD during a crisis situation will establish a number of crisis action teams (CATs) or joint interagency working groups (JIAWGs). These groups or teams are formed to address specified subjects or topics in support of the warfighter or decisionmaker. The groups will focus on the crisis and normally publish their products on a homepage. Analysts with online access can participate in a number of ways. This includes passively accessing the homepages to study the products, sending queries to the identified POCs, or by having their organization join the CAT or JIAWG. Table 2-2 shows some examples of collaborative tools.

Requirements Management

2-82. The RM system provides a mechanism for submitting RFIs, tasking, and managing ISR assets. Analysts who are trained and familiar with the RM process and the various tasking procedures can leverage its systems for refined information. Each intelligence discipline has established procedures for requesting specific information. It is therefore advantageous to have someone familiar with each discipline's procedures to participate in synchronizing the ISR effort. For example, an analyst receives a HUMINT-based intelligence information report, which provides information on an event or subject. If he requires more refined data or clarification, he should submit a time-sensitive requirement or a source-directed requirement. This creates a two-way communication from the field to the collectors without creating additional standing requirements. The more familiar the analyst is with the RM process, the better he can leverage it.

2-83. The intelligence reach component of RM includes the ability of an intelligence officer at any level to request information, which is beyond what is available at his location. The normal procedure for obtaining intelligence or information not obtainable through the use of available ISR assets is to submit an RFI to the next higher echelon. Users enter RFIs into the RM system. See ST 2-33.5 for more information on intelligence reach operations and FM 2-01 for more information on RM.

Table 2-2. Examples of Collaborative Tools.

Tool	Description
Chat (Audio and Text)	Used to conduct conversations online.
Whiteboard	Permits real-time display of drawings, pictures, or documents for group discussion and comment. Participants can annotate in real time as well.
Bulletin Board	Used to post notices and facilitate discussions on any topic.
Video	Used at a desktop computer or a video teleconferencing (VTC) center to allow the person or group to see with whom they are communicating.
Discussion Groups (Newsgroups)	Topics are posted to a website for discussion and comment where participants can follow a line of discussion on a topic.
File-Sharing Tools	Virtual file cabinets allow information to be stored on web servers, and are available to anyone having access to the site and electronic permission to use the files.
Presentation Tools	Used in a virtual auditorium to allow lectures and briefings to be given to an audience.
Application Sharing	An entire team can use an application running on one computer to revise documents.
Text Tools	Allows live text input and editing by group members. Once complete, the text document can be copied into word processing software.
E-mail	Electronic mail.
Persistent Capability	The ability to preserve files, briefings, or other team or project material for future reference. Properly organized, it becomes an information management device and is invaluable to a long-term effort.
Instant Messaging	This allows real-time exchange of notes and messages.

PART TWO

Intelligence in Full Spectrum Operations

Part Two provides a primer of what and who constitutes both MI and the Intelligence BOS and the process that these entities use in order to provide warfighters and decisionmakers with the intelligence they require in order to accomplish their respective missions.

Chapter 3 discusses the role of MI and the Intelligence BOS within full spectrum operations. It provides an overview of intelligence readiness, particularly the intelligence requirements associated with force projection. The concept of MI asset technical control is discussed as a complement to, not a replacement of, the Army's command and support relationships.

Chapter 4 presents the Intelligence Process—the Intelligence BOS methodology that accomplishes the primary focus of intelligence in full spectrum operations, which provides the warfighter with effective intelligence.

Chapter 3

Fundamentals in Full Spectrum Operations

FULL SPECTRUM OPERATIONS

3-1. Army commanders at all echelons may combine different types of operations simultaneously and sequentially to accomplish missions. Full spectrum operations include offensive, defensive, stability, and support operations. Missions in any environment require ARFOR that are prepared to conduct any combination of these operations:

- **Offensive operations** aim to destroy or defeat an enemy. Their purpose is to impose US will on the enemy and achieve decisive victory.
- **Defensive operations** defeat an enemy attack, buy time, economize forces, or develop conditions favorable for offensive operations. Defensive operations alone normally cannot achieve a decision. Their purpose is to create conditions for a counteroffensive that allow ARFOR to regain the initiative.
- **Stability operations** promote and protect US national interests by influencing the threat, political, and information dimensions of the OE through a combination of peacetime developmental, cooperative activities, and coercive actions in response to crises. Regional security is supported by a balanced approach that enhances regional stability

and economic prosperity simultaneously. ARFOR presence promotes a stable environment.

- **Support operations** employ ARFOR to assist civil authorities, foreign or domestic, as they prepare for or respond to crises and relieve suffering. Domestically, ARFOR respond only when the President or Secretary of Defense direct. ARFOR operate under the lead federal agency and comply with provisions of US law, to include the Posse Comitatus and Stafford Acts. (See FM 3-0.)

3-2. Intelligence supports the commander across full spectrum operations. It helps the commander decide when and where to concentrate sufficient combat power to overwhelm the enemy. ISR is essential for the commander to preclude surprise from the enemy, maintain the initiative on the battlefield, and win battles. Commanders at all levels synchronize intelligence with the other BOSs to maximize their ability to see and strike the enemy simultaneously throughout the AO.

3-3. Every soldier in the command is responsible for detecting and reporting enemy activities, dispositions, and capabilities. This task is critical because the environment we operate in is characterized by violence, uncertainty, complexity, and asymmetric methods by the threat. The increased situational awareness that soldiers develop through personal contact and observation is a critical element of that unit's ability to more fully understand the OE. However, soldiers collect information, they are not intelligence collectors. While medical personnel cannot be assigned ISR tasks due to their Geneva Convention category status, medical personnel who gain information through casual observation of activities in plain view while discharging their humanitarian duties will report the information to their supporting intelligence element.

OFFENSIVE OPERATIONS

3-4. Offensive operations at all levels require effective intelligence to help the commander avoid the enemy's main strength and to deceive and surprise the enemy. During offensive operations, intelligence must provide the commander with the composition, disposition, limitations, employment characteristics, and anticipated enemy actions in a timely enough manner for the commander to significantly affect the enemy commander's decision cycle. It ensures commanders have the intelligence they need to conduct offensive operations with minimum risk of surprise.

DEFENSIVE OPERATIONS

3-5. The immediate purpose of defensive operations is to defeat an enemy attack. Commanders defend to buy time, hold key terrain, hold the enemy in one place while attacking in another, or destroy enemy combat power while reinforcing friendly forces. Intelligence should determine the enemy's strength, COAs, and location of enemy follow-on forces. The defending commander can then decide where to arrange his forces in an economy-of-force role to defend and shape the battlefield. Intelligence support affords him the time necessary to commit the striking force precisely.

3-6. Intelligence in area defensive operations identifies, locates, and tracks the enemy's main attack and provides the commander time to allocate sufficient combat power to strengthen the defense at the point of the enemy's main effort. Intelligence should also identify where and when the commander can most decisively counterattack the enemy's main effort or exploit enemy vulnerabilities.

3-7. Although the battlefield is normally organized as decisive, shaping, and sustaining operations, commanders conducting defensive operations may delineate intelligence in battlefield organizational terms of deep area, close area, and rear area.

STABILITY OPERATIONS

3-8. The environment is often much more complex during stability operations and as a result intelligence is often more complex. In fact, intelligence is even more important a factor (or operational multiplier) during stability operations. As a result, the commander must be even more involved in and knowledgeable of ISR (to include ISR operations the commander controls and other higher level ISR operations that may be occurring within his AO) during stability operations. For example, the commanders must understand the complex details of HUMINT and special access program (SAP) operations.

3-9. The commander requires the appropriate intelligence and IPB products in order to determine how best to influence the threat, political and information dimensions of the operational environment, and enhance regional stability. The identification and analysis of characteristics of the terrain and weather, politics, infrastructure, health status, civilian press, attitudes, and culture of the local populace and all possible threats are important in conducting stability operations. A lack of knowledge concerning local politics, customs, and culture could lead to US actions which attack inappropriate targets or which may offend or cause mistrust among the local population.

SUPPORT OPERATIONS

3-10. Support operations are usually nonlinear and noncontiguous. Commanders designate the decisive, shaping, and sustaining operations necessary for mission success. However, determining the intelligence requirements that drive the ISR effort intending to identify any potential threat's centers of gravity and decisive points may require a more complex and unorthodox thought process than that used in offensive and defensive operations. The G2/S2 may have to define the enemy differently. In support operations, the adversary is often the effects of disease, hunger, or disaster on a civilian population. US forces conducting support operations must also fully understand the organization and identity of key figures or groups within the region where they are operating, as these figures may influence greatly the actions of the population—both civilian and military.

ELEMENTS OF COMBAT POWER

3-11. Combat power is the ability to fight. It is the total means of destructive or disruptive force, or both, that a military unit or formation can apply against the adversary at a given time. Commanders combine the elements of combat power to meet constantly changing requirements and defeat an enemy. The elements of combat power are—

- **Manuever** is the employment of forces, through movement combined with fire or fire potential, to achieve a position of advantage with respect to the enemy to accomplish the mission. Maneuver is the means by which commanders concentrate combat power to achieve surprise, shock, momentum, and dominance.
- **Firepower** provides the destructive force essential to overcoming the enemy's ability and will to fight. Firepower magnifies the effects of maneuver by destroying enemy forces and restricting his ability to counter friendly actions; maneuver creates the conditions for the effective use of firepower.
- **Leadership** is the most dynamic element of combat power. Confident, audacious, and competent leadership focuses on the other elements of combat power and serves as the catalyst that creates conditions for success. Leadership provides purpose, direction, and motivation in all operations.
- **Protection** is the preservation of fighting potential of a force so the commander can apply maximum force at the decisive time and place. Protection is neither timidity nor risk avoidance.
- **Information** enhances leadership and magnifies the effects of maneuver, firepower, and protection. Today, ARFOR use information collected to increase their situational understanding before engaging the enemy. Information from the COP and running estimates, transformed into situational understanding, allows commanders to combine the elements of combat power in new ways.

THE FOUNDATIONS OF ARMY OPERATIONS

3-12. Understanding the principles of war and tenets of Army operations is fundamental to operating successfully across the range of military operations. The principles of war and tenets of Army operations form the foundation of Army operational doctrine. (Refer to FM 3-0 for a full description of the Principles of War and Tenets of Army Operations.)

- **The Principles of War** provide general guidance for conducting war and military operations other than war at the strategic, operational, and tactical levels. The principles are the enduring bedrock of Army doctrine.
- **The Tenets of Army Operations** build upon the principles of war. They further describe the characteristics of successful operations. These tenets are essential to victory.

THE OPERATIONAL FRAMEWORK

3-13. The operational framework consists of the arrangement of friendly forces and resources in time, space, and purpose with respect to each other and the enemy or situation. It consists of the **AO**, **battlespace**, and the **battlefield organization**. The framework establishes an area of geographic and operational responsibility and provides a way for commanders to visualize how to employ forces against the enemy. Commanders design an operational framework to accomplish their mission by defining and arranging its three components. They use the framework to focus combat power. The operational framework provides a mechanism through which the commander can focus the Intelligence BOS effort. Understanding the operational framework, AOIR, and intelligence coordination line (ICL) and their relationship to each other is key to planning and executing ISR operations.

- AO is an operational area defined by the JFC for land and naval forces. AOs do not typically encompass the entire operational area of the JFC but should be large enough for component commanders to accomplish their missions and protect their forces. AOs should also allow component commanders to employ their available systems to the limits of their capabilities. The AO is the basic control measure for all operations that defines the geographical area for which a particular unit is responsible. The commander—
  - Assumes responsibility for intelligence, maneuver, fires, terrain management, security, and movement within his AO.
  - Establishes control measures within his AO to assign responsibilities, coordinate intelligence, fires, and maneuver, and to control other activities.
- **Battlespace** is the environment, factors, and conditions that must be understood to successfully apply combat power, protect the force, or complete the mission. This includes air, land, sea, space, and the included enemy and friendly forces; facilities; weather; terrain; the EMS; and the information environment within the operational areas and AOIs. The G2/S2 performs IPB and synchronizes ISR activities throughout the battlespace as determined by the commander's METT-TC considerations.
  - Area of influence is the geographical area in which a commander can directly influence operations by maneuver of FS systems normally under the commander's command or control. Areas of influence surround and include the associated AO.
  - AOI is that area of concern to the commander, including the area of influence, areas adjacent thereto, and extending into enemy territory to the objectives of current or planned operations. This area also includes areas occupied by enemy forces who could jeopardize the accomplishment of this mission. (See JP 1-02.)
  - Information environment refers to a commander's battlespace that encompasses information activity affecting the operation. To envision that part of the information environment that is within their battlespace, commanders determine the information activities that affect their own operational capabilities and opposing C2 and information systems.
  - Force Projection Bases are the intermediate staging bases and power projection platforms.
  - Home Stations are the permanent locations of active and RC units. To a significant degree, events occurring at home station affect the morale and performance of deployed forces. Thus, the commander's battlespace encompasses all home station functions.

- **Battlefield Organization** is the allocation of forces in the AO by purpose. It consists of three all-encompassing categories of operations:
  - Decisive Operations are those that directly accomplish the task assigned by higher headquarters. Decisive operations conclusively determine the outcome of major operations, battles, and engagements.
  - Shaping Operations are operations at any echelon that create and preserve conditions for the success of decisive operations.
  - Sustaining Operations are operations at any echelon that enable shaping and decisive operations by providing CSS, rear area and base security, movement control, terrain management, and infrastructure development.

3-14. AOIs extend into enemy territory, to the objectives of current or planned operations. They include areas occupied by enemy forces that could jeopardize the accomplishment of the mission. AOIs also serve to focus intelligence development and IO directed at factors outside the AO that may affect the operation. (FM 3-0, para 4-79) The scheme of support includes the coordination of reconnaissance and surveillance missions and AOIs with a joint force or higher headquarters and lateral units to answer the intelligence requirements within the AOI. The G2/S2 monitors enemy activities within the AOI and provides intelligence on enemy activities that may affect the unit.

3-15. The AOIR is an area allocated to a commander in which the commander is responsible for the provision of intelligence within the means at the commander's disposal. It is an area allocated to the commander in which the commander is responsible for the collection of information concerning the threat and the environment and the analysis of that information in order to produce intelligence. Higher headquarters also ensure through intelligence handovers, collection management, and deconfliction that problems with duplication, confliction, and command and control do not occur in the AOIR. (FM 2-19.402/FM 34-80-2) They include the available ISR assets, capability of the G2/S2 section, available intelligence architecture, and METT-TC considerations. The AOIR cannot extend beyond a unit's AOI; however, it can be smaller than its AO as well as vary (expand or contract) during an operation. An example of when a unit's AOIR is smaller than its AO is when a higher headquarters ISR effort covers an area within the unit's AO.

3-16. ICLs designate the boundary between AOIRs. The G2/S2 establishes ICLs to facilitate coordination between higher, lateral, and subordinate units; coordinates with the G3/S3 to direct subordinates to track enemy units and HPTs in their areas; and hands over intelligence responsibility for areas of the battlefield. The establishment of ICLs ensures that there are no gaps in the collection effort; that all echelons are aware of the location, mission, and capabilities of other assets. The G2/S2 keeps abreast of collection activities in progress (all echelons) and battlefield developments through the ICLs. ICLs help—

- Facilitate coordination between higher, lateral, and subordinate units.

- Ensure higher, lateral, and subordinate units share information and intelligence as enemy entities and HPTs move into, within, or transition between AOIRs.
- Specify intelligence responsibility for areas of the battlefield.

ARMY CAPABILITIES

3-17. Commanders combine AC and RC ARFOR—consisting of different types of units with varying degrees of modernization—with multinational forces and civilian agencies to achieve effective and efficient unified action. A broad range of organizations makes up the institutional Army that supports the field Army. Institutional Army organizations design, staff, train, and equip the force. The institutional Army assists in effectively integrating Army capabilities. It does this through leadership and guidance regarding force structure, doctrine, modernization, and budget. (See FM 100-11.)

TASK ORGANIZATION

3-18. The Army supports JFCs by providing tailored force packages to accomplish joint missions and dominate enemies and situations on land. Trained and equipped AC and RC units comprise these force packages. Within these force packages, Army commanders organize groups of units for specific missions. They reorganize for subsequent missions when necessary. This process of allocating available assets to subordinate commanders and establishing their command and support relationships is called task organizing. A temporary grouping of forces designed to accomplish a particular mission is a task organization. The ability of ARFOR to tailor (select forces based upon a mission) and task organize (temporarily organize units to accomplish a tactical mission) gives them extraordinary agility. It allows operational and tactical level commanders to organize their units to make the best use of available resources. The ability to task organize means ARFOR can shift rapidly among offensive, defensive, stability, and support operations.

COMBINED ARMS

3-19. The fundamental basis for the organization and operations of ARFOR is combined arms. Combined arms is the synchronized or simultaneous application of several arms—such as infantry, armor, field artillery, engineers, air defense, and aviation—to achieve an effect on the enemy that is greater than if each arm was used against the enemy separately or sequentially. The ultimate goal of Army organization for operations remains success in joint and combined arms warfare. Its combined arms capability allows commanders to form Army combat, combat support (CS), and CSS forces into cohesive teams focused on common goals.

COMMAND AND SUPPORT RELATIONSHIPS

3-20. Establishing clean command and support relationships is fundamental in organizing for all operations. These relationships can achieve clear responsibilities and authorities among subordinate and supporting units. The commander designates command and support relationships within his authority to weigh the decisive operation and support his scheme of maneuver. Some forces available to a commander are given command or support relationships that limit his authority to prescribe additional relationships. Command and support relationships carry with them varying responsibilities to the subordinate unit by parent and gaining units. By knowing the inherent responsibilities of each command and support relationship, a commander may organize his force to establish clear relationships.

3-21. Command relationships establish the degree of control and responsibility commanders have for forces operating under their tactical control (TACON). When commanders establish command relationships, they determine if the command relationship includes administrative control (ADCON). Figure 3-1 shows command and support relationships.

3-22. Support relationships define the purpose, scope, and effect desired when one capability supports another. Support relationships establish specific responsibilities between supporting and supported unit. Army support relationships are direct support (DS), general support (GS), general support-reinforcing (GSR), and reinforcing.

3-23. While not an actual C2 function, technical control often affects certain intelligence operations. Technical control ensures adherence to existing policies or regulations and provides technical guidance for MI activities, particularly HUMINT, SIGINT, and CI operations. Commanders direct operations but often rely on technical expertise to plan, prepare, execute, and assess portions of the unit's collection effort. Technical control also involves translating ISR tasks into the specific parameters used to focus highly technical or legally sensitive aspects of the ISR effort. Technical control includes, but is not limited to—

- Defining, managing, or guiding the employment of specific ISR assets.
- Identifying critical technical collection criteria such as technical indicators.
- Recommending collection techniques, procedures, or assets.
- Conducting operational reviews.
- Conducting operational coordination.
- Conducting specialized training for specific MI personnel or units.

3-24. An example of technical control is the Prophet control team converting the PIR and SOR sets from the MDMP process and assigning times and anticipated enemy frequencies for subordinate Prophets to collect.

	IF RELATIONSHIP IS:	INHERENT RESPONSIBILITIES ARE: Has Command Relationship with:	May Be Task Organized by:	Receives CSS from:	Assigned Position or AO By:	Provides Liaison To:	Establishes/ Maintains Communications with:	Has Priorities Established by:	Gaining Unit Can Impose Further Command or Support Relationship of:
COMMAND	Attached	Gaining unit	Gaining unit	Gaining unit	Gaining unit	As required by gaining unit	Unit to which attached	Gaining unit	Attached; OPCON; TACON; GS; GSR; R; DS
COMMAND	OPCON	Gaining unit	Parent unit and gaining unit; gaining unit may pass OPCON to lower HQ. Note 1	Parent unit	Gaining unit	As required by gaining unit	As required by gaining unit and parent unit	Gaining unit	OPCON; TACON; GS; GSR; R; DS
COMMAND	TACON	Gaining unit	Parent unit	Parent unit	Gaining unit	As required by gaining unit	As required by gaining unit and parent unit	Gaining unit	GS; GSR; R; DS
COMMAND	Assigned	Parent unit	Parent unit	Parent unit	Gaining unit	As required by parent unit	As required by parent unit	Parent unit	Not Applicable
SUPPORT	Direct Support (DS)	Parent unit	Parent unit	Parent unit	Supported unit	Supported unit	Parent unit; Supported unit	Supported unit	Note 2
SUPPORT	Reinforcing (R)	Parent unit	Parent unit	Parent unit	Reinforced unit	Reinforced unit	Parent unit; reinforced unit	Reinforced unit: then parent unit	Not Applicable
SUPPORT	General Support Reinforcing (GSR)	Parent unit	Parent unit	Parent unit	Parent unit	Reinforced unit and as required by parent unit	Reinforced unit and as required by parent unit	Parent unit; then reinforced unit	Not Applicable
SUPPORT	General Support (GS)	Parent unit	Parent unit	Parent unit	Parent unit	As required by parent unit	As required by parent unit	Parent unit	Not Applicable

NOTE 1. In NATO, the gaining unit may not task organize a multinational unit (see TACON).

NOTE 2. Commanders of units in DS may further assign support relationships between their subordinate units and elements of the supported unit after coordination with the supported commander.

Figure 3-1. Army Command and Support Relationships and Inherent Responsibilities.

Chapter 4

Intelligence Process in Full Spectrum Operations

THE INTELLIGENCE PROCESS

4-1. Commanders use the operations process of plan, prepare, execute, and assess to continuously design and conduct operations (see Figure 4-1). The commander cannot successfully accomplish the activities involved in the operations process without information and intelligence. The design and structure of intelligence operations support the commander's operations process by providing him with intelligence regarding the enemy, the battlefield environment, and the situation.

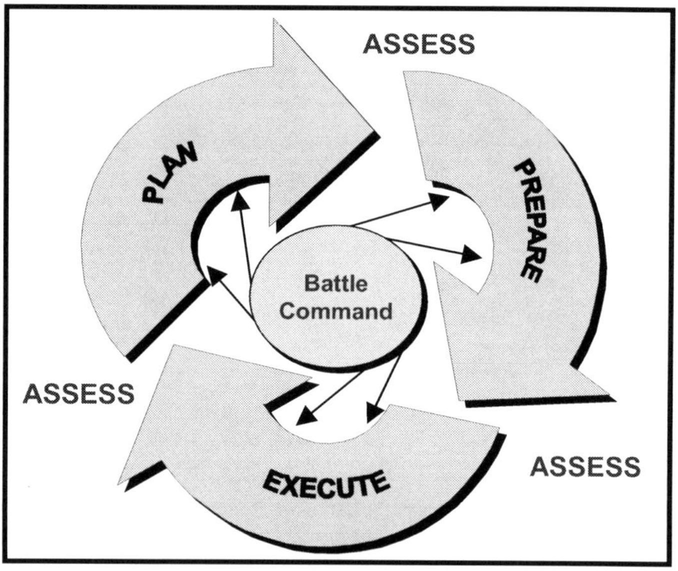

Figure 4-1. The Operations Process.

4-2. Intelligence operations consist of the functions that constitute the intelligence process: **plan, prepare, collect, process, produce**, and the three common tasks of **analyze, disseminate,** and **assess**. Just as the activities of the operations process overlap and recur as circumstances demand, so do the functions of the intelligence process. Additionally, the analyze, disseminate, and assess functions of the intelligence process occur continuously throughout the intelligence process.

4-3. The operations process and the intelligence process are mutually dependent. The commander, through the operations process, provides the guidance and focus through CCIRs and PIRs that drives the intelligence process; the intelligence process provides the continuous intelligence essential to the operations process. Intelligence about the enemy, the battlefield environment, and the situation allows the commander and staff to develop a plan, seize and retain the initiative, build and maintain momentum, and exploit success (see Figure 4-2). The intelligence process is just one of the mechanisms that provides input to build the COP and facilitate the commander's situational understanding.

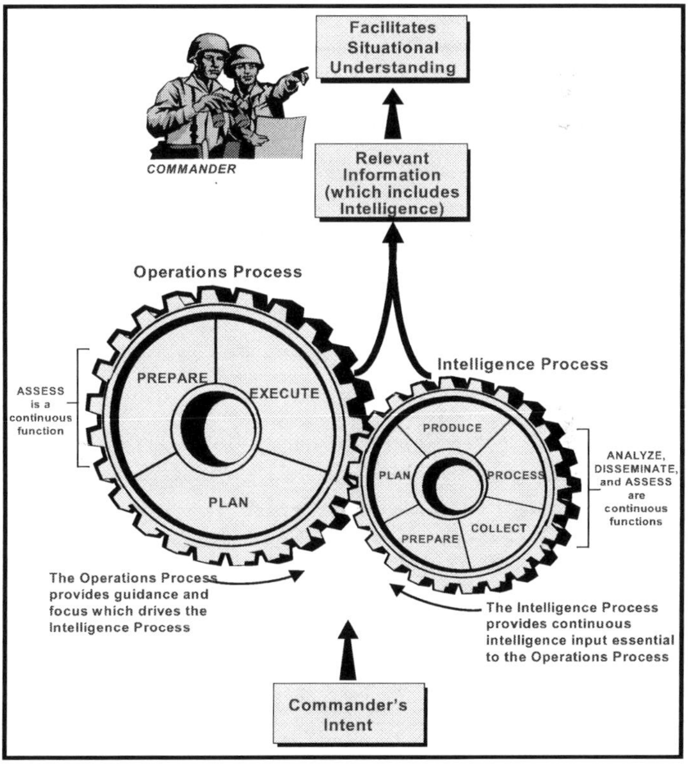

Figure 4-2. The Relationship Between the Operations and Intelligence Processes.

PLAN

4-4. The planning step of the intelligence process consists of activities that include assessing the situation, envisioning a desired outcome (also known as setting the vision), identifying pertinent information and intelligence requirements, developing a strategy for ISR operations to satisfy those requirements, directing intelligence operations, and synchronizing the ISR effort. The commander's intent, planning guidance, and CCIRs (PIRs and FIRs) drive the planning of intelligence operations. Planning, managing, and coordinating these operations are continuous activities necessary to obtain information and produce intelligence essential to decisionmaking.

COORDINATE

4-5. Staff and leaders coordinate with various elements, units, and organizations to ensure the necessary resources, linguist support (see Appendix B), information, intelligence, training, and procedures are in place to facilitate effective intelligence operations.

- Coordination for Movement of ISR Assets. All ISR assets at one time or another will move through or near another unit's AO. To avoid fratricide, ISR elements must coordinate with units, G3/S3, G2/S2, and each other, as well as coordinate with the fire support officer (FSO) to establish no-fire areas and/or other control measures around ISR assets and the air defense officer (ADO) in reference to aerial ISR assets in order to establish the appropriate weapons control status.
- Coordination for Information and Intelligence. The intelligence staff must prepare and practice coordination with personnel from all MI units, non-MI units, other service components, and multinational organizations that may contribute to or facilitate the collection effort. This coordination enables the G2/S2 to share and update databases, information, and intelligence and ensures connectivity with those organizations. All units are sources of relevant information regarding the enemy and the operational environment.
- Liaison. In order to accomplish the mission, exchange information and intelligence, move through certain areas and ensure FP, it may be necessary to coordinate with many different elements, organizations, and local nationals of the country in which friendly forces are conducting operations. Local nationals include police, town officials, foreign military forces, and political and other key figures within the AO. Operations may also necessitate coordination with other US and multinational forces; for example, the International Police Task Force (IPTF), Joint Commission Observers (JCO), Organization for Security and Cooperation in Europe (OSCE), Allied Military Intelligence Battalion (AMIB), and Defense HUMINT Service (DHS).
- Movement. Coordination with the G3/S3 ensures ease of movement and safe passage of friendly forces through an area. Coordinating movement also helps avoid fratricide.

INTELLIGENCE BOS CONSIDERATIONS

4-6. The Intelligence BOS is a unified system that anticipates and satisfies intelligence needs. Commanders ensure its proper employment by clearly articulating intent, designating CCIRs (PIRs and FFIRs), and prioritizing targets. Commanders must, however, understand the limitations of the Intelligence BOS to preclude unrealistic expectations of the system. The following are Intelligence BOS limitations:

- Intelligence only reduces uncertainty on the battlefield; it does not eliminate it entirely. The commander will always have to determine the presence and degree of risk involved in conducting a particular mission.
- The Intelligence BOS is composed of finite resources and capabilities. Intelligence systems and soldiers trained in specific ISR skills are limited in any unit. Once lost to action or accident, these soldiers and systems are not easily replaceable; for some, it may not be possible to replace them during the course of the current operation. The loss of soldiers and equipment can result in the inability to detect or analyze enemy actions. The loss of qualified language-trained soldiers, especially soldiers trained in low-density languages or skills, could adversely affect intelligence operations as well.
- The Intelligence BOS cannot effectively and efficiently provide intelligence without adequate communications equipment, capacity, and connectivity. Commanders and G2/S2s must ensure communications support to intelligence has the appropriate priority.
- Commanders and G2/S2s cannot expect that higher echelons will automatically send them everything they need. While intelligence reach is a valuable tool, the push of intelligence products from higher echelons does not relieve subordinate staffs from conducting detailed analysis and focusing the efforts of higher headquarters. Nor can they expect products pushed to them to be always at the level of detail they require. Commanders and G2/S2s must focus higher echelons by clearly articulating and actively pursuing intelligence requirements. By providing higher echelons with a clear picture of the required intelligence products, commanders can also narrow the flow of intelligence and information and preclude being overwhelmed by too much information.

4-7. Commanders should be aware that intelligence collection is enabled by, and subject to, laws, regulations, and policies to ensure proper conduct of intelligence operations. While there are too many to list here specifically, categories of these legal considerations include United States Codes (USCs), Executive Orders, National Security Council Intelligence Directives (NCSIDs), Army Regulations, United States Signal Intelligence Directives (USSIDs), SOFAs, ROE, and other international laws and directives.

PREPARE

4-8. The prepare step includes those staff and leader activities which take place upon receiving the OPORD, OPLAN, WARNO, or commander's intent to improve the unit's ability to execute tasks or missions and survive on the battlefield. These activities include—

- Effecting necessary coordination in accordance with the OPORD, METT-TC, unit SOP.
- Establishing and testing the intelligence architecture. This activity includes complex and technical issues like hardware, software, communications, COMSEC materials, network classification, technicians, database access, liaison officers (LNOs), training, funding, and TTP.
- Establishing an intelligence team attitude. This activity includes knowing different unit's and organization's capabilities, training the necessary collective skills, establishing effective relationships with different units and organizations, developing mutual battle rhythms and TTP, and leveraging the right architectures and collaboration tools.
- Coordinating effective analytic collaboration. Effective analytic collaboration is necessary to maximize the complementary analytic capabilities of different units and organizations that produce intelligence within the same theater of operations. Coordinating this collaboration is an effort-intensive activity that requires careful mutual planning, division of labor, defined responsibilities, and procedures for adapting to changing circumstances as they develop.
- Establishing reporting procedures.
- Conducting IPB.
- Producing Intelligence Estimates.
- Presenting briefings.
- Ensuring staff and personnel are trained. If personnel are not adequately trained at this point, they must be trained or the leader must evaluate the risk they bring to the operation.
- Planning refinement, brief-backs, SOP reviews, rehearsals, and coordinating with various elements and organizations.
- Establishing other troop-leading procedures (TLPs) or coordination, as necessary, in accordance with METT-TC factors.

G2/S2 PREPARATION ACTIVITIES

4-9. The G2/S2 takes numerous steps before mission execution to ensure intelligence operations run smoothly and effectively. These steps include, but are not limited to, the following:

- Conduct rehearsals.
- Conduct communication rehearsals.
- Review and update available databases and IPB products.
- Review applicable SOPs, Army Regulations, DA Pamphlets, Field Manuals, and ROE for guidance in conducting intelligence operations.
- Plan and practice actions supporting likely contingencies, or the branches or sequels to an operation.
- Ensure coordination measures are still in effect.
- Ensure training (individual and collective).

- Verify communications protocols with theater and higher headquarters and subordinate and lateral units.
- Update intelligence databases.
- Update the forces with the most recent intelligence on the AO immediately before mission execution.

VERIFICATION

4-10. Coordination for or requesting provisions or services is only the first step in acquiring them. It is crucial that staff and leaders check to verify that procedures, personnel, equipment, and services are in place and ready for mission execution.

REHEARSALS

4-11. Rehearsals help units prepare for operations by either verifying that provisions and procedures are in place and functioning or identifying inadequacies, which staff and leaders must remedy. They allow participants in an operation to become familiar with and to translate the plan into specific actions that orient them to their environment and other units when executing the mission. They also imprint a mental picture of the sequence of key actions within the operation and provide a forum for subordinate and supporting leaders and units to coordinate. (FM 6-0)

REPORTING PROCEDURES

4-12. The timely and accurate reporting of CCIRs (PIRs and FFIRs) and IRs is key to successful operations. All assets should know when, how often, and what format to use when reporting. The G2/S2 must verify the frequencies, alternate frequencies, and reactions during jamming, as well as the LTIOV for specific information to be reported. Unit SOPs provide the proper reporting procedures.

4-13. The G2/S2 coordinates with the unit staff, subordinate and lateral commands, and higher echelon units to ensure that specific reporting assets, personnel, equipment (especially communications), and procedures are in place. The G2/S2 requests or establishes the appropriate message addresses, routing indicators, mailing addresses, and special security office (SSO) security accreditation for units.

COMMUNICATIONS

4-14. Staff and leaders must work closely with the G6/S6 or signal officer (SIGO) to coordinate for the required communication links. The unit may require classified and unclassified network connections for their equipment. If elements of the unit will be working outside the range of the unit's communications systems, then it is necessary to coordinate for global or extended range communications. Leaders must obtain the required type and amount of communications equipment and related components as well as the latest fills and frequencies. They must possess and be familiar with all the instructions, passwords, policies, regulations, and directives conducive to OPSEC. They must also ensure soldiers are trained in the use and procedures involved in operating communications equipment.

SITUATION UPDATES

4-15. Each staff section and element conducts activities to maximize the operational effectiveness of the force. Coordination and preparation are just as important, if not more important, as developing the plan. Staff preparation includes assembling and continuously updating estimates. For example, continuous IPB provides accurate situational updates for commanders.

INTELLIGENCE HANDOFF

4-16. A well-executed intelligence hand-off will ensure a smooth and seamless transition between units. It is important that the incoming unit becomes familiar with the operation as soon as possible to avoid compromising the intelligence production and flow of the mission. The following are points to consider during a mission hand-off:

- Briefings and reports (learn what briefings are required and when as well as report formats and requirements).
- Past, present, and planned activities within the AOI.
- Established SOPs (know procedures for reporting; intelligence contingency funds [ICFs] and incentive use if applicable; emplacement and use of ISR equipment).
- Key personalities (introductions are required; establish rapport and a good working relationship with all key personalities).
  - Key personnel on the base or camp (their responsibilities; how to contact them).
  - Key personnel in other US and multinational service components (coordinate for exchange of information and intelligence).
  - Key personalities from surrounding towns (local figures).
  - Key national level political and military figures.
- Supporting units (know where to go for provisions, information, or assistance and POCs within those organizations).
- Current attitudes (understand current attitudes and perspectives of the local populace).
- Equipment operation and idiosyncrasies (equipment may run on different applications; personnel may need to train on specific equipment and procedures).
- Area familiarization (identify NAIs, key terrain, minefields, and boundaries; know camp locations, routes and route names, checkpoints, towns, and troubled resettlement areas).

RULES OF ENGAGEMENT

4-17. Although ROE training was presented during the plan function of the intelligence process, leaders at all levels can take the opportunity during the prepare function to ensure their subordinates completely understand the ROE. It is also during this function that commanders may need to consider exceptions to, or modifications of, the ROE to facilitate HUMINT and CI collection or to enable the placement of ISR assets.

COLLECT

4-18. Recent ISR doctrine necessitates that the entire staff, especially the G3/S3 and G2/S2, must change their reconnaissance and surveillance mindset to conducting ISR. The staff must carefully focus ISR on the CCIR (PIR and FFIR) but also enable the quick retasking of units and assets as the situation changes. This doctrinal requirement ensures that the enemy situation not just our OPLAN "drives" ISR operations. Well-developed procedures and carefully planned flexibility to support emerging targets, changing requirements, and the need to support combat assessment is critical. The G2/S2 and G3/S3 play a critical role in this challenging task that is sometimes referred to as "fighting ISR" because it is so staff intensive during planning and execution (it is an operation within the operation). Elements of all units on the battlefield obtain information and data about enemy forces, activities, facilities, and resources as well as information concerning the environmental and geographical characteristics of a particular area.

ISR TASKS

4-19. ISR tasks are the actions of the intelligence collection effort. ISR tasks consists of three categories:

- Intelligence.
- Surveillance.
- Reconnaissance.

4-20. Intelligence tasks are included in Annex B of the OPORD under Scheme of Intelligence. They include the following:

- **Intelligence Production**. Intelligence production includes analyzing information and intelligence and presenting intelligence products, conclusions, or projections regarding the OE and enemy forces in a format that enables the commander to achieve situational understanding.
- **Request for Information**. Submitting an RFI to the next higher or lateral echelon is the normal procedure for obtaining intelligence information not available through the use of available ISR assets. Users enter RFIs into an RFI management system where every other user of that system can see it. Hence, an echelon several echelons above the actual requester becomes aware of the request and may be able to answer it. A G2/S2 who receives an RFI from a subordinate element may use intelligence reach to answer RFIs.
- **Intelligence Reach**. Intelligence reach allows the commander to access the resources of national, joint, foreign, and other military organizations and units. Requestors can acquire information through push and pull of information, databases, homepages, collaborative tools, and broadcast services. (See Chapter 2 for more information on intelligence reach.)

4-21. For information on reconnaissance or surveillance tasks, refer to FM 7-15 and FM 3-55.

SPECIAL RECONNAISSANCE

4-22. Special reconnaissance (SR) is the complementing of national and theater intelligence collection assets and systems by obtaining specific, well-defined, and time-sensitive information of strategic or operational significance. It may complement other collection methods where there are constraints of weather, terrain, hostile countermeasures, and/or other systems availability. SR places US or US-controlled personnel conducting direct observation in hostile, denied, or politically sensitive territory when authorized. SOF may conduct these missions unilaterally or in support of conventional operations. (See JP 3-05 and FM 101-5-1.)

4-23. Army Special Operations Forces (ARSOF) elements conduct SR missions to obtain information not available through other means. SR operations encompass a broad range of collection activities to include reconnaissance, surveillance, and TA. SR complements national and theater collection systems that are more vulnerable to weather, terrain masking, and hostile countermeasures. SR missions provide intelligence or information that is often not available through other means. Typical SR missions include—

- TA and surveillance of hostile C2 systems, troop concentrations, deep-strike weapons, lines of communication (LOCs), WMD systems, and other targets.
- Location and surveillance of hostage, prisoner of war, or political prisoner detention facilities.
- Post-strike reconnaissance for BDA.
- Meteorologic, geographic, or hydrographic reconnaissance to support specific air, land, or sea operations.

4-24. For more information on special reconnaissance, see FM 3-05.102.

REPORTING

4-25. The most critical information collected is worthless if not reported in a timely manner. Collectors may report information via verbal, written, graphic, or electronic means. Unit SOPs must clearly state the transmission means of different types of reports (for example, sent by voice frequency modulated [FM] radios or by automated means). In general, the transmission of reports for enemy contact and actions, CCIRs, exceptional information, and NBC reports is by voice FM, and then followed up with automated reports. Commanders and staffs must remember that timely reporting, especially of enemy activity, is critical in fast-moving operations. Collectors must report accurate information as quickly as possible. Commanders and staff must not delay reports for the sole purpose of editing and ensuring the correct format. This is particularly true for reporting information or intelligence that answers the PIR.

TIME-SENSITIVE REPORTING

4-26. Intelligence and time-sensitive combat information that affects the current operation is disseminated immediately upon recognition. Combat information is unevaluated data, gathered by or provided directly to the tactical commander which, due to its highly perishable nature or the criticality of the situation, cannot be processed into tactical intelligence in time to satisfy the user's tactical intelligence requirements. Thus combat information is provided directly to the tactical commander (see JP 1-02). The routing of combat information proceeds immediately in two directions: directly to the commander and through routine reporting channels, which include intelligence analysis and production elements.

4-27. Time-sensitive information usually includes reports concerning enemy contact and actions and CCIRs.

PROCESS

4-28. The process function converts relevant information into a form suitable for analysis, production, or immediate use by the commander. Processing also includes sorting through large amounts of collected information and intelligence (multidiscipline reports from the unit's ISR assets, lateral and higher echelon units and organizations, and non-MI elements in the battlespace). Processing identifies and exploits that information which is pertinent to the commander's intelligence requirements and facilitates situational understanding. Examples of processing include developing film, enhancing imagery, translating a document from a foreign language, converting electronic data into a standardized report that can be analyzed by a system operator, and correlating dissimilar or jumbled information by assembling like elements before the information is forwarded for analysis.

4-29. Often collection assets must collect and process their data prior to disseminating it. MI systems have their own reporting and processing systems, the details of which are in the appropriate MI system manuals and technical manuals. Some collection assets, particularly air reconnaissance and ground scouts, can report relevant information that is immediately usable by the tactical commander (for example, for targeting purposes). However, the personnel in the reporting chain still process these reports by evaluating their relevancy and accuracy. In many cases, the output of a collection asset is data, or information of limited immediate use to a commander. Also, in certain situations ROE dictate a requirement for target confirmation by other sources.

4-30. The intelligence staff processes information collected by the unit's assets as well as that received from higher echelons. The intelligence staff processes many types of information and data from intelligence reach, unmanned aerial vehicle (UAV) imagery, radar imagery, mobile target indicators (MTIs), and HUMINT and SIGINT reports.

PRODUCE

4-31. In the production step, the G2/S2 integrates evaluated, analyzed, and interpreted information from single or multiple sources and disciplines into finished intelligence products. Like collection operations, the G2/S2 must ensure the unit's information processing and intelligence production are prioritized and synchronized to support answering the CCIRs (PIRs and FFIRs).

4-32. Intelligence products must be timely, relevant, accurate, predictive, and usable. The accuracy and detail of every intelligence product has a direct effect on how well the unit plans and prepares for operations. However, the G2/S2 and unit must use intelligence (no matter what form the intelligence is in) that meets the requirements but might not be as detailed or refined as possible or in a better form. A good answer on time is better than a more refined answer that is late.

4-33. The G2/S2 produces intelligence for the commander as part of a collaborative process. The commander drives the G2/S2's intelligence production effort by establishing intelligence and IRs with clearly defined goals and criteria. Differing unit missions, environments, and situations impose numerous and varied production requirements on the G2/S2 and his staff.

4-34. The G2/S2 must employ collaborative analysis techniques and procedures that leverage intelligence production capability of higher and subordinate echelons to meet these requirements. Proficiency in these techniques and procedures enables the G2/S2 to answer the commander's and staff's requirements regardless of the mission, environment, and situation. The G2/S2 and staff intelligence products enable the commander to—

- Plan operations and employ maneuver forces effectively.
- Recognize potential COAs.
- Employ effective tactics and techniques.
- Take appropriate security measures.
- Focus ISR.

COMMON INTELLIGENCE PROCESS TASKS

ANALYZE

4-35. Analysis occurs at various stages throughout the intelligence process. Personnel conducting intelligence operations at all levels analyze intelligence, information, and problems to produce intelligence, solve problems and, most importantly, answer the PIRs. Leaders at all levels conduct analysis to assist in making many types of decisions. An example is a HUMINT collector analyzing an intelligence requirement in order to determine the best possible collection strategy to use against a specific source.

4-36. Analysis in RM is critical to ensuring the IRs receive the appropriate priority for collection. The intelligence staff analyzes each requirement to determine its feasibility, whether or not it supports the commander's intent, and to determine the best method of satisfying the IRs. The staff also analyzes collected information to determine if it satisfies requirements.

4-37. During the produce function, the intelligence staff analyzes information from multiple sources to develop all-source intelligence products. The intelligence staff analyzes information and intelligence to ensure the focus, prioritization, and synchronization of the unit's intelligence production is in accordance with the PIRs.

4-38. In situation development, the intelligence staff analyzes information to determine its significance relative to predicted ECOAs and the CCIRs (PIRs and FFIRs). Through predictive analysis, the staff attempts to identify enemy activity or trends that represent opportunities or risks to the friendly force.

They use the indicators developed for each ECOA and CCIRs (PIRs and FFIRs) during the MDMP as the basis for their analysis and conclusions.

DISSEMINATE

4-39. Successful operations at the tactical and operational levels require an increased ability to synchronize fires, have faster access to intelligence, and enhance situational understanding and effective FP. Timely and accurate dissemination of intelligence is key to the success of these and other operations. Commanders must receive combat information and intelligence products in time and in an appropriate format to support decisionmaking. Additionally, sharing the most current all-source information and intelligence at all echelons is essential for commanders to maintain situational understanding.

4-40. To achieve this, it is imperative that the commander and staff establish and support a seamless intelligence architecture—including an effective dissemination plan—across all echelons to ensure information and intelligence flow in a timely manner to all those who need them. Intelligence and communications systems continue to evolve in their sophistication, application of technology, and accessibility to the commander. Their increasing capabilities also create an unprecedented volume of information available to commanders at all echelons. Finally, the commander and staff must have a basic understanding of these systems and how they contribute to the Intelligence BOS. A dissemination plan can be a separate product, or integrated into existing products such as the ISR synchronization plan or ISM, the decision support template (DST), or decision support matrix (DSM).

Dissemination Procedures

4-41. The G2/S2 and intelligence personnel at all levels assess the dissemination of intelligence and intelligence products.

4-42. Disseminating intelligence simultaneously to multiple recipients is one of the most effective, efficient, and timely methods. This can be accomplished through various means; for example, push, broadcast. However, within the current tactical intelligence architecture, reports and other intelligence products move along specific channels. The staff helps streamline information distribution within these channels by ensuring dissemination of the right information in a timely manner to the right person or element. There are three channels through which commanders and their staffs communicate: command, staff, and technical.

4-43. **Command Channel**. The command channel is the direct chain-of-command link that commanders, or authorized staff officers, use for command-related activities. Command channels include command radio nets (CRNs), video teleconferences (VTC), and the Maneuver Control System (MCS).

4-44. **Staff Channel**. The staff channel is the staff-to-staff link within and between headquarters. The staff uses the staff channel for control-related activities. Through the staff channel, the staff coordinates and transmits intelligence, controlling instructions, planning information, provides early warning information, and other information to support C2. Examples of staff channels include the operations and intelligence radio net, telephone, the staff huddle, VTC, and the BOS-specific components of the Army Battle Command System (ABCS).

4-45. **Technical Channel.** Staffs typically use technical channels to control specific combat, CS, and CSS activities. These activities include fire direction and the technical support and sensitive compartmented information (SCI) reporting channels of intelligence and ISR operations. The SIGINT tasking and reporting radio net, intelligence broadcast communications, and the wide area networks (WANs) supporting single intelligence discipline collection, processing, and production are examples of technical channels.

Presentation Techniques and Procedures

4-46. The staff's objective in presenting information is to provide the commander with relevant information. Table 4-1 lists the three general methods that the staff uses to present information and meet its information objective. Systems within the ABCS contain standard report formats, maps, and mapping tools that assist the staff in presenting information in written, verbal, and graphic form. Audio and video systems such as large format displays and teleconferencing systems enable the staff to use a combination of the methods in multimedia presentations.

Table 4-1. Presentation Methods and Products.

Method	Products
Written Narrative	Reports, Estimates, and Studies
Verbal Narrative	Briefing (information, decision, mission, and staff)
Graphic	Charts, Overlays, and Electronic Displays

Intelligence Communications Architecture

4-47. The intelligence communications architecture transmits intelligence and information to and from various ISR elements, units, and agencies by means of automation and communication systems. With the continued development of sensors, processors, and communications systems, it is increasingly important to understand the requirements of establishing an effective communications architecture. The G2/S2 must identify the Intelligence BOS specific requirements of the unit's overall communications architecture. Refer to FM 2-33.5 for more information on intelligence communications reach. The following are some (but not all) of the questions which the staff must answer in order to establish the intelligence communications architecture:

- Where are the unit's collectors?
- What and where are the unit's processors?
- Where are the unit's intelligence production elements?
- Where are the unit's decisionmakers?
- How does the unit disseminate information from its producers to its decisionmakers and/or consumers?

- Are the systems which the unit's collectors, producers, processors, and consumers use compatible with each other? If not, what is the plan to overcome this challenge?
- How can the unit access databases and information from higher and other agencies?

ASSESS

4-48. Assessment is the continuous monitoring—throughout planning, preparation, and execution—of the current situation and progress of an operation, and the evaluation of it against criteria of success to make decisions and adjustments. Assessment plays an integral role in all aspects of the intelligence process. Assessing the situation and available information begins upon receipt of the mission and continues throughout the intelligence process. The continual assessment of intelligence operations and ISR assets, available information and intelligence, the various aspects of the battlefield environment, and the situation are critical to—

- Ensure the CCIRs (PIRs and FFIRs) are answered.
- Ensure intelligence requirements are met.
- Redirect collection assets to support changing requirements.
- Ensure operations run effectively and efficiently.
- Ensure proper use of information and intelligence.
- Identify enemy efforts at deception and denial.

4-49. During planning, the intelligence staff conducts a quick initial assessment of the unit's intelligence posture and holdings, status of intelligence estimates, and any other available intelligence products. From this assessment the commander issues his initial guidance and a WARNO.

4-50. While the majority of the unit is engaged in preparation, the ISR effort should already have begun. It is during this period when the prepare and execute activities of the operations process overlap, that the G2/S2 assesses the current situation as well as the progress of ISR operations.

4-51. During execution the intelligence staff continues assessing the effectiveness of the ISR effort while at the same time assessing the results and products derived from the collection effort. The critical aspects of assessment at this point include determining whether the PIRs have been answered, will be answered with the current ISR operations, or which ISR operations to adjust in order to answer the CCIRs (PIRs and FFIRs). This type of assessment requires sound judgment and a thorough knowledge of friendly military operations, characteristics of the AO and AOI, and the threat situation, doctrine, patterns, and projected future COAs.

PART THREE

Military Intelligence Disciplines

Part Three provides a more detailed explanation of the intelligence disciplines introduced in Part One of this manual.

Chapter 5 defines and discusses the roles and fundamentals of the all-source intelligence discipline.

Chapter 6 defines and discusses the roles, fundamentals, and generic organization of the HUMINT discipline.

Chapter 7 defines and discusses the roles, fundamentals, and generic organization of the IMINT discipline.

Chapter 8 defines and discusses the roles, fundamentals, and generic organization of the SIGINT discipline.

Chapter 9 defines and discusses the roles, fundamentals, and generic organization of the MASINT intelligence discipline.

Chapter 10 defines and discusses the roles, fundamentals, and generic organization of the TECHINT discipline.

Chapter 11 defines and discusses the roles, fundamentals, and generic organization of the CI discipline.

Chapter 5

All-Source Intelligence

DEFINITION

5-1. All-source intelligence is defined as the intelligence products, organizations, and activities that incorporate all sources of information and intelligence, including open-source information, in the production of intelligence. All-source intelligence is a separate intelligence discipline, as well as the name of the function used to produce intelligence from multiple intelligence or information sources.

ROLE

5-2. The operational environment provides an ever-growing volume of data and information available from numerous sources, from which the commander can use to achieve situational understanding. His situational understanding enables him to make decisions in order to influence the outcome of the operation; prioritize and allocate resources; assess and take risks; and understand the needs of the higher and subordinate commanders. The commander depends upon a skilled G2/S2 working within his intent to support his ISR effort and provide all-source intelligence analysis conclusions and projections of future conditions or events.

FUNDAMENTALS

5-3. All-source intelligence production satisfies intelligence requirements. It provides an overall picture of the adversary and the battlespace. It reduces the possibility of error, bias, and misinformation through the use of multiple sources of information and intelligence.

PLAN

5-4. The utilization of all-source intelligence facilitates the development of accurate and concise contingency-specific plans, orders, and intelligence products. Additionally, the G2/S2 must retrieve, update, or develop any required intelligence databases. The most important all-source products are the modified combined obstacle overlay (MCOO), event templates, ECOA sketches, and the HPT list provided during initial IPB in support of the plan function. Through all-source intelligence, the commander can make informed decisions about COAs and adversary capabilities.

5-5. The intelligence staff coordinates its efforts with other elements such as the engineer terrain team, the unit surgeon, the Air Force weather team, and other assets or elements that can support the analytical effort.

5-6. IPB plays a significant role in the planning phase of the intelligence process. It is a systematic, continuous process of analyzing the threat and environment and is designed to support staff estimates and the MDMP. The IPB is led by the G2/S2, with participation by the entire staff. IPB allows the commander and staff to make informed decisions, develop COAs, and focus ISR efforts where they are most needed. IPB continuously assists the commander and staff in focusing ISR assets on the appropriate targets at critical points in time and space.

5-7. The time available for the IPB process may not permit the luxury of conducting each step in detail. Overcoming time limitations requires the commander to identify those parts of IPB that are most important to the commander in planning and executing his mission. Applying the specific steps or degree of detail performed varies according to METT-TC. There are four steps in the IPB process: define the battlefield environment, describe the battlefield's effects, evaluate the threat, and determine ECOAs.

Define the Battlefield Environment

5-8. In this step, the G2/S2—

- Identifies characteristics of the battlefield that will influence friendly and threat operations including terrain (mobility), weather, hydrological data, infrastructure, and civilian demographics. Produces accurate, timely, and predictive IPB products that depict the aspects of the battlefield.

- Identifies the limits of the command's AO and AOI.
- Identifies gaps in current intelligence holdings, identifies IRs, and recommends CCIRs (PIRs and FFIRs).

5-9. Defining the battlefield environment includes identifying enemy forces (their location, mobility, general capabilities, and weapon ranges) and all other aspects of the environment that could have an effect on the unit's ability to accomplish the mission. Depending on the situation, these considerations may also include—

- Geography, terrain, and weather of the area.
- Information environment to include but not limited to computer and communications systems and capabilities, data acquisition systems and capabilities, media access and distribution, areas prone to electromagnetic interferences, and systems to generate electromagnetic interference.
- Population demographics (ethnic groups, religious sects, age distribution, health status, income groups).
- Medical threat to include but not limited to endemic and epidemic diseases, occupational and environmental health hazards, and poisonous and toxic plants and animals.
- Political or socio-economic factors, including factions, clans, and gangs.
- Infrastructures, such as transportation or telecommunications, and critical decisionmaking infrastructures and supporting information systems.
- ROE or legal restrictions, such as international treaties or agreements.

Describe the Battlefield's Effects

5-10. This IPB step deals with the effects of the battlefield environment on the current operations and potential enemy and friendly COAs. It begins with the assessment of existing and projected conditions of the battlefield environment, which the staff accomplishes through terrain analysis, weather analysis, and analysis of other characteristics of the battlefield. To conduct terrain analysis, the staff uses maps, reconnaissance, and other specialized terrain products (maps, overlays, databases, software). These products address such factors as wet or dry cross-country mobility, transportation systems (road and bridge information), vegetation type and distribution, surface drainage and configuration, surface materials (soils), ground water, manmade structures, and obstacles. The results of evaluating the terrain's effects should be expressed by identifying areas of the battlefield that influence each COA.

Evaluate the Threat

5-11. Evaluating the threat involves determining the threat force capabilities and the doctrinal principles and TTP that threat forces prefer to employ. The result of this evaluation produces a threat model, which portrays how threat forces normally execute operations and how they have reacted to similar situations in the past. The threat model includes an evaluation of the threat's strengths, weaknesses, and vulnerabilities, including an evaluation of typical HVTs.

Determine ECOAs

5-12. Determining ECOAs involves the identification and development of likely ECOAs that will influence accomplishing the friendly mission. Developing ECOAs is a form of predictive intelligence analysis and production. The procedures for this step include—

- Identifying the enemy's likely objectives and predicting the desired end-state.
- Identifying the full set of COAs available to the enemy.
- Predicting the enemy's most likely and most dangerous COAs.
- Evaluating and prioritizing each COA.
- Developing each COA in the amount of detail that time permits.
- Identifying initial ISR and collection requirements.

5-13. After conducting the initial IPB, the staff, primarily the G2/S2, identifies gaps in the available intelligence, develops the initial PIRs and IRs, and develops the initial input to the ISR plan based on the commander's guidance.

5-14. Indicators are the basis for situation development. Indicators are activities that will confirm or deny the event specified in an intelligence requirement. They are any positive or negative evidence of enemy activity or characteristic of the AO that points toward enemy capabilities, vulnerabilities, or intentions. Individual indicators cannot stand alone. Each indicator is integrated with other factors and indicators before patterns are detected and enemy intentions established. Indicators are developed by the analysts in the G2/S2 section. The event matrix shows the threat activities, or indicators, to look for in each NAI, and the timelines during which each NAI should be active. All indicators are developed to answer the commander's PIRs and IRs. The analyst uses indicators to correlate particular events or activities with probable ECOAs and to determine what events or activities must occur for an enemy to follow a particular COA. The ability to read indicators (including recognition of enemy deception indicators) contributes to the success of friendly operations. The analyst integrates information from all sources to confirm indicators of enemy activities. As indicators are detected and confirmed, PIRs are answered.

PREPARE

Activities

5-15. All-source activities during the prepare function include—

- Conducting rehearsals.
- Conducting communications rehearsals and verifiying communications protocols with higher, lateral, and subordinate units.
- Planning and practicing actions that support likely contingencies, branches, or sequels.
- Reviewing and updating available databases and IPB products.
- Ensuring control and coordination measures are still in effect.

- Updating the force with the most recent intelligence before mission execution.

Intelligence Estimates

5-16. The intelligence estimate is a continuous process that is the product of all actions the intelligence staff performs throughout the MDMP. The intelligence estimate provides a timely and accurate evaluation of the enemy and the AO (and often the AOIR) at a given time. It provides the background the G2/S2 uses to portray enemy actions during COA analysis.

5-17. The G2/S2 must clearly understand the weather and terrain effects and the ability to visualize the battlespace before producing the intelligence estimate. This understanding facilitates accurate assessments and projections regarding the enemy: enemy situation (including strengths and weaknesses), enemy capabilities and an analysis of those capabilities (COAs available to the enemy), and conclusions drawn from that analysis. The intelligence estimate's conclusion identifies the enemy's most likely COA and most dangerous COA. ECOAs must include a sketch of the COA with a task and purpose for the enemy's actions. Any gaps in the intelligence estimate are identified as information requirements.

5-18. The intelligence estimate may be written or oral. At the tactical level, especially during operations and exercises, intelligence estimates are usually delivered orally, supported by graphic displays and other decision support tools. During contingency planning, especially at corps level and above, estimates are usually written. During deliberate planning at joint headquarters, estimates are always written (see JP 5-00.2). However, the intelligence estimate should always be prepared as thoroughly as time and circumstances permit. A comprehensive intelligence estimate considers both the tangible (quantifiable) and the intangible aspects of the enemy's operations. It translates enemy strengths, weapon systems, training, morale, and leadership into combat capabilities and projections of future enemy actions.

5-19. Different sections of the intelligence estimate receive more emphasis during different activities within the operations process.

- During planning, the most important decision the commander makes is selecting a COA on which to base the plan. Thus, the sections of the intelligence estimate that focus on the commander's selecting the most appropriate COA are the most important at that time.
- During preparation, the intelligence estimate must focus on the commander's decisions that affect the ability of the unit to execute the upcoming operation. The intelligence estimate—and functions—that supports these decisions must focus on answering the CCIRs (PIRs and FFIRs) and guiding the ISR effort.
- During execution, the intelligence estimate focus is to support the anticipated command decisions. The most important action of the intelligence estimate—and the ISR effort— is to answer the PIR. However, it is also during the execution phase that the intelligence estimate must look ahead: anticipating branches or sequels to the

current operation, transition to other operations, changes in the ECOAs, and required adjustments to the ISR effort.

5-20. Refer to FM 2-01 for examples of intelligence estimates.

COLLECT

5-21. The collect intelligence process function within the all-source intelligence discipline is limited to gathering the necessary information, intelligence, and intelligence products required to perform analysis. Thus, all-source intelligence depends upon the other intelligence disciplines to perform collection. This is also the primary reason why OSINT is defined as a source of information and intelligence and not a separate intelligence discipline.

PROCESS

5-22. The process function converts relevant information into a form suitable for analysis, production, or immediate use by the commander. Processing also includes sorting through large amounts of collected information and intelligence to identify and exploit the information which is pertinent to the commander's intelligence requirements and facilitates situational understanding.

PRODUCE

5-23. All-source intelligence analysis and production—

- Drives collection to answer the PIR.
- Provides the enemy situation.
- Provides INTSUMs and other intelligence reports.
- Supports situational understanding.
- Provides predictive estimates of enemy actions; specifically, ECOAs.
- Provides continuously updated IPB.
- Provides all-source target packages (or folders).

5-24. As previously stated, the intelligence estimate is a continuous process. The commander and staff constantly collect, process, store, display, and disseminate information. Staff members update their estimates as they receive new information, such as—

- When they recognize new facts.
- When they replace assumptions with facts or find their assumptions invalid.
- When they receive changes to the mission or when changes are indicated.

5-25. Technological advances and NRT information allow estimates to be continuously updated; the running estimate and its intelligence component, the intelligence running estimate, exemplify this.

Running Estimate

5-26. Running estimates are continuously updated estimates based on new information as the operation proceeds. They serve as a staff technique to support the commander's visualization and decisionmaking, as well as the staff's tool for assessing during preparation and execution. In the running estimate, staff officers continuously update their conclusions and recommendations as they evaluate the impact of new facts. Current doctrine emphasizes the COP as the primary tool that provides the current situation and, when merged with the running estimates, facilitates commanders achieving situational understanding. All staff sections (BOSs) provide their respective input to the COP. The COP as used today, combined with the running estimate, is predictive and enhances our current ability to collect, process, store, display, and disseminate information.

5-27. The portion of the COP that depicts the enemy situation is currently limited to displaying the locations and dispositions of enemy and threat forces in a relatively static manner, sometimes referred to as snapshots in time. The enemy situation portion of the COP requires analysis to provide the required level of detail. The COP and running estimates sufficiently provide effective support for battle command of knowledge-based organizations.

5-28. The main differentiation between the running estimate and the old staff estimates is the emphasis on not only continuously updating the facts of the estimate but also continuously updating the conclusions and recommendations while including projections of future conditions of the entire battlespace. While the COP is primarily a display of current intelligence and information, the running estimate requires the merging of the staff's cognitive processes with automation applications. The primary focus of the staff's cognitive process is to present predictive or anticipatory intelligence in support of the commander's decisionmaking or situational understanding. See Figure 5-1.

5-29. The running estimate is a product of the entire battle staff. Just as all BOSs contribute to the COP, they will also contribute their portion of the running estimate. The running estimate integrates the running estimates from each BOS. The Intelligence BOS input to the running estimate is the intelligence running estimate.

Intelligence Running Estimate

5-30. The intelligence running estimate is a continuous flow and presentation of relevant information and predictive intelligence that, when combined with the other staff running estimates, enables the decisionmaker's visualization and situational understanding of the AOI in order to achieve information superiority. The intelligence running estimate requires constant verification to support situational understanding of the current situation as well as predictive assessments for future operations.

5-31. While it is possible to employ the concept of the intelligence running estimate today, the true seamless and continuous update feature will be achieved with technological enablers that are not yet present in the force. In the future, technology should allow the information from the running estimate to become implanted as a part of the COP. However, there will always be a place for some separate context, orally or written, separate from the COP display to add fidelity and assist in the commander's visualization and decisionmaking. Additional training must be implemented not just within MI but also throughout the force in order to institute the running estimate concept.

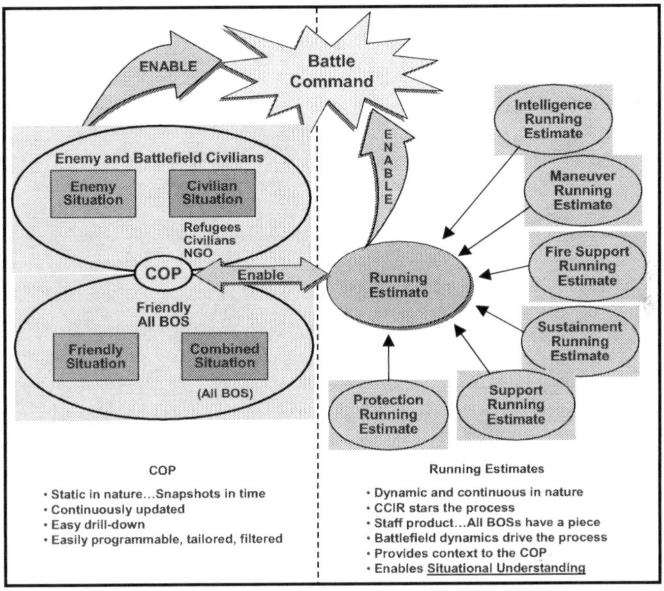

Figure 5-1. Example of the COP and a Running Estimate.

ANALYZE

5-32. Analysis is key in converting combat information and all intelligence from each discipline into all-source intelligence products and targeting information. From the multiple sources of information received, the staff analyzes and identifies critical information; determines the relevancy and accuracy of the information; and reaches conclusions about the information. These conclusions are either immediately disseminated or are used to form the basis of intelligence analysis products.

DISSEMINATE

5-33. The intelligence staff extracts and disseminates pertinent intelligence from products developed as a result of all-source analysis. The intelligence staff—

- Rapidly disseminates time-sensitive, all-source analysis intelligence and intelligence products to higher, lateral, and subordinate commands in order to keep all commanders abreast of current developments in the situation and battlespace in accordance with unit SOPs.

- Disseminates all-source analysis intelligence to other G2/S2 intelligence activities for additional processing or detailed analysis and exploitation in accordance with unit SOPs.

5-34. For intelligence reach operations all-source intelligence products are available and disseminated in a variety of forms. It is incumbent on the requestor to ensure that the all-source product can be transmitted over the available communications systems. This includes verifying the appropriate security level of the communication system.

ASSESS

5-35. The commander, the intelligence staff, and the intelligence consumers assess the analysis, production, and dissemination of all-source intelligence. The staff identifies intelligence gaps, emerging operational requirements, duplication of effort, or new targets or threats and redirects all-source activities as appropriate to meet intelligence requirements. Throughout the assessment process, the G2/S2, the intelligence staff, and intelligence users should be alert for evidence of possible enemy denial and deception efforts. The staff assesses intelligence and information for—

- Consistency with the current situation and threat trends.
- Accuracy and confirmation by other intelligence sources or disciplines.
- Source reliability and credibility.
- Pertinence to PIRs, intelligence requirements, RFIs, and other intelligence tasks or requests.

Chapter 6

Human Intelligence

DEFINITION

6-1. HUMINT is the collection by a trained HUMINT Collector of foreign information from people and multimedia to identify elements, intentions, composition, strength, dispositions, tactics, equipment, personnel, and capabilities. It uses human sources as a tool and a variety of collection methods, both passively and actively, to gather information to satisfy the commander's intelligence requirements and cross-cue other intelligence disciplines.

ROLE

6-2. The role of HUMINT Collectors is to gather foreign information from people and multiple media sources to identify adversary elements, intentions, composition, strength, dispositions, tactics, equipment, personnel, and capabilities. It uses human sources and a variety of collection methods to gather information to satisfy the commander's intelligence requirements, and cross-cue other intelligence disciplines.

HUMINT FUNCTIONS

6-3. HUMINT functions are interrelated, mutually supporting, and can be derived from one another. No single function or technical capability can provide a full understanding of our adversaries. HUMINT functions are defined below and shown in Figure 6-1.

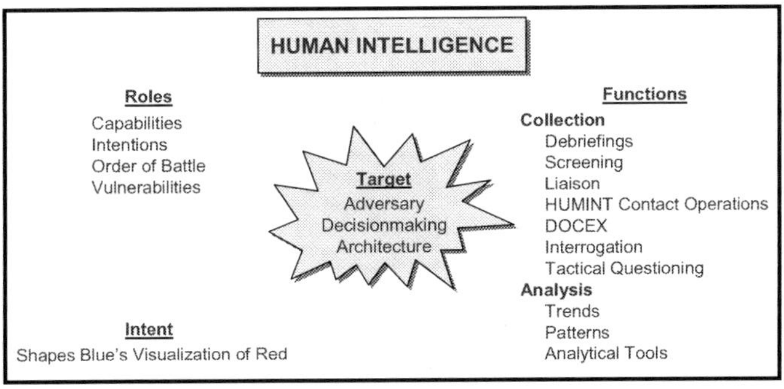

Figure 6-1. HUMINT Roles and Functions.

COLLECTION

6-4. HUMINT collection methods and operations include—

- **Debriefings.** Debriefings are the systematic questioning of individuals to procure information to answer specific collection requirements by direct and indirect questioning techniques. Sources for debriefings are categorized as friendly forces, US and non-US civilians to include refugees, displaced persons, and local inhabitants.
- **Screening.** Screening is the process of evaluating and selecting human and media sources for the prioritized collection of information in support of commander's PIRs and IRs. Screening categorizes and prioritizes sources based on the probability of a particular source having priority information and the degree of difficulty of extracting that information from the source. While screening is not in itself an information collection technique, it is vital to the rapid collection of information.
- **Liaison.** HUMINT elements conduct liaison with US, multinational, and HN military and civilian agencies, to include NGOs. Liaison is conducted to obtain information of interest and to coordinate or deconflict HUMINT activities.
- **HUMINT Contact Operations.** HUMINT contact operations are tactically oriented, overt collection activities that use human sources to identify attitude, intentions, composition, strength, dispositions, tactics, equipment, target development, personnel, and capabilities of those elements that pose a potential or actual threat to US and multinational forces. These forces provide early warning of imminent danger to US and multinational forces and contribute to the MDMP.
- **Document Exploitation (DOCEX).** DOCEX is the systematic extraction of information from all media formats in response to collection requirements.
- **Interrogation.** Interrogation is the systematic effort to procure information to answer specific collection requirements by direct and indirect questioning techniques of persons who are in the custody of the forces conducting the questioning.

UNIT SUPPORT TO HUMINT COLLECTION

6-5. Small units contribute to HUMINT collection through a number of different ways.

Tactical Questioning

6-6. Tactical questioning can provide critical information for situational understanding. Tactical questioning is the expedient initial questioning for information of immediate tactical value. Soldiers conduct tactical questioning based on the unit's SOP, ROE, and the order for that mission. Unit leaders must include specific guidance for tactical questioning in the order for appropriate missions. The unit S3 and S2 must also provide specific guidance down to the unit level to help guide tactical questioning. For a more detailed discussion of tactical questioning, see ST 2-91.6.

Combat Patrolling

6-7. Through observing and interacting with the local environment during the conduct of missions, handling EPWs/detainees, and handling captured documents, soldiers serve as the commander's "eyes and ears" whether—

- Performing traditional offensive or defensive operations.
- Performing a patrol in a stability operation.
- Manning a TCP or a roadblock in a support operation.
- Occupying an observation post.
- Passing through an area in a convoy.
- Performing any operation that involves observing and reporting on elements of the environment and activities in the AO.

ISR Operations

6-8. The information that the soldier reports as a result of completed missions up the chain of command forms a vital part of planning and operations. Careful and quick handling of EPWs/detainees and documents also helps the ISR effort. Unit headquarters must ensure that the reports they receive and forward contain all the subtle detail provided by the small unit. For tactical operations, there are four levels of reporting:

- Immediate reporting of information of critical tactical value, based either on predetermined criteria or common sense.
- Normal reporting, submitted before the unit S2 section performs the debriefing.
- Debriefing performed by the S2 section.
- Follow-up reporting, submitted after the unit S2 section performs the debriefing.

6-9. The four levels of reporting facilitate the unit S2 section capturing all the fine details of the small unit's activities for all-source analysis and future planning. The unit S2 must ensure that information of HUMINT and CI value is reported to the G2X.

ANALYSIS

6-10. IPB, all-source, and single-source analysis are used to template adversarial OB. HUMINT elements analyze operational taskings to determine the best collection methods to employ in satisfying CCIRs (PIRs and FFIRs) and IRs. As information is collected, it is analyzed for completeness and accuracy to identify significant facts for subsequent interpretation and inclusion in intelligence products, and passing to other collectors. This analysis also helps to identify collection gaps and focus or refocus collection efforts. Analysis provides the commander with situational understanding of the battlespace to execute operations. Raw information, open-source material, and finished intelligence products are analyzed in response to local and national requirements. Analysis occurs at all levels from tactical to strategic.

- At the tactical level, HUMINT teams focus their efforts on supporting mission requirements and contributing to the all-source COP.

- Operational analysis is used to assess adversary intentions, capabilities, dispositions, and their regional impacts.
- Strategic analysis supports national and Army programs through compilation of local and regional analysis into global assessments; identifies technology development trends and patterns in military activities and capabilities.

6-11. HUMINT products consist of, but are not limited to, target nomination, input to threat and vulnerability assessments, intelligence estimates, and intelligence information reports. Finalized intelligence derived from HUMINT activities is incorporated into joint and national intelligence databases, assessments, and analysis products. HUMINT products are also incorporated into the COP to support situational awareness. HUMINT production takes place at all levels.

- Operational and tactical production includes tactical alerts, spot reports, and current intelligence; input to threat and/or vulnerability assessments tailored to specific activities, units, installations, programs or geographic areas, and target studies to support contingency planning and major exercises; studies of military activities and capabilities.
- Strategic products include assessments supporting national and Army information requirements on foreign technology development; worldwide assessments of the organization, location, funding, training, operations capabilities and intentions of terrorist organizations; analyses of the capabilities of international narcotics trafficking organizations.

OPERATIONAL EMPLOYMENT

6-12. Army HUMINT must support the full spectrum of military operations. Commanders require a well-trained HUMINT force consisting of AC, RC, civilian government, and contractor personnel. The HUMINT force is focused on and dedicated to the collection of data and information relevant to the commander's PIRs and IRs. Effective employment of Army HUMINT elements in all phases of operations and at all levels from tactical to strategic will remain paramount to ensuring that commanders have the best pictures possible of their adversaries. HUMINT Collectors use the following collection methods.

LEVELS OF EMPLOYMENT

6-13. **Strategic and Departmental.** Strategic and departmental operations will be conducted by HUMINT elements supporting national, DOD, and DA required missions (for example, support to NATO and special operations and missions). The Army strategic HUMINT force resides in DHS. Trained Army HUMINT professionals must be available to augment DOD HUMINT efforts and coordinate between Army and Joint or DOD elements during crisis and war. Strategic HUMINT activities are generally carried out by departmental level HUMINT elements or Army HUMINT assets directly assigned to DOD and Joint positions. The Army overt HUMINT capability at the strategic level will be deployed and under operational control (OPCON) of the Army Service Component Commander (ASCC).

6-14. **Operational.** HUMINT operational missions must support combatant commanders, generally in geographic AORs. Operational HUMINT assets provide capabilities to support theater requirements for HUMINT collection. Operational HUMINT elements will focus on threat identification and capabilities. As with all levels of employment, HUMINT activities and functions will include DOCEX, collection, operations, analysis, and production. HUMINT elements must be capable of quickly transitioning from a peacetime mission to crisis operations to support combatant commander requirements. Operational elements may also be deployed to support or reinforce tactical forces in CONOPS.

6-15. **Tactical.** Army HUMINT play a crucial role in supporting tactical forces. HUMINT teams conduct operations throughout the battlespace during CONOPS. CONOPS support activities include debriefings, screenings, contact operations, and interrogations. HUMINT activities in CONOPS focus on the threat and assisting the senior intelligence officer (SIO) and commander in understanding the threat's decisionmaking process. During peacetime, organic tactical HUMINT teams conduct activities pursuant with approved regulations and command guidance.

SUPPORT TO CONTINGENCY OPERATIONS

6-16. **Initial Phase.** The initial phase of operations from peacetime military engagement (PME) to major theater war (MTW) lays the foundation of future team operations. In general, the priority of effort focuses inward on security of operating bases, areas of troop concentration, and C2 nodes to identify the collection threat to US forces that could be used by adversarial elements to plan hostile acts against US activities and locations.

6-17. **Continuation Phase.** Once security of the operating bases has been established, the operational focus of HUMINT teams shifts outside the operating base to continue to detect, identify, and neutralize the collection threat to US forces as well as to provide I&W of hostile acts targeting US activities. The HUMINT team uses several collection methods, to include HUMINT contact operations, elicitation, and liaison, to answer the supported commander's requirements.

6-18. **HUMINT Team.** A key element to the HUMINT team's success is the opportunity to spot, assess, and develop relationships with potential sources of information. Operating as independent teams, without being tied to combat assets, enables the HUMINT team's maximum interaction with the local population, thereby maximizing the pool of potential sources of information. A second key element of a HUMINT team's success is its approachability to the local population. A soft posture enables a HUMINT team to appear as non-threatening as possible. Experience has shown that the local population in general is apprehensive of fully and openly armed patrols and soldiers moving around population centers. During some operations civilian attire or non-tactical vehicles may be used to lower the

HUMINT team profile. NOTE: In some special situations, these measures are taken to make the operation less visible to the casual observer. Also, in some cultures, sharing food and beverages among friends is expected; exceptions to restrictions or general orders should be considered to facilitate successful HUMINT team operations, many of which are geared towards developing relationships with potential sources of information.

SUPPORT TO INSTALLATIONS AND OPERATING BASES

6-19. Commanders may restrict personnel to base camps and installations during the initial stages of operations (PME to MTW), when the operational environment is being assessed, or as a temporary expedient when the threat level exceeds the ability to provide reasonable FP. Operational restrictions minimize the risk to the HUMINT team, but minimizing its collection potential may increase the risk to the force as a whole. While confined to an installation or a base camp, the HUMINT team can maintain a limited level of information collection by—

- Screening locally employed personnel.
- Debriefing combat and ISR patrols.
- Debriefing friendly force personnel who are in contact with the local population.
- Conducting limited local open-source information collection.
- Contributing to the threat and vulnerability assessments of the base camp.

TACTICS, TECHNIQUES, AND PROCEDURES

6-20. At the HUMINT team level, team members conduct mission analysis and planning specific to their AO. Backwards planning and source profiling are used extensively to choose HUMINT targets. To verify adequate area coverage, the HUMINT team may periodically develop and use HUMINT target overlays and other HUMINT analytical tools that illustrate the HUMINT situation, identify gaps, and help refocus the collection effort.

6-21. The HUMINT team is also in constant contact with the supported S2 and the other ISR assets (Scouts, PSYOP, CA, and MPs) in order to coordinate and deconflict operations and to cross-check collected information. The supported unit S2, with the help of the HUMINT team, regularly and systematically debriefs all ISR assets.

6-22. The HUMINT team must be integrated into the supported unit's ISR plan. The HUMINT operational management team (OMT) chief will advise the supported unit on the specific capabilities and requirements of the team to maximize mission success.

OPERATIONAL RISK MITIGATION

6-23. The employment of HUMINT teams includes varying degrees of contact with the local population. As the degree of contact with the population increases, both the quantity and quality of collection increases. In many instances, however, there is a risk to the team inherent with increased exposure to the local population. The decision at what level to employ a team is METT-TC dependent. The risk to HUMINT assets must be balanced with the need to collect priority information and to protect the force as a whole. ROE, SOFA, direction from higher headquarters, and the overall threat level may also restrict the deployment and use of HUMINT teams. The commander should consider exceptions to the ROE to facilitate HUMINT collection.

6-24. Risks are minimized through the situational awareness of HUMINT team members. They plan and rehearse to readily react to any situation and carry the necessary firepower to disengage from difficult situations. If it becomes necessary to call for assistance, adequate and redundant communications equipment is critical. These scenarios and actions should be trained prior to deployment into a contingency area and rehearsed continuously throughout the deployment.

6-25. A supported unit commander is often tempted to keep the HUMINT team "inside the wire" when the threat condition (THREATCON) level increases. The supported commander must weigh the risk versus potential information gain when establishing operational parameters of supporting HUMINT teams. This is necessary especially during high THREATCON levels when the supported unit commander needs as complete a picture as possible of the threat arrayed against US and multinational forces.

6-26. When it is not expedient to deploy the HUMINT team independently due to threat levels or other restrictions, the team can be integrated into other ongoing operations. The HUMINT team may be employed as part of a combat, ISR, or MP patrol or used to support CA, PSYOP, engineer, or other operations. This method reduces the risk to the team while allowing a limited ability to collect information. It has the advantage of placing the team in contact with the local population and allowing it to spot, assess, and interact with potential sources of information. However, this deployment method restricts collection by subordinating the team's efforts to the requirements, locations, and timetables of the unit or operation into which it is integrated and does not allow for the conduct of sensitive source operations. **This method of employment should be considered a last resort.**

HUMINT EQUIPMENT

6-27. Basic C2, transportation, and weapons requirements do not differ significantly from most soldier requirements and are available as unit issue items. However, HUMINT teams have unique communications, collection, processing, and mission-specific requirements.

COMMUNICATIONS

6-28. Dedicated and secure long-range communications are key to the success of the HUMINT team mission. HUMINT team operations require a secure, three-tiered communications architecture consisting of inter/intra-team radios, vehicle-based communications, and a CI and HUMINT base station.

6-29. The HUMINT team must have access to existing communications networks such as the tactical local area network (LAN). The HUMINT team must also be equipped with its own COMSEC devices. It is imperative that the HUMINT team acquire access to the public communication system of the HN. This can be in the form of either landlines or cellular telephones. Such access enables the HUMINT team to develop leads which can provide early indicators to US forces.

Interoperability

6-30. Communications systems must be equipped with an open-ended architecture to allow for expansion and compatibility with other service elements, government organizations, NGOs, and multinational elements to effectively communicate during CONOPS. All ISR systems must be vertically and horizontally integrated to be compatible across all BOSs and with Legacy and Interim Force elements.

Satellite Communications On-The-Move (SOTM)

6-31. To provide real time or NRT information reporting, HUMINT elements must have the capability to transmit voice, data, imagery, and video while on the move. HUMINT teams must be able to transmit while geographically separated from their parent unit while operating remotely. This broadband requirement can only be achieved through a SATCOM capability and must be achievable while mobile.

HUMINT COLLECTION AND PROCESSING SYSTEMS

6-32. The HUMINT team must rely on automation to achieve and maintain information dominance in a given operation. With time, effective collection planning and management at all echelons, the HUMINT team can collect a wealth of information. The sorting and analysis of this information in a timely and efficient manner is crucial to operations. Automation helps the HUMINT team to report, database, analyze, and evaluate the collected information quickly and to provide the supported unit with accurate data in the form of timely, relevant, accurate, and predictive intelligence.

6-33. Automation hardware and software must be user friendly as well as interoperable among different echelons and services. They must interface with the communications equipment of the HUMINT team as well as facilitate the interface of audiovisual devices. Technical support for hardware and software must be available and responsive.

6-34. The demand for accurate and timely HUMINT reporting, DOCEX, and open-source information has grown tremendously. Biometric (physiological, neurological, thermal analysis, facial and fingerprint recognition) technologies will allow rapid identification; coding and tracking of adversaries and human sources; and cataloging of information concerning enemy prisoners of war (EPWs), detainees, and civilians of HUMINT interest on the battlefield. Biometrics will also provide secure authentication of individuals seeking network or facility access.

6-35. HUMINT teams work with multinational forces, and other foreign nationals, and require the ability to communicate in their respective languages. Often HUMINT personnel have little or no training in the target language, and lack of skilled interpreters can hinder HUMINT activities. HUMINT teams require textual and voice translation devices, source verification, and deception detection machines (biometrics) to improve collection capability and accuracy.

6-36. HUMINT teams require dynamic machine language translation (MLT) tools that provide both non-linguists and those with limited linguist skills a comprehensive, accurate means to conduct initial screenings and basic interviews in a variety of situations. HUMINT elements will focus on in-depth interviews and communications with persons of higher priority. MLT tools will minimize reliance on contract linguists and will allow soldiers to concentrate on mission accomplishment.

MISSION SPECIFIC

6-37. The HUMINT team may conduct night operations and must be equipped with NVDs for its members and photographic and weapons systems. The HUMINT team also may operate in urban and rural areas, where the threat level can vary from semi-hostile to hostile. The safety of the HUMINT team can be enhanced with equipment that can detect, locate, suppress, illuminate, and designate hostile optical and E-O devices. In addition, high power, gyro-stabilized binoculars, which can be used from a moving vehicle, also increase the survivability of the HUMINT team. It also gives the team another surveillance and collection device.

6-38. Some of the HUMINT team missions may require the documentation of incidents. The teams can use the following equipment in their open-source collection efforts.

- Small, rugged, battery-operated digital camcorders and cameras which are able to interface with the collection and processing systems as well as communication devices.
- GPSs that can be mounted and dismounted to move in the AO efficiently.
- Short-range multichannel RF scanning devices that can also identify frequencies which enhance their security.

6-39. In some cases HUMINT teams require a stand-off, high resolution optical surveillance and recording capability that can provide target identification at extended ranges to protect the intelligence collector while avoiding detection by the adversary target. An advanced optical capability provides intelligence collectors the ability to locate and track adversarial targets (passive and hostile) for identification, collection, and target exploitations.

INTEGRATION OF LINGUISTS

6-40. Integrating linguists into the HUMINT team should take place as soon as possible. Security clearances and contractual agreements will help the team determine the level of integration.

6-41. Along with the basic briefing of what is expected of the civilian linguists as interpreters, HUMINT teams should be informed about the civilians' chain of command and the scope of their duties beyond interpreting. The HUMINT team leader must ensure that linguists are trained and capable of completing all tasks expected of them.

BATTLE HAND-OFF

6-42. HUMINT teams are always engaged. A good battle hand-off is critical to smooth transition and mission success. The battle hand-off can directly contribute to mission success or failure of the outgoing team, but especially of the incoming team. The battle hand-off begins the first day the HUMINT team begins to operate in an AO. Regardless of how long the team believes it will operate within the AO, it must ensure there is a seamless transition to an incoming team, other US unit, or agency. The HUMINT team accomplishes this transition by establishing procedures for source administration, database maintenance, and report files.

6-43. Teams must plan and implement a logical and systematic sequence of tasks to enable an incoming team to assume the operations in the AO. Adequate time must be allotted for an effective battle hand-off. In some environments, a few weeks may be necessary to accomplish an effective battle hand-off. Introductions to sources of information, especially HUMINT contact operations sources, are critical and teams must prioritize their time. During this time the outgoing HUMINT team must familiarize the new HUMINT team with all aspects of the operation, which includes past, present, and planned activities within the AO. Area orientation is critical. These include major routes, population centers, potential hot spots, and other points of interest (such as police stations, political centers, and social centers).

ORGANIZATION

6-44. HUMINT activities require a complex C2 relationship to ensure that the requirements of the supported commander are fulfilled while balancing the need for strict integrity and legality of HUMINT operations. This complex relationship balances the role of the SIO as the requirements manager and the *2X as the mission manager with the MI commander as the asset manager. ("*2X" indicates 2X functions at all levels.)

COMMAND VERSUS CONTROL

6-45. ARFOR will normally deploy as part of a joint and/or multinational operation. In all cases, commanders at each echelon will exercise command over the forces assigned to their organization. Command includes the authority and responsibility for effectively using resources, planning for and employment of forces, and ensuring that forces accomplish assigned missions. Leaders and staffs exercise control to facilitate mission accomplishment.

6-46. While the MI commander supervises subordinates and produces reports, the *2X synchronizes activities between intelligence units and provides single-source processing and limited analysis. While the MI commander takes care of the operators executing missions, the *2X obtains the data and reports from higher echelons required to execute the missions.

STAFF RESPONSIBILITIES AND FUNCTIONS

*2X Staff

6-47. The *2X staff is responsible for the integration, correlation, and fusion of all Human Sensor information into the Intelligence BOS within the *2X

AOIR. The *2X is also responsible for analyzing adversary collection, terrorist and sabotage activities, developing countermeasures to defeat threat collection activities, identifying and submitting collection requirements to fill collection gaps, and providing input to the all-source picture regarding adversary intelligence activities.

*2X Staff Officer

6-48. The *2X Staff Officer provides CI and HUMINT collection expertise. The *2X—

- Is the single focal point for all matters associated with CI and HUMINT in the AOIR.
- Is the CI and HUMINT advisor to the G2 and commander.
- Is an extension of the collection manager and ensures that the best asset or combinations of assets are used to satisfy information requirements.
- Along with his subordinate elements (CICA, HOC, OSC, CI Analysis Cell [CIAC], and HAC), exercises technical control over his assigned Army CI and HUMINT elements in the designated AOIR.
- Is the principal representative of the G2 and the commander when coordinating and deconflicting CI and HUMINT activities with national or theater agencies operating in the AOIR.
- Supports specific RM efforts in conjunction with the requirements manager through—
  - Planning and coordinating CI and HUMINT operations.
  - Reviewing and validating HUMINT requirements.
  - Recommending assignment of tasks to specific collectors.
  - Conducting liaison with non-organic HUMINT collection. This liaison includes national level and multinational force assets for source deconfliction and special activities outside the *2X AOIR.
- Will provide OMTs with capability to reach back to current database information, technical information and guidance, and source deconfliction necessary to monitor the collection activities of the HUMINT teams.

The HUMINT Analysis Cell (HAC)

6-49. The HAC is the single-source fusion point for all HUMINT reporting and operational analysis. It determines gaps in reporting and coordinates with other analysis teams and technical controllers to cross-cue other collection sensor systems. The HAC—

- Uses analytical tools to develop long-term collection plans and provides reporting feedback that supports all HUMINT and CI entities in the supported command's AOIR.
- Produces and disseminates HUMINT products and provides input to intelligence summaries.
- Uses analytical tools found at the ACE or JISE to develop long-term analyses and provides reporting feedback that supports the J/G/S2X, HUMINT operations section, OMTs, and HUMINT teams.
- Produces country and regional studies tailored to HUMINT collection.

- Compiles target folders to assist J/G/S2X assets in focusing collection efforts.
- Analyzes and reports on trends and patterns found in HUMINT reporting.
- Analyzes source reliability and credibility as reflected in reporting and communicating that analysis to the collector.
- Develops and maintains databases specific to HUMINT collection activities that directly support the collection efforts of HUMINT teams and are directly accessible by HUMINT teams.
- Provides collection requirements input to the HOC.
- Supports RM through the development of HUMINT SIRs based on command PIRs.
- Answers HUMINT-related RFIs.

6-50. For intelligence reach operations, HUMINT products are available and disseminated in a variety of forms. It is incumbent on the requestor to ensure that the HUMINT product can be transmitted over the available communications systems. This includes verifying the appropriate security level of the communications systems.

HUMINT Operations Cell (HOC)

6-51. The HOC in the *2X coordinates and synchronizes all HUMINT activities in the AOIR. The HOC exercises technical control over all HUMINT entities in the designated AOIR and deconflicts HUMINT activities with higher, lower, and adjacent HUMINT elements. The HOC accomplishes all responsibilities through coordination with the operational units and the CICA and operations support cell (OSC). The HOC tracks all HUMINT activities in the AOR. The J/G2X uses this information to advise the SIO on all HUMINT activities conducted within the AO. The HOC—

- Exercises technical control of all HUMINT assets and coordinates and deconflicts HUMINT activities in the deployed AO.
- Establishes and maintains a HUMINT source database.
- Coordinates and supervises HUMINT FP source operations conducted by all services and components in the AO.
- Develops and manages collection requirements for HUMINT in coordination with the requirements manager.
- Develops and provides the HUMINT portion of the intelligence synchronization plan to the J/G/S2X and requirements manager for inclusion in the intelligence synchronization plan.
- Coordinates the activities of HUMINT collectors assigned or attached to interrogations and debriefing facilities.
- Expedites preparation of intelligence information reports and their distribution to consumers at all levels.
- Performs liaison with HN and US national HUMINT organizations.

Operations Support Cell

6-52. The OSC in the *2X staff maintains the source registry for all HUMINT activities in the designated AOIR. The OSC provides management of intelligence property book operations, source incentive programs, and ICFs.

6-53. Integrating linguists and DOD emergency-essential civilians, such as technical support contractors, into the HUMINT team should take place as soon as possible. Security clearances and contractual agreements will help the team determine the level of integration.

6-54. Along with the basic briefing of what is expected of the civilian linguists as interpreters and the emergency-essential civilians as support personnel, HUMINT teams should be informed about the civilians' chain of command and the scope of their duties beyond interpreting and technical support. The HUMINT team leader must ensure that linguists and emergency-essential civilians are trained and capable of completing all tasks expected of them.

Counterintelligence Coordinating Authority

6-55. The CICA coordinates all CI activities for a deployed force. There can be only one CICA in a theater of operations. When multiple echelons exist, the highest echelon has the CICA and subordinate G2X offices have a CICA. Depending on the size and scope of the operation, the CICA could be the unified command's CI staff officer; the CI or HUMINT staff officer from corps or division; or a senior warrant officer or branch CI officer designated by the unified command CI staff officer or task force commander. For more information about the CICA functions, see Chapter 11.

HUMINT TEAM STRUCTURE

Operational Management Team

6-56. The OMT is a four-person team consisting of a warrant officer (WO), two noncommissioned officers (NCOs), and a junior enlisted soldier. Civilians may be inserted into this structure when appropriate. Rank structure and standards of grade for OMTs vary depending upon the skill sets required and mission focus. HUMINT OMTs provide operational guidance for two to four HUMINT teams, depending on mission focus and operational tempo. When two or more HUMINT teams are deployed in DS of a maneuver element, an OMT also deploys to provide technical control. The OMT works closely with the supported S2 and analysis and control team (ACT) to furnish current threat information and answer the supported commander's PIRs and IRs. OMTs coordinate with the supported 2X and manage subordinate HUMINT team echelons to—

- Provide guidance and technical control of operational activity.
- Provide the collection and operational focus for HUMINT teams.
- Provide quality control and dissemination of reports for subordinate HUMINT teams.
- Conduct single-discipline HUMINT analysis, and assist in mission analysis for the supported commander.

- Act as a conduit between subordinate HUMINT teams, the HOC, and supported unit headquarters.
- Provide administrative support for subordinate HUMINT teams, to include reporting mission and equipment status to the HOC and the supported unit headquarters.
- Educate the supported commander on the capabilities of the HUMINT teams.
- Integrate the HUMINT teams directly into the maneuver commander's ISR planning.

HUMINT Team

6-57. The HUMINT team is a four-person team consisting of two NCOs and two junior enlisted personnel. Civilians may be inserted into this structure when appropriate. Rank structure and standards of grade for HUMINT teams vary depending upon the skill sets required and the mission focus. HUMINT teams are trained to execute the full range of HUMINT functions as defined in **HUMINT FUNCTIONS** section. However, they may be assigned to mission-focused elements such as DOCEX, interrogation, debriefing, or contact operations.

Chapter 7

Imagery Intelligence

DEFINITION

7-1. IMINT is intelligence derived from the exploitation of imagery collected by visual photography, infrared, lasers, multi-spectral sensors, and radar. These sensors produce images of objects optically, electronically, or digitally on film, electronic display devices, or other media.

ROLE

7-2. The role of imagery is to assist the commander in focusing and protecting his combat power. Imagery often enhances the commander's situational understanding of the battlespace. Other than direct human observation, imagery is the only intelligence discipline that allows the commander to see the battlefield in real time as the operation progresses. In those cases where maps are not available, digital imagery in hardcopy or softcopy can be used as a substitute. Imagery can also be used to update maps or produce grid-referenced graphics. Detailed mission planning often requires imagery, to include three-dimensional stereo images, in order to provide the degree of resolution necessary to support such specialized planning.

FUNDAMENTALS

7-3. Some imagery assets are very responsive to the individual commander's intelligence requirements. Some imagery systems can directly transmit imagery into the tactical operations center (TOC); examples include imagery from UAVs and the Joint Surveillance Target Attack Radar System (JSTARS). This direct downlink enables the G2/S2 to use the imagery as soon as possible instead of having to wait for finished imagery products. A note of caution is required, however, if not a trained imagery analyst because the G2/S2 could incorrectly interpret the imagery.

7-4. Imagery-related equipment has undergone a reduction in size as well as a reduction in the time it takes to provide products, particularly softcopy imagery. The modularity and size reduction of imagery analysis, processing, and display systems make transport easier; they also allow the commander to bring lesser amounts than in the past, while still retaining those systems (or subsystems) required to complete the mission. Additionally, data compression allows faster transmission of imagery products directly to the warfighter.

SOURCES OF IMAGERY

7-5. There are three general sources of imagery: national, civil, and commercial. National imagery traditionally refers to imagery collected by DOD imagery systems. However, there are other sources of imagery provided by non-national sources such as handheld photography (film, digital, and video), UAV imagery, and gun-camera images.

National

7-6. National systems are developed specifically for supporting the President of the United States, the Secretary of Defense, other national agencies, and US military forces. These systems respond to the needs of the nation and those of the combatant commands.

Civil

7-7. Civil imagery systems are usually government funded in terms of building, launching, and operating the system. In many, but not all, cases the agencies operating these civil imagery systems also process, distribute, and archive the imagery data or images.

Commercial

7-8. Commercial companies build, launch, and operate imagery systems for profit. In times of crises, license agreements with the US Government obligate US commercial satellite imaging systems to provide data only to the US Government at the market value. This protects information concerning US operations from adversarial exploitation from commercial systems such as the ACE Imagery Company. However, foreign commercial imagery systems are not bound to this arrangement, and thus may be used by our nation's adversaries. Commercial imagery has become increasingly valuable for many reasons:

- Due to its unclassified nature, civil and commercial imagery is useful in an open environment, may be released to multinational partners, and can be made available to the press. They are especially useful for geospatial products. Their use allows national systems more time to focus on other intelligence functions.
- Civil and commercial imagery offer radar and multi-spectral. Some offer large area collection useful for broad area coverage purposes.
- Commercial satellite imagery resolution varies from less than one meter to several kilometers.

7-9. The Central Imagery Tasking Office (CITO) is responsible for ordering commercial imagery. The Commercial Satellite Imagery Library is available to research DOD purchased commercial imagery. The G2/S2 should consult the CITO when forming commercial imagery requests. NGA will deliver the imagery primarily on CD-ROM media via courier or mail service. Limited digital or electronic delivery is available as well.

TYPES OF IMAGERY SENSORS

7-10. There are four types of imagery sensors: visible (optical), infrared, radar, and multi-spectral. Each sensor has a unique capability, with distinct advantages and disadvantages. The G2/S2 must understand each sensor's capability in order to select the best sensor for the mission and thus enable the user to better understand the intelligence received. Certain sensors are better suited for military operations than others. (See Table 7-1.)

Table 7-1. Sensor Characteristics Matrix.

SENSORS	ADVANTAGES	DISADVANTAGES
Visible (Optical) Best tool for daytime, clear weather, detailed analysis. Includes video and electro-optical.	• Affords a familiar view of a scene. • Offers system resolution that cannot be achieved in other optical systems or in thermal images and radars. • Preferred for detailed analysis and mensuration. • Offers stereoscopic viewing.	• Restricted by terrain and vegetation. • Limited to daytime use only. • Reduced picture size.
Infrared Best tool for nighttime, clear weather, detailed analysis. Includes Overhead Non-Imaging Infrared (ONIR).	• A passive sensor and is impossible to jam. • Offers camouflage penetration. • Provides good resolution. • Nighttime imaging capability.	• Not effective during thermal crossover periods. • Product not easily interpretable. • Requires skilled analysis. • Cannot penetrate clouds.
Radar Useful for detecting presence of objects at night and in bad weather. Includes synthetic aperture radar (SAR), coherent change detection (CCD), and MTI.	• All weather; can penetrate fog, haze, clouds, smoke. • Day or night use. • Does not rely on visible light nor thermal radiation. • Good standoff capability. • Large area coverage. • Allows moving target detection. • Foliage and ground penetration.	• Product not easily interpretable. • Requires skilled analysis. • Terrain masking inhibits use.
Multi-Spectral Imagery (MSI) Best tool for mapping purposes and terrain analysis.	• Large database available. • Band combinations can be manipulated to display desired requirements. • Images can be merged with other digital data to provide higher resolution.	• Product not easily interpretable. • Requires skilled analysis. • Computer manipulation requires large amounts of memory and storage; requires large processing capabilities.

PLAN

7-11. The first step in planning for IMINT is determining the need for IMINT products based on the PIRs and the initial IPB. The G2/S2 should research targets using online imagery databases early and request those imagery products that are not perishable for contingency planning. National and COCOM imagery databases may hold recently imaged targets that could meet the commander's immediate needs instead of requesting new imagery. The staff must clearly articulate their intelligence requirements to include communicating what the mission is and how the requested product will aid in mission accomplishment. The G2/S2 should submit the imagery requirement using established procedures such as those in the unit's SOP or as established by the COCOM.

7-12. The G2/S2 must also determine the specific imagery requirements so as not to burden the system with unnecessary requests. The desire for imagery products often exceeds the capabilities of the IMINT system. Therefore, it is imperative that the G2/S2 consider what type of analysis they require, and request only that which they require. The specifications of the request for IMINT products often affect the timeliness of the response. For example, determining if vehicles are tanks takes less time and requires less resolution than determining if a tank is a T-64 or a T-72.

7-13. Here are some of the roles IMINT can perform that the G2/S2 may consider when determining IMINT requirements:

- Imagery can detect and/or identify and locate specific unit types, equipment, obstacles, potential field fortifications, etc., from which intelligence analysts are able to analyze enemy capabilities and develop possible COAs.
- Imagery can also update maps and enhance the interpretation of information from maps. Detailed mission planning uses imagery to include stereo images for three-dimensional viewing of the terrain and many other geospatial uses.
- Imagery can be used as a substitute for maps when maps are not available. The most common application of this technique is by constructing imagery mosaics: a combination of two or more overlapping photographic prints that form a single picture.
- MTI displays or products can provide an NRT picture of an entity's movement by indicating its speed, location, and direction of travel. MTI systems do not differentiate friendly from enemy. Imagery assets, particularly MTI systems, are useful in cueing other ISR systems.
- Imagery can be used to support protection of the force by helping the commander visualize how his forces look—including their disposition, composition, and vulnerabilities—as exploited by enemy IMINT systems.
- Imagery analysts use combat assessment imagery to confirm destruction, determine the percentage of destruction, or whether the target was unaffected.

PREPARE

7-14. The G2/S2 IMINT-related actions during the prepare function of the intelligence process include establishing or verifying the portion of the intelligence communications architecture that supports IMINT display and analysis functions properly. Additionally, the G2/S2 must ensure that required IMINT analytical assets and resources are prepared to provide support or are available through intelligence reach. Lastly, the G2/S2 must also ensure IMINT reporting and dissemination channels and procedures are in place and rehearsals are conducted with all pertinent IMINT elements to ensure interoperability.

COLLECT

7-15. As previously mentioned, there are four types of imagery sensors. Depending on the type of sensor, it can record hardcopy or softcopy single frame or continuous (video). A given imagery target will not necessarily receive continuous coverage due to the possible conflict between the number and priority of targets and the number and availability of imagery assets. However, a commander may decide to have continuous surveillance of certain targets, for specified periods of time, usually using his own imagery assets (for example, UAV) even though this detracts from the commander's ability to use these assets for other imagery targets within his AOI.

PROCESS

7-16. The process function regarding IMINT involves converting imagery data into a form that is suitable for performing analysis and producing intelligence. Examples of IMINT processing include developing film, enhancing imagery, converting electronic data into visual displays or graphics, and constructing electronic images from IMINT data.

PRODUCE

7-17. The IMINT producer must ensure the IMINT product satisfies the associated intelligence requirements and that the product is in the required format. The quality and resolution of the product is highly dependent upon the type of sensor, the time of day, and the weather conditions, as well as the imagery analyst's ability to identify items, vehicles, equipment, and personnel within the images. Specific IMINT products are discussed in the ANALYZE section below.

ANALYZE

7-18. Timeliness is critical not only to IMINT collection but also to IMINT analysis and reporting. It is difficult to separate IMINT reporting from IMINT analysis in this discussion. This is demonstrated by the three phases of IMINT reporting presented below; all are dependent upon the timeliness requirements. Each phase represents a different degree of analysis and period of time available to accomplish the exploitation of the imagery.

- First Phase imagery analysis is the rapid exploitation of newly acquired imagery and reporting of imagery-derived information within a specified time from receipt of imagery. This phase satisfies priority requirements of immediate need and/or identifies changes or activity of immediate significance. First Phase imagery analysis results in an Initial Phase Imagery Report (IPIR).
- Second Phase imagery analysis is the detailed exploitation of newly acquired imagery and the reporting of imagery-derived intelligence and information while meeting the production and timeliness requirements. Other intelligence discipline source material may support phase two imagery as appropriate. Second phase imagery analysis results in a Supplemental Imagery Report (SUPR).
- Third Phase imagery analysis is the detailed analysis of all available imagery pertinent to an SIR, and the subsequent production and reporting resulting from this analysis within a specified time. This

phase provides an organized detailed analysis of an imagery target or topic, using imagery as the primary data source but incorporating data from other sources as appropriate.

7-19. The two types of imagery exploitation are national and direct support (DS).

- National exploitation is imagery exploitation that supports presidential requirements, National Security Council (NSC) requirements, congressional requirements, or requirements of a common concern to the intelligence community.
- DS exploitation is imagery exploitation that supports assigned missions of a single agency, department, or command (the warfighter).

7-20. IMINT assets will complete DS exploitation in order to satisfy (First Phase) requirements and report the results as soon as possible, but not later than 24 hours after receipt of the imagery. Collectors will complete national exploitation in order to satisfy (Second and Third Phases) requirements and report the results within the time specifications of each individual requirement.

DISSEMINATE

7-21. IMINT products are distributed or disseminated in hardcopy, softcopy, or direct viewing such as through remote terminals. The distribution of hardcopy products will be via couriers or other type of mail system. The dissemination of softcopy products will be either as hardcopy products (for example, CD-ROM and 3.5-inch disks) or electronically. The requestor must ensure that the requested product is transmittable over the available communications systems.

ASSESS

7-22. The requestor should immediately assess the imagery product upon receipt for accuracy and relevance to the original request. The requestor must then notify the producer and inform him of the extent to which the product answered the PIR. Providing feedback to the producer regarding the product helps ensure the producer will provide the required information in the correct format. The following are some of the questions which the requestor should consider when providing feedback to the producer.

- Is the format of the product acceptable?
- Is additional information needed on the product or future products?
- Is excess information included on the product?

Chapter 8

Signals Intelligence

DEFINITION

8-1. SIGINT is a category of intelligence comprising either individually or in combination all COMINT, ELINT, and FISINT, however transmitted; intelligence is derived from communications, electronics, and foreign instrumentation signals. SIGINT has three subcategories:

- COMINT – The intelligence derived from foreign communications by other than the intended recipients.
- ELINT – The technical and geo-location intelligence derived from foreign non-communications electromagnetic radiations emanating from other than nuclear detonations or radioactive sources.
- FISINT – Technical information and intelligence derived from the intercept of foreign electromagnetic emissions associated with the testing and operational deployment of non-US aerospace, surface, and subsurface systems. Foreign instrumentation signals include but are not limited to telemetry, beaconry, electronic interrogators, and video data links. (See JP 1-02)

8-2. SIGINT provides intelligence to the commander based upon intercepted communications and provides transmitter location data.

ROLE

8-3. SIGINT provides intelligence on threat capabilities, disposition, composition, and intentions. In addition, SIGINT provides targeting information for the delivery of lethal and non-lethal effects.

FUNDAMENTALS

8-4. It is important that the G2/S2 understand how SIGINT assets are organized not only within the Army but also throughout the DOD. The majority of SIGINT assets are located at EAC. SIGINT assets from all the armed services, combined with national SIGINT assets, work together to support commanders from the tactical to the strategic level. Only by understanding the SIGINT structure that transcends traditional service component boundaries can the G2/S2 understand how to use SIGINT effectively.

TECHNICAL CONTROL AND ANALYSIS ELEMENT (TCAE) ARCHITECTURE

8-5. The TCAE architecture supports Army SIGINT collectors and analysts by providing SIGINT-specific intelligence and guidance to SIGINT personnel. The SIGINT technical architecture complements existing C2 relationships; it does not replace the commander's authority or chain of command.

8-6. There is a TCAE at each Army operational echelon. The TCAE at each echelon provides a single POC and resource dedicated to supporting the commander at that level. TCAEs or SIGINT elements at varying echelons draw information from each other in order to provide a more complete and detailed intelligence picture for their respective commanders.

ARMY TCAE (ATCAE)

8-7. The ATCAE, established at the national level, plays a significant role in TCAE operations by providing technical support oversight and providing access to national databases—in coordination with the NSA and other intelligence organizations. Additionally, the ATCAE works closely with NSA's IO element assisting in operations, monitoring technical capabilities, and providing liaison with Army's 1st Information Operations Command (Land). (This unit was previously called Land Information Warfare Activity.)

8-8. The ATCAE supports the Army's Quick Reaction Capability (QRC) SIGINT and signals research and target development (SRTD) operations. The QRC systems are responsive to the ground force commander's requirements and allow the ability to conduct SIGINT operations against modern communications systems.

REGIONAL TCAE (RTCAE)

8-9. The Army established RTCAEs to allow supported units at all echelons access to SIGINT regarding their respective AOI. The Army placed an RTCAE within each Regional Security Operations Center (RSOC). An RSOC is a joint service facility established by the Director, NSA, to conduct continuous security operations on selected targets in support of national and warfighter intelligence requirements using remoting technologies. The RSOC embeds a focus on the tactical commander's SIGINT requirements at a national level. It also creates a framework for enhanced interoperability among service SIGINT activities, especially within the context of a deployed JTF.

8-10. In addition to each RSOC, the RTCAE works closely with Theater TCAEs as well as Corps and Division SIGINT elements with the same regional focus. It provides analytic support and tailored products and answers RFIs. During peacetime and real-world operations RTCAEs facilitate intelligence reach support. They provide tailored technical support packages, support for surge and survey operations, analytic expertise, and signal data access and digital file transfers of live collection for Army tactical units.

THEATER TCAE

8-11. The Theater TCAE performs SIGINT and EW technical control and analysis and management. It provides SIGINT technical support for assigned, OPCON, and lower echelon SIGINT resources deployed in the theater. This includes mission tasking, processing, analyzing, and reporting of SIGINT data, information, and intelligence. The TCAE provides direction for the Theater Collection and Exploitation Battalion's SIGINT mission and for other theater tactical SIGINT assets.

SIGINT THEATER INTELLIGENCE BRIGADE/GROUP (TIB/TIG)

8-12. The TIB/TIG conducts operational level multidiscipline SIGINT operations. It provides timely intelligence to the commander throughout the full spectrum of operations. The TIB/TIG is structured to provide ground and aerial SIGINT support. It executes the full range of SIGINT missions or tasks.

SIGINT AT ECHELONS CORPS AND BELOW

8-13. There is a varying mixture of SIGINT assets within each corps. A reference listing the types of SIGINT assets found at echelons corps and below is listed in ST 2-50.

PLAN

8-14. An important SIGINT planning consideration is that if at all possible, SIGINT collection should be employed in conjunction with another intelligence discipline collection system. SIGINT is often used to cue, and be cued by, other ISR assets.

8-15. During planning, the G2/S2 should retrieve, update, or develop any required SIGINT databases. This includes effecting coordination with other SIGINT assets or elements that can support the operation.

8-16. SIGINT ground collection assets are usually placed in proximity to enemy signal sources due to the limited height of collector antennas, the low power output of threat or enemy emitters, and the line-of-sight (LOS) constraints imposed by terrain. Ground-based SIGINT teams are most effective when positioned to—

- Maximize threat emitter interception. This allows teams to overcome the constraints of threat emitter characteristics usually allowed by proximity to the threat.
- Minimize system receiver interference. This increases the potential capability of the team to acquire threat emitters in a timely manner.
- Optimize overlapping areas of intercept coverage. This ensures coverage of the AOIR and allows targets to be handed off from team to team if necessary.

8-17. Aerial SIGINT assets have additional planning requirements. Coordination must be effected with Army Airspace Command and Control (A2C2) for ingress and egress routes, restricted operating zones (ROZs) where the aircraft can travel on an intercept track, as well as with the supporting air operations center to determine the availability of asset types and times.

PREPARE

8-18. The G2/S2 ensures the SIGINT unit and asset leaders have effected all necessary coordination and conducted rehearsals. This includes establishing or verifying the operation of the SIGINT technical architecture. The G2/S2 also ensures all required SIGINT assets and resources are available, SIGINT reporting and dissemination channels and procedures are in place, and connectivity and interoperability exist with all pertinent SIGINT elements.

8-19. SIGINT OPCON is the authoritative direction of SIGINT activities, including tasking and allocation of effort, and the authoritative prescription of those uniform techniques and standards by which SIGINT information is collected, processed, and reported.

8-20. SIGINT operational tasking is the authoritative operational direction of and direct levying of SIGINT information needs by a military commander on designated SIGINT resources. These requirements are directive, regardless of other priorities, and are conditioned only by the capability of those resources to produce such information. Operational tasking includes authority to deploy all or part of the SIGINT resources for which SIGINT operational tasking authority (SOTA) has been delegated.

8-21. SOTA is the military commander's authority to operationally direct and levy SIGINT requirements on designated SIGINT resources; it includes authority to deploy and redeploy all or part of the SIGINT resources for which SOTA has been delegated.

COLLECT

8-22. SIGINT performs two major collection activities: signals intercept and direction finding (DF).

Signals Intercept

8-23. Signals intercept are those SIGINT actions used to search for, intercept, and identify threat electromagnetic signals for the purpose of immediate threat recognition. Signals intercept provides information required to answer PIRs, and other intelligence requirements in support of the ISR effort.

Direction Finding

8-24. Even when threat radio operators use COMSEC procedures, SIGINT teams can often intercept and approximate the location of the threat's signals. Specifically, SIGINT teams can use DF to determine—

- Movement of threat personnel or equipment.
- Locations of emitters associated with weapon systems and units.
- New emitter locations and confirm known emitter locations.
- Possible friendly targets the enemy intends to attack (lethal and non-lethal).

8-25. In addition to finding threat forces, DF operations can assist the (radio-equipped) friendly force by—

- Locating and vectoring assets or units during limited visibility.
- Locating downed aircraft and personnel radio beacons.
- Conducting signal security assessments.
- Locating sources of communication interference and jamming.

PROCESS

8-26. SIGINT processing involves converting intercepts of SIGINT into written and verbal reports, automated message, graphic displays, recordings, and other forms suitable for analysis and intelligence production. Since US forces routinely conduct operations against adversaries who speak languages other than English, SIGINT processing often also includes translation of these intercepts.

8-27. Due to the complexity of many SIGINT systems, automated processing may occur several times before SIGINT data or information receives any human interaction.

PRODUCE

8-28. The SIGINT producer must ensure the SIGINT product satisfies the associated intelligence requirements and that the product is in the required format. The quality, fidelity, and timeliness of SIGINT products are highly dependent upon the type of intercept, the collection system, the system's position in relation to the threat emitter, the weather (including space-based weather), as well as the SIGINT operator's ability to identify the appropriate threat signal activity.

8-29. SIGINT production results in some of the reports and formats mentioned in the process section above; however, the objective for SIGINT is to be used in an all-source analytical approach.

ANALYZE

8-30. The intelligence staff analyzes intelligence and information about the enemy's communications capabilities to determine appropriate SIGINT collection strategies. Conversely, a corresponding analysis of the friendly forces' SIGINT capabilities must be conducted to ensure the continued effectiveness of, or to improve upon, SIGINT collection.

8-31. SIGINT analysts also sort through large amounts of SIGINT and information and intelligence to identify and use only that which is pertinent to the CCIRs (PIRs and FFIRs).

DISSEMINATE

8-32. SIGINT of critical importance to the force, including answers to the CCIRs (PIRs and FFIRs), is disseminated via the most expeditious means possible. Due to the usually highly perishable nature of SIGINT, the most expeditious reporting means is often immediately augmented with a follow-up report or augmented by a report transmitted through additional means, enhancing the probability of receipt. Sometimes the most expeditious means of reporting critical SIGINT information to the commander is face to face.

8-33. For intelligence reach operations, SIGINT products are available and disseminated in a variety of forms: hardcopy, softcopy, direct viewing, or listening (television or radio). It is incumbent on the requestor to ensure that the SIGINT product can be transmitted over the available communications systems. This includes verifying the appropriate security level of the communications system.

ASSESS

8-34. The primary goal of the assess function when applied to SIGINT is to determine whether the results of SIGINT collection meet the requirements of the unit's ISR effort. SIGINT producers must assess all facets of SIGINT operations, from receipt of the ISR task to the dissemination of SIGINT, in an effort to determine their effectiveness. This assessment is not only directed at each SIGINT asset individually but also throughout the supporting SIGINT architecture and the unit's entire ISR effort.

8-35. The G2/S2 immediately assesses SIGINT products upon receipt for accuracy and relevance. He must inform the SIGINT producer of the extent to which the product answered the PIR or intelligence requirement. Providing feedback to the SIGINT producer—and collector—helps improve the effectiveness and efficiency of SIGINT.

Chapter 9

Measurement and Signatures Intelligence

DEFINITION

9-1. MASINT is technically derived intelligence that detects, locates, tracks, identifies, and/or describes the specific characteristics of fixed and dynamic target objects and sources. It also includes the additional advanced processing and exploitation of data derived from IMINT and SIGINT collection.

9-2. MASINT collection systems include but are not limited to radar, spectroradiometric, E-O, acoustic, RF, nuclear detection, and seismic sensors, as well as techniques for gathering NBC and other material samples.

9-3. It requires the translation of technical data into recognizable and useful target features and performance characteristics. Computer, communication, data, and display processing technologies now provide MASINT in support of commanders throughout the full spectrum of operations.

9-4. The subdisciplines within MASINT include, but are not limited to, the following:

- **Radar Intelligence (RADINT).** The active or passive collection of energy reflected from a target or object by LOS, bistatic, or over-the-horizon radar systems. RADINT collection provides information on radar cross-sections, tracking, precise spatial measurements of components, motion and radar reflectance, and absorption characteristics for dynamic targets and objectives. A SAR system, coupled with advanced MASINT processing techniques—
  - Provides a high resolution, day and night collection capability.
  - Can produce a variety of intelligence products that identify or provide change detection, terrain mapping, underwater obstacles, dynamic sensing of targets in clutter, and radar cross-section signature measurements.
- **Frequency Intelligence.** The collection, processing, and exploitation of electromagnetic emissions from a radio frequency weapon (RFW), an RFW precursor, or an RFW simulator; collateral signals from other weapons, weapon precursors, or weapon simulators (for example, electromagnetic pulse signals associated with nuclear bursts); and spurious or unintentional signals.
  - *Electromagnetic Pulses.* Measurable bursts of energy that result from a rapid change in a material or medium, resulting in an explosive force, produces RF emissions. The RF pulse emissions associated with nuclear testing, advanced technology devices, power and propulsion systems, or other impulsive events can be used to detect, locate, identify, characterize, and target threats.

- *Unintentional Radiation Intelligence (RINT).* The integration and specialized application of MASINT techniques against unintentional radiation sources that are incidental to the RF propagation and operating characteristics of military and civil engines, power sources, weapons systems, electronic systems, machinery, equipment, or instruments. These techniques may be valuable in detecting, tracking, and monitoring a variety of activities of interest.

- **E-O Intelligence.** The collection, processing, exploitation, and analysis of emitted or reflected energy across the optical portion (ultraviolet, visible, and infrared) of the EMS. MASINT E-O provides detailed information on the radiant intensities, dynamic motion, spectral and spatial characteristics, and the materials composition of a target. E-O data collection has broad application to a variety of military, civil, economic, and environmental targets. E-O sensor devices include radiometers, spectrometers, non-literal imaging systems, lasers, or laser radar (LIDAR).
  - *Infrared Intelligence (IRINT).* A subcategory of E-O that includes data collection across the infrared portion of the EMS where spectral and thermal properties are measured.
  - *LASER Intelligence (LASINT).* Integration and specialized application of MASINT E-O and other collection to gather data on laser systems. The focus of the collection is on laser detection, laser threat warning, and precise measurement of the frequencies, power levels, wave propagation, determination of power source, and other technical and operating characteristics associated with laser systems—strategic and tactical weapons, range finders, and illuminators.
  - *Hyperspectral Imagery (HSI).* A subcategory of E-O intelligence produced from reflected or emitted energy in the visible and near infrared spectrum used to improve target detection, discrimination, and recognition. HSI can detect specific types of foliage—supporting drug-crop identification; disturbed soil—supporting the identification of mass graves, minefields, caches, underground facilities or cut foliage; and variances in soil, foliage, and hydrologic features—often supporting NBC contaminant detection.

- **Spectroradiometric Products.** Include E-O spectral (frequency) and radiometric (energy) measurements. A spectral plot represents radiant intensity versus wavelength at an instant in time. The number of spectral bands in a sensor system determines the amount of detail that can be obtained about the source of the object being viewed. Sensor systems range from multispectral (2 to 100 bands) to hyperspectral (100 to 1,000 bands) to ultraspectral (1,000+ bands). More bands provide more discrete information, or greater resolution. The characteristic emission and absorption spectra serve to fingerprint or define the makeup of the feature that was observed. A radiometric plot represents the radiant intensity versus time. An example is the radiant intensity plot of a missile exhaust plume as the missile is in flight. The intensity or brightness of the object is a function of several conditions including its temperature, surface properties or material, and how fast it is moving. For each point along a time-intensity radiometric plot, a spectral plot can be generated based on the number of spectral bands in the collector.

- **Geophysical Intelligence.** Geophysical MASINT involves phenomena transmitted through the earth (ground, water, atmosphere) and manmade structures including emitted or reflected sounds, pressure waves, vibrations, and magnetic field or ionosphere disturbances.
  - *Seismic Intelligence.* The passive collection and measurement of seismic waves or vibrations in the earth surface.
  - *Acoustic Intelligence.* The collection of passive or active emitted or reflected sounds, pressure waves or vibrations in the atmosphere (ACOUSTINT) or in the water (ACINT). ACINT systems detect, identify, and track ships and submarines operating in the ocean.
  - *Magnetic Intelligence.* The collection of detectable magnetic field anomalies in the earth's magnetic field (land and sea). An example is a Remotely Emplaced Battlefield Surveillance System (REMBASS) sensor detection indicating the presence and direction of travel of a ferrous object.
- **Nuclear Intelligence (NUCINT).** The information derived from nuclear radiation and other physical phenomena associated with nuclear weapons, reactors, processes, materials, devices, and facilities. Nuclear monitoring can be done remotely or during onsite inspections of nuclear facilities. Data exploitation results in characterization of nuclear weapons, reactors, and materials. A number of systems detect and monitor the world for nuclear explosions, as well as nuclear materials production.
- **Materials Intelligence.** The collection, processing, and analysis of gas, liquid, or solid samples. Materials intelligence is critical to collection against NBC warfare threats. It is also important to analyzing military and civil manufacturing activities, public health concerns, and environmental problems. Samples are both collected by automatic equipment, such as air samplers, and directly by humans. Samples, once collected, may be rapidly characterized or undergo extensive forensic laboratory analysis to determine the identity and characteristics of the sources of the samples.

ROLE

9-5. MASINT provides intelligence to the commander throughout the full spectrum of operations to facilitate situational understanding. MASINT can thwart many of the camouflage, concealment, and deception techniques currently used to deceive ISR systems.

9-6. MASINT is perceived as a "strategic" discipline with limited "tactical" support capabilities. But, by application of real-time analysis and dissemination, MASINT has a potential ability to provide real-time situation awareness and targeting not necessarily available to the classic disciplines. Specifically, MASINT "sensors" have unique capabilities to detect missile launch, detect and track aircraft, ships, and vehicles; do non-cooperative target identification (NCTI), combat assessment, and BDA; and detect and track fallout from nuclear detonations. Often, these contributions are the first indicators of hostile activities. For example, two EXOCET-equipped

Mirage F-1s were shot down during the Operation DESERT STORM due to MASINT collection and analysis. As evidenced by Operation IRAQI FREEDOM (OIF), MASINT will play a decisive role in the targeting of smart munitions with the signatures (fingerprint) of the targets they are seeking (for example, infrared signatures).

9-7. The MASINT systems most familiar on today's battlefield are employed by ground surveillance and NBC reconnaissance elements.

9-8. MASINT spans the entire EMS and its capabilities complement, rather than compete with, the other intelligence disciplines. MASINT provides, to varying degrees, the capability to—

- Use automatic target recognition (ATR) and aided target recognition (AiTR).
- Penetrate manmade and/or natural camouflage.
- Penetrate manmade and/or natural cover, including the ability to detect subterranean anomalies or targets.
- Counter stealth technology.
- Detect recently placed mines.
- Detect natural or manmade environmental disturbances in the earth's surface not discernible through other intelligence means.
- Provide signatures (target identification) to munitions and sensors.
- Enhance passive identification of friend or foe.
- Detect the presence of NBC agents to include prior to, during, or after employment.
- Detect signature anomalies that may affect target-sensing systems.

FUNDAMENTALS

9-9. Before discussing the functions of the intelligence process within a MASINT, the following paragraph provides an overview of organizational structure of MASINT.

9-10. Within DOD, the DIA provides central coordination for MASINT collection efforts through the Central MASINT Office. Each service, in turn, has a primary command or staff activity to develop requirements and coordinate MASINT effort. Army responsibility currently resides with INSCOM. Army weapons systems programs that require MASINT information to support system design or operations submit requests through the Army Reprogramming Analysis Team (ARAT) or INSCOM channels for data collection and processing. The S&TI community also performs MASINT collection and processing primarily to support R&D programs and signature development. Every S&TI center has some involvement in MASINT collection or production that reflects that center's overall mission (for example, NGIC does work on armored vehicles and artillery). Service R&D centers such as the Communications-Electronics Command (CECOM) Research, Development, and Engineering Center (RDEC), and Night Vision and Electronic Systems Laboratory are also involved in developing sensor systems for collecting and processing MASINT.

9-11. In addition to supporting the S&TI mission, INSCOM units also execute limited ground-based operational collection to support Theater and ASCC PIRs. This capability will expand upon the standup of INSCOM TIBs and TIGs.

PLAN

9-12. Some MASINT sensors can provide extremely specific information about detected targets, whereas other sensors may only be capable of providing an indication that an entity was detected. Additionally, there are varying capabilities of detection, identification, and classification among MASINT sensors. It is these varying capabilities that require synchronizing the employment of MASINT sensors both within the MASINT discipline and within the ISR effort as a whole. See FM 2-01 for more specific information on ISR synchronization.

9-13. As previously mentioned, there are many types of MASINT sensors. Depending on the type of sensor employed, a given MASINT collection target or NAI may not necessarily receive continuous coverage due to the possible conflict between the number and priority of targets and the number and availability of MASINT assets. However, a commander may decide to have continuous surveillance of certain targets by using his own MASINT assets (for example, REMBASS or Improved-REMBASS).

9-14. Another consideration when planning MASINT missions is whether to use active, passive, or a combination of both when planning MASINT coverage.

PREPARE

9-15. The G2/S2 MASINT related actions during the prepare function of the intelligence process include properly establishing or verifying the MASINT portion of the intelligence communications architecture functions. Additionally, the G2/S2 must ensure that required MASINT analytical assets and resources are prepared to provide support or are available through intelligence reach. Since the products of MASINT are not as well known as products from other intelligence disciplines, the G2/S2 must be aware of the types of MASINT products available to support the operation, and then educate the rest of his unit's staff on the use of these MASINT products. Lastly, the G2/S2 must also ensure MASINT reporting and dissemination channels and procedures are in place and rehearsals are conducted with all pertinent MASINT elements to ensure interoperability.

COLLECT

9-16. MASINT provides information required to answer PIRs and other intelligence requirements in support of the ISR effort. As stated earlier in this chapter, MASINT collection must not only be synchronized within its own discipline but also be synchronized and integrated into the unit's overall ISR effort in order to be effective.

9-17. MASINT sensors are employed throughout the full spectrum of operations from a variety of platforms—sub-surface, ground, marine, and aerospace.

PROCESS

9-18. Just as in the other intelligence disciplines, MASINT involves dealing with huge volumes of data that have to be processed before beginning analysis and production. The process function regarding MASINT involves converting esoteric data into a form that is suitable for performing analysis and producing intelligence. MASINT processing can include relatively simple actions such as converting a REMBASS sensor activation into a report, to a complex task such as processing HSI into a report identifying the composition and concentrations of carcinogenic chemicals present in the emissions from a factory upwind from a US forces encampment.

PRODUCE

9-19. Effective and timely MASINT requires personnel with diverse skill sets. The MASINT producer must ensure the MASINT product satisfies the associated intelligence requirements and that the product is in the required format. The quality, fidelity, and timeliness of MASINT products are highly dependent upon the type of target, the collection system, the system's position in relation to the target or NAI, and the weather, as well as the MASINT system operator's ability to identify the appropriate threat activity.

9-20. The objective of MASINT production is to be used in an all-source analytical approach.

ANALYZE

9-21. The intelligence staff analyzes intelligence and information about the enemy's equipment, doctrine, and TTP to determine appropriate MASINT collection strategies. Conversely, a corresponding analysis of the friendly force's MASINT capabilities must be conducted to ensure the continued effectiveness of, or to improve upon, MASINT collection.

DISSEMINATE

9-22. MASINT of critical importance to the force, including answers to the PIRs, is disseminated via the most expeditious means possible.

9-23. For intelligence reach operations, MASINT products are available and disseminated in a variety of forms. The requestor must ensure that the MASINT product can be transmitted over the available communications systems. This includes verifying the appropriate security level of the communications system.

ASSESS

9-24. The primary goal of the MASINT assess function is to determine whether the results of MASINT collection and production meet the requirements of the unit's ISR effort. MASINT producers must assess all facets of MASINT operations, from receipt of the ISR task to the dissemination of MASINT, in an effort to determine the effectiveness of MASINT. This assessment is not only directed at each MASINT asset individually but also throughout the supporting intelligence communications architecture, to include intelligence reach and the unit's entire ISR effort.

9-25. Also, the G2/S2 immediately assesses MASINT products upon receipt for accuracy and relevance. He must inform the MASINT producer of the extent to which the product answered the PIR or intelligence requirement. Providing feedback to the MASINT producer—and collector—helps improve the effectiveness and efficiency of MASINT.

Chapter 10

Technical Intelligence

DEFINITION

10-1. TECHINT is intelligence derived from the collection and analysis of threat and foreign military equipment and associated materiel.

ROLE

10-2. The strength of the US military lies, in part, to the diversity and extent of its technology base. While the US aspires to be the leader in integrating technology, the threat can achieve temporary technological advantage in certain areas by acquiring modern systems or capabilities. The world arms market is willing to provide these advanced systems to countries or individuals with the resources to pay for them. A concerted TECHINT program is vital to providing precise direction and purpose within the US R&D process to ensure quick and efficient neutralization of this advantage.

10-3. The role of TECHINT is to ensure that the warfighter understands the full technological capabilities of the threat. With this understanding, the warfighter can adopt appropriate countermeasures, operations, and tactics.

10-4. TECHINT has two goals within its role:

- To ensure the US armed forces maintain technological advantage against any adversary.
- To provide tailored, timely, and accurate TECHINT support to the warfighter throughout the entire range of military operations. This includes providing US forces intelligence, information, and training on foreign weapons systems to an extent that allows their use of CEE.

FUNDAMENTALS

10-5. The G2/S2 must understand how TECHINT assets are organized in order to properly apply the intelligence process, as the majority of TECHINT assets are located at EAC.

DEFENSE INTELLIGENCE AGENCY

10-6. DIA manages and reviews overall TECHINT activities. The S&TI Directorate within DIA is the action element for TECHINT. This directorate coordinates with external TECHINT agencies on non-policy matters concerning the production of S&TI. The following organizations provide TECHINT support under the control of DIA:

- **Armed Forces Medical Intelligence Center (AFMIC).** AFMIC, based at Fort Detrick, MD, is a DOD intelligence production center under DIA control. AFMIC is responsible for exploiting foreign medical materiel. The director supports the Army Foreign Materiel Exploi-tation

Program (FMEP) and Army medical R&D requirements. The director coordinates planning, programming, and budgeting with the Army Deputy Chief of Staff, Intelligence (DCS, G2).

- **Missile and Space Intelligence Center (MSIC).** MSIC, based at Redstone Arsenal, AL, is a DOD intelligence production center under DIA control and supports the FMEP. The MSIC acquires, produces, maintains, and disseminates S&TI pertaining to missile and space weapons systems, subsystems, components, and activities. The S&TI produced at MSIC also covers foreign state-of-the-art technology and research applicable to missiles.
- **Defense HUMINT Service.** DHS conducts worldwide HUMINT operations in support of foreign materiel acquisition (FMA) and foreign materiel exploitation (FME).

10-7. The organizations and agencies discussed below constitute the Army TECHINT structure.

ARMY DCS, G2

10-8. The Army DCS, G2 exercises general staff responsibility for all Army TECHINT activities. The Army DCS, G2 forms policies and procedures for S&TI activities, supervises and carries out the Army S&TI program, coordinates DA staff and MSC requirements for TECHINT, and is responsible for the Army Foreign Materiel Program (FMP).

INSCOM

10-9. Under the direction of Headquarters, Department of the Army (HQDA), INSCOM is responsible for peacetime TECHINT operations. HQ, INSCOM, fulfills its responsibilities through its TECHINT oversight function and manages the Army's Foreign Materiel for Training (FMT) Program and FMEP. It provides the interface with strategic S&TI agencies in support of FME and organizes, trains, and equips EAC TECHINT organizations during peacetime. TECHINT exploitation within INSCOM is performed by the following elements:

- **National Ground Intelligence Center.** HQ, INSCOM, exercises direct OPCON over the NGIC. NGIC produces and maintains intelligence on foreign scientific developments, ground force weapons systems, and associated technologies. NGIC analysis includes but is not limited to military communications electronics systems, types of aircraft used by foreign ground forces, NBC systems, and basic research in civilian technologies with possible military applications.
- **203d Military Intelligence Battalion.** The 203d MI Battalion is a multi-component unit headquartered at Aberdeen Proving Ground, MD, and is the Army's sole TECHINT battalion. It performs the following functions:
  - Conducts TECHINT collection and reporting in support of validated S&TI objectives.
  - Acts as the HQDA executive agent for foreign materiel used for training purposes.

- Conducts TECHINT training for DOD analysts and RC TECHINT personnel.
- Supports INSCOM's FMA and FME operations as directed.
- Analyzes and exploits foreign captured enemy documents (CEDs), equipment, weapon systems, and other war materiel.
- Reports on the capabilities and limitations of enemy combat materiel.
- Provides reports alerting the command to the tactical threat posed by technical advances in new or recently discovered foreign or enemy materiel.
- Provides countermeasures to any enemy technical advantage.
- Provides foreign or enemy equipment for troop familiarization and training.
- Provides recommendations on the reuse of CEM.
- Supervises evacuating items of TECHINT interest.
- Provides task-organized battlefield TECHINT teams to support a subordinate command's TECHINT effort.

US ARMY MATERIEL COMMAND (AMC)

10-10. AMC plays a significant support role in TECHINT. Among AMC elements are a series of RDECs, the Army Research Laboratory System, and the Test and Evaluation Command (TECOM). Each conducts highly technical evaluations of foreign equipment. In peacetime, the AMC conducts FME on equipment purchased by each laboratory and RDEC for the intelligence community and for DOD as part of the International Materiel Evaluation Program (IMEP). AMC elements include—

- The Foreign Ordnance Exploitation Team, which is located at the Fire Support Armaments Center (FSAC) in the Picatinny Arsenal. This team exploits foreign ground ordnance and develops render safe procedures (RSPs) for foreign ordnance. It also prepares detailed intelligence reports to support explosive ordnance disposal (EOD), MI, and US munitions developers.
- The Science and Technology Center Europe and the Science and Technology Center Far East, which have the responsibility of collecting information on foreign technical developments by attending arms shows and technology exhibitions.
- The Soldier Biological Chemical Defense Command, which is the headquarters for the US Army Technical Escort Unit (TEU). The mission of the TEU is to collect and escort (transport) chemical, biological, radiological, and nuclear (CBRN) samples for testing and evaluation.

10-11. There are many other agencies with TECHINT responsibilities within the DOD. Refer to FM 34-54 for more information on TECHINT.

PLAN

10-12. TECHINT collection usually begins when an organization or individual reports the recovery or acquisition of unusual, new, or newly employed threat materiel. However, there are often indications that the threat may be using materiel not yet associated with the threat among the myriad intelligence products available. Conversely, it may be known that the threat is using a particular item, the capabilities of which are unknown to US forces. It is in these cases that a unit may receive a TECHINT related ISR task. It is conceivable that such a task may be reflected in the PIR. An example of such a task is the identification or verification of suspected external modifications on a particular model of an enemy's main battle tank; the results of which are linked to a commander's decision.

10-13. TECHINT related ISR tasks should include a mission or target folder. At a minimum, this folder should include a description of the item, with its associated major combat systems, as well as handling instructions, reporting instructions, and a photograph or sketch of the item if available.

10-14. Additionally, the G2/S2 is responsible for ensuring the staff coordinates and establishes a plan for evacuating the desired materiel.

PREPARE

10-15. The G2/S2 must ensure that required TECHINT analytical assets, resources, and evacuation means are prepared to provide support. This includes verifying coordination effected with the task-organized battlefield TECHINT teams from the 203d MI Battalion. The G2/S2 must also ensure the means to report and disseminate TECHINT results to the unit and its soldiers are in place so that they can immediately adopt appropriate countermeasures, operations, or tactics in order to enhance their survival and mission accomplishment.

COLLECT

10-16. TECHINT collection includes capturing, reporting, and evacuating CEM. TECHINT collection begins when an organization or individual reports the recovery or acquisition of threat materiel. An item of materiel is exploited at each level, and continues on through succeeding higher levels until an appropriate countermeasure to neutralize the item's capabilities is identified or developed.

10-17. Army personnel (soldiers and civilian) and units will normally safeguard CEM and report it through intelligence channels to the first TECHINT element in the reporting chain. The location of this TECHINT element will be in accordance with the METT-TC factors; however, there will usually be TECHINT representation at the Corps G2 or the COCOM J2. The TECHINT representative or element will verify if the type of materiel is of intelligence value and determine its further disposition in conjunction with the unit's staff.

PROCESS

10-18. TECHINT processing starts (simultaneously with collection) with the capture of a piece of equipment of TECHINT value. This confirms that the enemy is indeed employing this materiel. In accordance with METT-TC factors, a TECHINT team may move to the location of the item at the capture site or wait until the item is evacuated before conducting a hasty exploitation. After hasty exploitation, the team decides if further processing is required. If it is, the items are sent to the first (or nearest) Captured Materiel Exploitation Center (CMEC). If the item is deemed to yield no immediate tactical intelligence value, it may still be evacuated to the S&TI centers in CONUS for further analysis if the systems represent a change in the technological posture of an enemy.

PRODUCE

10-19. Battlefield TECHINT teams normally report initial and secondary examinations of CEM using either a preliminary technical report or a complementary technical report.

- A preliminary technical report—
  - Includes a general description of the item reported and recommended RSP.
  - Alerts others to information that can be used immediately by tactical units.
- A complementary technical report is more in-depth and—
  - Follows a secondary or an in-depth initial examination.
  - Allows the CMEC to compare new information with intelligence holdings.

10-20. At each successive echelon of exploitation, TECHINT analysts add to the overall body of information on an item by either adding to previous reports or by preparing new reports. The CMEC or other national level S&TI activities prepare more advanced technical reports and analyses. These reports include—

- Detailed technical reports.
- Translation reports.
- Special technical reports.

10-21. Other TECHINT products include—

- CMEC publications such as operator manuals, maintenance manuals, TECHINT bulletins, and tactical user bulletins.
- S&TI analysis bulletins.
- Foreign materiel exploitation reports.

ANALYZE

10-22. TECHINT analysts use checklists established by S&TI agencies and the CMECs to analyze each type of the adversary's equipment for which requirements exist. Analysis always begins with what is, and what is not, known about the piece of equipment. TECHINT units maintain procedures and plans for sampling, analyzing, and handling materiel.

DISSEMINATE

10-23. TECHINT of critical importance to the force, including answers to the PIR, is disseminated via the most expeditious means possible.

10-24. Routine TECHINT reports and products are usually transmitted through the unit's existing intelligence communications architecture in the format of an intelligence information report (IIR) format. For intelligence reach operations, TECHINT products are available and disseminated in a variety of forms. The requestor must ensure that the TECHINT product can be transmitted over the available communications systems. This includes verifying the appropriate security level of the communications systems.

ASSESS

10-25. The primary goal of the TECHINT assess function is to determine whether the results of TECHINT production meet the unit's PIR or intelligence requirements. The G2/S2 immediately assesses TECHINT products upon receipt for accuracy and relevance. He must inform the TECHINT producer of the extent to which the product answered the PIR or intelligence requirement. Providing feedback to TECHINT analysts helps improve the effectiveness and efficiency of TECHINT.

10-26. The G2/S2 also assesses the success of the unit's ISR effort in accomplishing any TECHINT associated ISR task and shares his assessment with the staff and the pertinent units or personnel.

Chapter 11

Counterintelligence

DEFINITION

11-1. CI counters or neutralizes intelligence collection efforts through collection, CI investigations, operations, analysis and production, and functional and technical services. CI includes all actions taken to detect, identify, exploit, and neutralize the multidiscipline intelligence activities of friends, competitors, opponents, adversaries, and enemies. It is the key intelligence community contributor to protect US interests and equities. Figure 11-1 shows the CI overview.

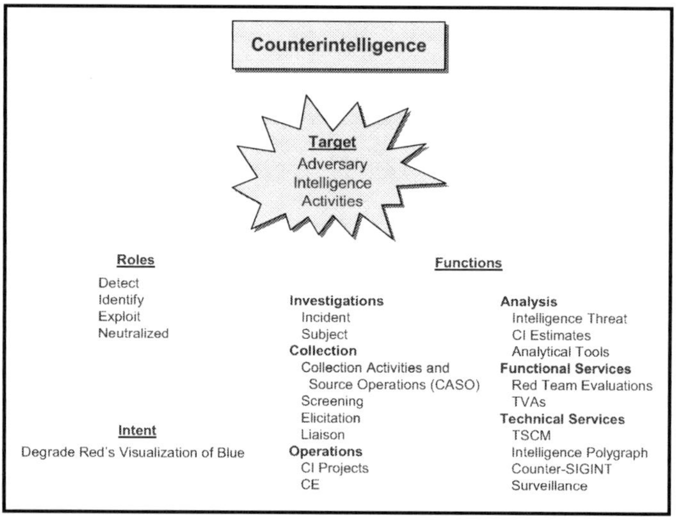

Figure 11-1. Counterintelligence Overview.

ROLE

11-2. The role of CI is to detect, identify, exploit, and neutralize all adversary intelligence entities targeting US and multinational interests. CI will focus on countering adversary intelligence collection activities targeting information or material concerning US personnel, activities, operations, plans, equipment, facilities, publications, technology, or documents—either classified or unclassified—without official consent of designated US release authorities, for any purpose that could cause damage or otherwise adversely impact the interests of national security of the US ability to fulfill national policy and objectives. Adversary intelligence threats include but are not limited to any (US, multinational, friendly, competitor, opponent, adversary, or recognized enemy) government or NGOs, companies, businesses, corporations, consortiums, groups, agencies, cells or persons, terrorists, insurgents, guerrilla entities, and persons whose demonstrated actions, views, or opinions are inimical to US interests.

11-3. CI elements are instrumental in contributing to situational awareness in the AOI. CI elements may corroborate other intelligence discipline information as well as cue other intelligence assets through the CI core competencies and through CI technical services. The CI core competencies are collection, investigations of national security crimes within the purview of CI, operations, and analysis and production. CI technical services include computer network operations, technical surveillance countermeasures (TSCM), and polygraph. CI focuses on combating adversary intelligence activities targeting Army personnel, plans, operations, activities, technologies, and other critical information and infrastructure.

COUNTERINTELLIGENCE FUNCTIONS

11-4. CI functions are interrelated, mutually supporting, and can be derived from one another. No single function or technical capability can defeat adversary intelligence efforts to target US interests. CI functions are discussed below.

CI INVESTIGATIONS

11-5. Investigative activity is essential to countering the adversary intelligence threat to Army interests. CI places emphasis on investigative activity to support force, infrastructure and technology protection, homeland defense, information assurance, and security programs. CI investigations focus on resolving allegations of known or suspected acts that may constitute National Security Crimes under US law. The primary objective in any CI investigation is the detection, identification, exploitation, and/or neutralization of adversary intelligence threats directed against the US Army. CI investigations are also conducted to identify systemic security problems that may have damaging repercussions to Army operations and national security interests. All CI investigations are conducted within guidelines established in AR 381-10, AR 381-12, AR 381-20, applicable DOD policy and directives, and US laws.

CI OPERATIONS

11-6. CI operations are characterized as those activities that are not solely associated with investigative, collection, analysis, or production functions. CI operations can be either offensive or defensive in nature; they are derived from or transition to a collection or investigative activity depending on the scope, objective, or continued possibility for operational exploitation. CI operations fall into two categories: CI support operations and CI sensitive operations.

- **CI Support Operations.** These are defensive operations used to support ARFOR and technology protection, security projects, and programs. They include technical services support, support to acquisition, FP, special access, international security, foreign visitor or contact, treaty verification, information assurance, homeland defense, and other approved projects and programs.
- **CI Sensitive Operations.** These operations are generally offensive in nature and involve direct or indirect operations against a known adversary intelligence threat. These operations include counter-espionage (CE) and CI projects and are conducted by designated units.

CI COLLECTION

11-7. **Collection Activities.** CI elements conduct collection activities focused on adversary intelligence threats that target US and multinational interests. CI collection is conducted through the use of sources, assets, official contacts, and other human or multimedia sources to obtain information that impacts the supported unit. CI will not be used as a substitute for Army HUMINT collection. These activities are designed to collect specific information or develop leads concerning adversary intelligence collection requirements, capabilities, efforts, operations, structure, personalities, and methods of operation targeting US and multinational interests. CI collection can result from ongoing CI investigations or operations or serve to initiate CI investigations and/or operations.

11-8. **Liaison.** CI elements conduct liaison with US, multinational, and HN military and civilian agencies, to include NGO, for the purpose of obtaining information of CI interest and coordinating or deconflicting CI activities. Liaison activities are designed to ensure a cooperative operating environment for CI elements and to develop CI leads for further exploitation.

11-9. **CI Collection Activities and Source Operations (CASO).** CASO is used to collect information on direct threats to US Army installations, organizations, activities, and personnel. The CASO program is not intended to be used as a substitute for tactical HUMINT contact operations. CASO can be used to initiate CI investigations, identify potential leads for offensive operations, or develop additional CASO leads. **Only designated units, in accordance with applicable policy, will be authorized to pursue investigative and offensive operational leads from ongoing CASO operations.**

11-10. **Screening.** CI Special Agents work jointly with HUMINT Collectors during screening operations to identify civilians on the battlefield, EPWs, detainees, and other noncombatants who may have information of CI interest or to develop CI leads. CI screening is also conducted during the process of hiring HN citizens for Army and DOD employment. Information obtained during screening operations may be used to initiate CI investigations and operations or to cue other intelligence collection disciplines such as HUMINT, IMINT, SIGINT, and MASINT.

11-11. **Debriefing.** CI Special Agents conduct debriefings of friendly force, HN, or the local population who may have information of CI interest regarding adversary intelligence collection or targeting efforts focused on US and multinational interests.

11-12. **Functional Services.** CI elements conduct functional services to assist traditional CI activities of investigations, collection, operations, analysis, and production; they also provide tailored support to US, DOD, and Army protection and security programs for commanders at all echelons. CI elements may use one or several of the functional services simultaneously to provide tailored support to a particular CI mission or supported program. CI functional services consist of—

- CI threat vulnerability assessments (TVAs).
- Adversary intelligence simulation (Red Team Evaluation).
- Covering agent support.

11-13. **CI Technical Services.** CI technical services are used to assist the CI functions of investigations, collections, and operations or to support functional services conducted by CI elements. For additional information on CI technical services, refer to AR 381-20 (SECRET/NOFORN) and FM 34-5. CI technical services consist of—

- Surveillance.
- Intelligence polygraphs.
- TSCM.
- Computer Network Operations (CNO).
- IO.
- Counter-Signals Intelligence (C-SIGINT).

ANALYSIS

11-14. IPB, all-source, and single-source analysis are used to template adversary intelligence activities. Analysis is also used to make recommendations to the commander on how to counter adversary intelligence efforts and to refine CI activities to potentially neutralize and/or exploit those efforts and to continually focus the efforts of CI teams. CI analysis focuses on the multidiscipline adversary intelligence collection threat targeting information on US multinational personnel, operations, activities, technology, and intentions. Raw information, open-source material, and finished intelligence products are analyzed in response to local and national requirements. Analysis occurs at all levels from tactical to strategic.

- At the tactical level, CI teams focus their efforts on supporting mission requirements and contributing to the all-source COP.
- Operational analysis is used to assess how adversary intelligence views and targets US interests; identify US vulnerabilities that could be exploited or targeted; and to determine adversary intelligence targeting methods of operation (MOs).
- CI collection priorities; assessing adversary intelligence technical options for countering US weapons and intelligence systems; and

assessing the impact of technology transfer activities on the US military's technological overmatch.

PRODUCTION

11-15. CI products consist of, but are not limited to, target nomination, CI input to TVAs, CI estimates, and investigative and intelligence information reports. Finalized intelligence derived from CI activities are incorporated into joint and national intelligence databases, assessments, and analysis products. CI products are also incorporated into the COP to support battlefield situational awareness. CI production takes place at all levels.

- Operational and tactical production includes tactical alerts, spot reports, and current intelligence; CI threat and/or vulnerability assessments tailored to specific activities, units, installations, programs, or geographic areas; CI studies to support contingency planning and major exercises; studies of adversary intelligence organization, MO, personnel, activities, and intentions that pose a current or potential threat to the supported command.
- Strategic products include assessments supporting national and Army programs including SAPs and acquisition programs; worldwide assessments of the organization, location, funding, training, operations capabilities, and intentions of terrorist organizations; global trends in adversary intelligence MO; after-action studies of individual espionage cases; analyses of the intelligence collection capabilities of international narcotics trafficking organizations; and multimedia threat products to support Army CI awareness programs.

TECHNICAL SERVICES

11-16. CI organizations with technically trained CI Special Agents are chartered with providing unique technical capabilities to augment CI investigations, collection, and operations. These technical capabilities are not used as substitutes for CI activities, but support traditional CI techniques employed to counter and neutralize adversary intelligence activities targeting US interests.

11-17. Selected CI Special Agents are trained to identify human deception indicators during the course of investigations, operations, and collection missions. Required training and skills include the use of polygraph and emerging biometric technologies that can recognize indications of deception by human sources, contacts, and subjects of investigations, as well as analysis and reporting of results.

11-18. TSCM CI Special Agents are trained in sophisticated electronic and sensing equipment to identify technical collection activities carried out by adversary intelligence entities. The use of TSCM is critical to ensuring that sensitive and/or restricted areas are clear of any adversary intelligence-placed active or passive electronic sensing, or eavesdropping or collection devices. These areas include SCIF, secure working and planning areas, and C2 facilities.

11-19. CNO consist of computer network attack (CNA), computer network defense (CND), and computer network exploitation (CNE). CI Special Agents are specially trained in the areas of computer operation, network theory and administration, and forensics, along with IO in order to ensure US information dominance. The reliance on networked systems will result in greater emphasis being placed on information assurance. Specially selected CI agents will be trained in CNO in order to assist in protecting US information and information systems while exploiting and/or attacking adversary information and information systems.

11-20. For more information on technical services, refer AR 381-20 and FM 34-5.

OPERATIONAL EMPLOYMENT

11-21. Army CI supports the full spectrum of military operations. The Army requires a well-trained CI force consisting of AC, RC, civilian government employee, and contractor personnel. CI elements are focused on and dedicated to detecting, identifying, neutralizing, and/or exploiting adversary intelligence elements attempting to collect information on US forces. Effective employment of Army CI elements in all phases of operations and at all levels from tactical to strategic is paramount to countering any adversary intelligence threat to US interests and resources. Moreover, Army CI protects special programs that provide R&D, acquisition, and integration of technologies leading to future technological overmatch. CI elements have the capability and authority to conduct complex, nontraditional operations in this Information Age to CE, to both protect critical technologies and satisfy Army, DOD, and national level CI objectives.

LEVELS OF EMPLOYMENT

Strategic and Departmental

11-22. Strategic and departmental operations will be conducted by CI elements supporting national, DOD, and DA required missions (for example, support to NATO and special operations and missions). Strategic and departmental CI will conduct compartmented investigations and operations to affect the knowledge of adversary intelligence regarding CONOPS and defense information. Army CI will execute the full range of CI functions and missions at the strategic and departmental level including CI investigations and operations, CE, technology protection, SAP support, treaty verification, and technical CI services (polygraph, TSCM, and computer forensics). Strategic and departmental CI will also support SOF and special mission units (SMUs) within the scope of applicable national, DOD, and DA security policies and regulations.

Operational

11-23. Operational missions of CI elements will support combatant commanders, generally in geographic AOIR. Operational CI elements will focus on threat identification and countering regional adversary intelligence threats. Operational level CI activities and functions include CI investigations and operations, CE, technology protection, SAP support, treaty verification, and technical CI services (polygraph, TSCM, and computer forensics). CI elements must be capable of quickly transitioning from a peacetime mission to crisis operations to support combatant commander requirements. Theater CI assets will conduct unilateral to multinational operations in designated theaters. Operational elements may also be deployed to support or reinforce tactical forces in CONOPS.

Tactical

11-24. CI teams will conduct operations throughout the battlespace during CONOPS. CONOPS support activities include conduct of CASO, limited CI investigative capability, personnel security investigations, screenings, and debriefings. CI activities in CONOPS focus on countering the adversary intelligence threat and assisting in conducting TVAs. TVAs will be conducted in conjunction with MPs, Engineers, and Medical Service personnel to provide the commander with a comprehensive FP assessment. During peacetime, organic tactical CI teams conduct activities in accordance with approved regulations and command guidance.

SUPPORT TO CONTINGENCY OPERATIONS

11-25. The initial phase of operations from PME to MTW lays the foundation of future team operations. In general, the priority of effort focuses inward on security of operating bases, areas of troop concentration, and C2 nodes to identify the collection threat to US forces that could be used by adversary elements to plan hostile acts against US activities and locations.

11-26. Once security of the operating bases has been established, the operational focus of CI teams shifts outside the operating base to continue to detect, identify, and neutralize the collection threat to US forces as well as to provide I&W of hostile acts targeting US activities. The CI team uses several collection methods, to include CASO, elicitation, and liaison, to answer the supported commander's requirements. This is referred to as the continuation phase. The CI team conducts CI investigations to identify, neutralize, and exploit reported threat intelligence collection efforts.

11-27. A key element to the CI team's success is the opportunity to spot, assess, and develop relationships with potential sources of information. Operating as independent teams, without being tied to ISR or combat assets, enables the CI team's maximum interaction with the local population, thereby maximizing the pool of potential sources of information. Along with the opportunity to spot, assess, and interact with potential sources of information, a second key element of a CI team's success is its approachability to the local population. A soft posture enables a CI team to appear as non-threatening as possible. Experience has shown that the local population in general is apprehensive of fully and openly armed patrols and soldiers moving around population centers.

11-28. During some operations, civilian attire or nontactical vehicles may be used to lower the CI team profile. In some special situations, these measures are taken to make the operation less visible to the casual observer. Also, in some cultures, sharing food and beverages among friends is expected; exceptions to restrictions or general orders should be considered to facilitate successful CI team operations, many of which are geared towards developing relationships with potential sources of information.

SUPPORT TO INSTALLATIONS AND OPERATING BASES

11-29. CI teams, as part of a multi-agency team consisting of MPs, CA, medical, and EOD, support the conduct of TVAs of installations and operating bases to identify the intelligence threat to the operating locations. Detailed TVAs identify weaknesses in operational and physical security procedures and recommend countermeasures to mitigate intelligence collection on friendly forces limiting the ability to plan hostile acts on US activities and locations. CI activities supporting installations and operating bases include—

- Interviewing walk-in sources and locally employed personnel.
- Screening local national (LN) hires. Commanders, staff planners, and SIOs should always provide input to personnel assigned to establish and negotiate contracts using local national (LN) hires. This requirement ensures that LN hires can be screened, interviewed, and in some instances used as CI sources or assets in order to provide intelligence information that impacts the security of the base camp.
- Debriefing friendly force personnel who are in contact with the local population, such as—
  - ISR patrols.
  - MP patrols.
  - Combat patrols.
  - Liaison personnel.
  - CA and PSYOP teams.
- Conducting limited local open-source information collection.
- Providing support to TVAs of the base camp.

TACTICS, TECHNIQUES, AND PROCEDURES

11-30. At the CI team level, team members conduct mission analysis and planning specific to their AO. Backwards planning and source profiling are used extensively to choose CI targets. To verify adequate area coverage, the CI team may periodically develop and use CI target overlays and other CI analytical tools that illustrate the CI situation, identify CI gaps, and help refocus the collection effort.

11-31. The CI team is also in constant contact with the supported S2 and the other ISR assets (Scouts, PSYOP, CA, and MP) in order to coordinate and deconflict operations and to cross-check collected information. The supported unit S2, with the help of the CI team, regularly and systematically debriefs all ISR assets.

11-32. The CI team must be integrated into the supported unit's ISR plan. The CI OMT chief will advise the supported unit on the specific capabilities and requirements of the team to maximize mission success.

OPERATIONAL RISK MITIGATION

11-33. The employment of CI teams includes varying degrees of contact with the local population. As the degree of contact with the population increases, both the quantity and quality of CI collection increases. In many instances, however, there is a risk to the CI team inherent with increased exposure to the local population. The decision at what level to employ a CI team is METT-TC dependent. The risk to the CI assets must be balanced with the need to collect priority information and to protect the force as a whole. ROE, SOFA, direction from higher headquarters, and the overall threat level may also restrict the deployment and use of CI teams. The commander should consider exceptions to the ROE to facilitate CI collection.

11-34. Risks are minimized through the situational awareness of CI team members. They plan and rehearse to readily react to any situation and carry the necessary firepower to disengage from difficult situations. If it becomes necessary to call for assistance, adequate and redundant communications equipment is critical. These scenarios and actions should be trained prior to deployment into a contingency area and rehearsed continuously throughout the deployment.

11-35. A supported unit commander is often tempted to keep the CI team "inside the wire" when the THREATCON level increases. The supported commander must weigh the risk versus potential information gain when establishing operational parameters of supporting CI teams. This is necessary especially during high THREATCON levels when the supported unit commander needs as complete a picture as possible of the threat arrayed against US and multinational forces.

11-36. When it is not expedient to deploy the CI team independently due to threat levels or other restrictions, the team can be integrated into other ongoing operations. The CI team may be employed as part of a combat, ISR, or MP patrol or used to support CA, PSYOP, engineer, or other operations. This method reduces the risk to the team while allowing a limited ability to collect information. It has the advantage of placing the team in contact with the local population and allowing it to spot, assess, and interact with potential sources of information. However, this deployment method restricts collection by subordinating the team's efforts to the requirements, locations, and timetables of the unit or operation into which it is integrated and does not allow for the conduct of sensitive source operations. **This method of employment should be considered a last resort.**

COUNTERINTELLIGENCE EQUIPMENT

11-37. Basic C2, transportation, and weapons requirements do not differ significantly from most soldier requirements and are available as unit issue items. However, CI teams have unique communications, collection, processing, and mission-specific requirements.

COMMUNICATIONS

11-38. **Dedicated and Secure Long-Range Communications.** These are keys to the success of the CI team mission. CI team operations require a secure, three-tiered communications architecture consisting of inter/intra-team radios, vehicle-based communications, and a CI and HUMINT base station.

11-39. **Communications Network.** The CI team must have access to existing communications networks such as the tactical LAN. The CI team must also be equipped with its own COMSEC devices. It is imperative that the CI team acquire access to the public communication system of the HN. This can be in the form of either landlines or cellular telephones. Such access enables the CI team to develop leads which can provide early indicators to US forces.

11-40. **Interoperability.** Communications systems must be equipped with an open-ended architecture to allow for expansion and compatibility with other service elements, government organizations, NGOs, and multinational elements to effectively communicate during CONOPS. All ISR systems must be vertically and horizontally integrated to be compatible across all BOSs and with Legacy and Interim Force elements.

11-41. **SOTM.** To provide real-time and NRT information reporting, CI elements must have the capability to transmit voice, data, imagery, and video while on the move. CI teams must be able to transmit while geographically separated from their parent unit while operating remotely. This broadband requirement can only be achieved through a SATCOM capability and must be achievable while mobile.

CI COLLECTION AND PROCESSING SYSTEMS

11-42. The CI team must rely on automation to achieve and maintain information dominance in a given operation. With time, effective collection planning and management at all echelons, the CI team can collect a wealth of information. The sorting and analysis of this information in a timely and efficient manner is crucial to operations. Automation helps the CI team to report, database, analyze, and evaluate the collected information quickly and to provide the supported unit with accurate data in the form of timely, relevant, accurate, and predictive intelligence.

11-43. Automation hardware and software must be user friendly as well as interoperable among different echelons and services. They must interface with the communications equipment of the CI team as well as facilitate the interface of audiovisual devices. Technical support for hardware and software must be available and responsive.

11-44. The demand for accurate and timely CI reporting, DOCEX, and open-source information has grown tremendously. Biometric (physiological, neurological, thermal analysis, facial and fingerprint recognition) technologies will allow rapid identification, coding, and tracking of adversaries and human sources; as well as cataloging of information concerning EPWs, detainees, and civilians of CI interest on the battlefield. Biometrics will also provide secure authentication of individuals seeking network or facility access.

11-45. CI teams work with multinational forces and other foreign nationals and require the ability to communicate in their respective languages. Often CI personnel have little or no training in the target language, and lack of skilled interpreters can hinder CI activities. CI teams require textual and voice translation devices, source verification, and deception detection machines (biometrics) to improve collection capability and accuracy.

11-46. CI teams require dynamic MLT tools that provide both non-linguists and those with limited linguist skills a comprehensive, accurate means to conduct initial CI screenings and basic interviews in a variety of situations. CI elements will focus on in-depth interviews and communications with persons of higher priority. MLT tools minimize reliance on contract linguists and allow soldiers to concentrate on mission accomplishment.

MISSION SPECIFIC

11-47. The CI team may conduct night operations and must be equipped with NVDs for its members, and photographic and weapons systems. The CI team also may operate in urban and rural areas, where the threat level can vary from semi-hostile to hostile. The safety of the CI team can be enhanced with equipment that can detect, locate, suppress, illuminate, and designate hostile optical and E-O devices. In addition, high power, gyro-stabilized binoculars, which can be used from a moving vehicle, increases the survivability of the CI team and also gives the team another surveillance and collection device.

11-48. Some of the CI team missions may require the documentation of incidents. The CI teams can use the following equipment in their open-source collection efforts.

- Small, rugged, battery-operated digital camcorders and cameras which are able to interface with the collection and processing systems as well as communication devices.
- GPSs that can be mounted and dismounted to move in the AO efficiently.
- Short-range multichannel RF scanning devices that can also identify frequencies which enhance their security.
- In some cases CI teams require a stand-off, high resolution optical surveillance and recording capability that can provide target identification at extended ranges to protect the intelligence collector while avoiding detection by the adversary target. An advanced optical capability provides intelligence collectors the ability to locate and track adversary targets (passive and hostile) for identification, collection, and target exploitations.

INTEGRATION OF LINGUISTS

11-49. Integrating linguists into the CI team should take place as soon as possible. Security clearances and contractual agreements will help the team determine the level of integration.

11-50. Along with the basic briefing of what is expected of the civilian linguists as interpreters, CI teams should be informed about the civilians' chain of command and the scope of their duties beyond interpreting. The CI team leader must ensure that linguists are trained and capable of completing all tasks expected of them.

BATTLE HAND-OFF

11-51. CI teams are always engaged. A good battle hand-off is critical to smooth transition and mission success. The battle hand-off can directly contribute to mission success or failure of the outgoing team, but especially of the incoming team. The battle hand-off begins the first day the CI team begins to operate in an AO. Regardless of how long the team believes it will operate within the AO, it must ensure there is a seamless transition to an incoming team, other US unit, or agency. The CI team accomplishes this transition by establishing procedures for source administration, database maintenance, and report files.

11-52. Teams must plan and implement a logical and systematic sequence of tasks that enables an incoming team to assume the operations in the AO. Adequate time must be allotted for an effective battle hand-off. In some environments, a few weeks may be necessary to accomplish an effective battle hand-off. Introductions to sources of information, especially CASO sources, are critical, and teams must prioritize their time. During this time the outgoing CI team must familiarize the new CI team with all aspects of the operation, which include past, present, and planned activities within the AO. Area orientation is critical. These include major routes, population centers, potential hot spots, and other points of interest (such as police stations, political centers, and social centers).

ORGANIZATION

11-53. CI activities require a complex C2 relationship to ensure that the requirements of the supported commander are fulfilled while balancing the need for strict integrity and legality of CI operations. This complex relationship balances the role of the SIO as the requirements manager and the 2X as the mission manager with the MI commander as the asset manager.

COMMAND VERSUS CONTROL

11-54. ARFOR will normally deploy as part of a joint, multinational, and/or combined operation. In all cases, commanders at each echelon will exercise command over the forces assigned to their organization. Command includes the authority and responsibility for effectively using resources, planning for and employment of forces, and ensuring that forces accomplish assigned missions. Leaders and staffs exercise control to facilitate mission accomplishment.

11-55. While the MI commander supervises subordinates and produces reports, the *2X synchronizes activities between intelligence units and provides single-source processing and limited analysis. **(*2X " indicates 2X functions at all levels.)** While the MI commander takes care of the operators executing missions, the *2X obtains the data and reports from higher echelons required to execute the missions.

STAFF RESPONSIBILITIES AND FUNCTIONS

11-56. The *2X staff is responsible for the integration, correlation, and fusion of all Human Sensor information into the Intelligence BOS within the *2X AOIR. The *2X is also responsible for analyzing adversary intelligence collection, terrorist and sabotage activities, developing countermeasures to defeat threat collection activities, identifying and submitting collection requirements to fill CI collection gaps, and providing input to the all-source picture regarding adversary intelligence activities.

11-57. The *2X Staff Officer provides CI and HUMINT collection expertise. The *2X—

- Is the single focal point for all matters associated with CI and HUMINT in the AOIR.
- Is the CI and HUMINT advisor to the G2 and commander.
- Is an extension of the collection manager and ensures that the best asset or combinations of assets are used to satisfy information requirements.
- Along with his subordinate elements—CICA, HOC, OSC, CIAC, and HAC—exercises technical control over his assigned Army CI and HUMINT elements in the designated AOIR.
- Is the principal representative of the G2 and the commander when coordinating and deconflicting CI and HUMINT activities with national or theater agencies operating in the AOIR.
- Supports specific RM efforts in conjunction with the requirements manager through the planning and coordination of CI and HUMINT operations; the review and validation of CI requirements; the recommendation for assignment of tasks to specific collectors; and the conduct of liaison with non-organic HUMINT collection. This liaison includes national level and multinational force assets for source deconfliction and special activities outside the *2X AOIR.
- Will provide OMTs with capability to reach back to current database information, technical information and guidance, and source deconfliction necessary to monitor the collection activities of the CI teams.
  - *CICA*. The CICA is responsible for coordinating and synchronizing all CI activities in the designated AOIR. The CICA exercises technical control over all CI entities in the designated AOIR and deconflicts CI activities with higher, lower, and adjacent CI elements. The CICA accomplishes all responsibilities through coordination with the operational units and other *2X staff elements.
  - *OSC*. The OSC in the *2X staff maintains the source registry for all CI activities in the designated AOIR. The OSC provides management of intelligence property book operations, source incentive programs, and ICFs.
  - *CIAC*. The CIAC analyzes adversary intelligence collection capabilities. The CIAC leverages all intelligence discipline reporting and analysis to counter threat collection capabilities against the deployed force. CIAC analysis provides information and analysis to the COP.

11-58. The ACE and JISE CI analysis team analyzes threat intelligence collection and the intelligence collection efforts of foreign organizations involved in terrorism and sabotage in order to develop countermeasures against them. CI analysis cross-cues IMINT, SIGINT, MASINT, and TECHINT resources in addition to CI-related HUMINT reporting and analysis to counter threat collection capabilities against the deployed force. While the HAC supports the positive collection efforts of the force, the CI analysis team supports the "defend" aspects of the commander's FP program.

11-59. CI analysis is the analysis of the adversary's HUMINT, IMINT, SIGINT, and MASINT capabilities in support of intelligence collection, terrorism, and sabotage in order to develop countermeasures against them. It involves a reverse IPB process in which the analyst looks at US forces and operations from the threat's perspective. CI analytical products are an important tool in the COA development in the MDMP. This analytical tool supports the commander's FP program and facilitates the nomination of CI targets for neutralization or exploitation. (See FM 2-01.2 (FM 34-60) for more information on CI analysis.) Specifically, CI analysis—

- Produces and disseminates CI products and provides input to INTSUMs.
- Provides collection requirements input to the CICA.
- Analyzes source reliability and credibility as reflected in reporting and communicating that analysis to the collector.
- Nominates CI targets for neutralization or exploitation.
- Identifies and submits CI-related requirements to fill collection gaps.
- Assists HAC personnel in focusing the CI aspects of the HUMINT collection program.
- Presents CI analysis products such as CI estimates, target lists, reports, and graphics that support the commander.

11-60. For intelligence reach operations, CI products are available and disseminated in a variety of forms. It is incumbent on the requestor to ensure the CI product can be transmitted over the available communications systems. This includes verifying the appropriate security level of the communications systems.

CI TEAM STRUCTURE

11-61. **OMT.** The OMT is a four-person team consisting of a WO, two NCOs, and a junior enlisted soldier. (Civilians may be inserted into this structure as appropriate.) Rank structure and standards of grade for OMTs will vary depending upon the skill sets required and mission focus. CI OMTs will provide operational guidance for 1 to 4 CI teams, depending on mission focus and operational tempo. When two or more CI teams are deployed in a DS role, an OMT is also deployed to provide technical control. The OMT works closely with the supported S2 and ACT to furnish current threat information and to answer the supported commander's PIRs and IRs. OMTs coordinate with the supported 2X and manage subordinate CI teams to—

- Provide guidance and technical control of operational activity.
- Provide the collection and operational focus for CI teams.

- Provide quality control and dissemination of reports for subordinate CI teams.
- Conduct single-discipline CI analysis and assist in mission analysis for the supported commander.
- Act as a conduit between subordinate CI teams, the CICA, and supported unit headquarters.
- Provide administrative support for subordinate CI teams to include reporting mission and equipment status to the CICA and the supported unit headquarters.
- Educate the supported commander on the capabilities of the CI teams.
- Integrate the CI teams directly into the maneuver commander's ISR planning.

11-62. **CI Team.** The CI team is a four-person team consisting of two NCOs and two junior enlisted personnel. Rank structure and standards of grade for CI teams will vary depending upon the skill sets required and mission focus. CI teams are trained to execute the full range of CI functions; however, they may be assigned to mission-focused elements (for example, CE, CI projects). Assignment to a TSCM, polygraph, or information warfare team requires additional, specialized technical training.

Appendix A

Intelligence and Information Operations

THE INFORMATION ENVIRONMENT

A-1. The information environment is the aggregate of individuals, organizations, and systems that collect, process, store, display, and disseminate information; also included is the information itself. (FM 3-0) The information environment, it should be noted, is not an exclusively military one; in fact, the military applications of information are almost obscured in today's universal usage of the information spectrum by national, international, and non-state players.

THE COMMANDER AND INFORMATION

A-2. Information is facts, data, or instructions in any medium or form; the meaning that a human assigns to data by means of the known conventions used in their representation. (JP 1-02) Information provides the key to battlefield success in the 21st century. Commanders must have detailed information to command. Information is the medium that allows the commander's decisionmaking and execution cycle to function. Information gives direction to actions by the force, identifies the enemy's centers of gravity, provides COAs for force activity, and enables the force to accomplish its operational mission.

INFORMATION SUPERIORITY

A-3. Information superiority is the operational advantage derived from the ability to collect, process and disseminate an uninterrupted flow of information while exploiting or denying an adversary's ability to do the same. (FM 3-0) Relevant information drawn from intelligence supports the creation and development of situational understanding that contributes directly to information superiority during decisive operations. The requirement for information superiority is not new; what is new is that today's information technologies are creating a base of knowledge for military planning and execution that is unprecedented in scope, volume, accuracy, and timeliness. This means commanders receive accurate, timely information that enables them to make better decisions and act faster than their adversaries.

A-4. Information superiority, however, is neither a staple in today's battlespace nor is it necessary in military operations as a constant condition; rather, at specific times during operations, information superiority becomes a key enabler for assuring military success. At the operational level of predominant interest to the land component commander (LCC), information superiority is realized through the integration of operational level, interdependent ISR, information management, and IO to gain and maintain operational initiative and to achieve an operational advantage.

INFORMATION OPERATIONS

A-5. IO are the employment of core capabilities of electronic warfare, computer network operations, PSYOP, military deception, and operations security, in concert with specified supporting and related capabilities to affect or defend information and information systems, and to influence decisionmaking.... Information systems are...the equipment and facilities that collect, process, store, display, disseminate information. These include computers, hardware, software, and communications, as well as policies and procedures for their use. (FM 3-0). Offensive IO are capable of degrading an adversary's will to resist and ability to fight. Defensive IO measures passively and actively protect friendly information and C2 systems and limit their vulnerability.

A-6. Offensive and defensive IO are conducted as a fully coordinated effort to ensure the complementary, asymmetric, and reinforcing effects to attack enemy forces, influence others, and protect friendly forces. Relevant information assures that the right person has the right information at the right time for decisionmaking and execution.

THE ELEMENTS OF IO

A-7. Full-spectrum IO incorporates, integrates, and synchronizes traditionally independent capabilities and activities in support of the commander's mission. (See FM 3-0 and FM 3-13.) The Army's doctrinal view of full-spectrum IO as evolving core capabilities are—

- PSYOP.
- OPSEC.
- EW.
- Military deception.
- CNO.
  - CNA.
  - CND.
  - CNE.

A-8. Supporting IO capabilities are—

- Physical destruction.
- Physical security.
- Information assurance.
- Counterpropaganda.
- Counterdeception.
- Counterintelligence.

IO-RELATED ELEMENTS

A-9. The IO-related activities of PA and CMO are closely associated and integrated with the elements of IO as key contributors to information superiority.

INTELLIGENCE SUPPORT TO IO

A-10. All-source intelligence support, encompassing all of its forms, is the principal enabler for successful IO. The commander's capability to synchronize military operations can be heavily influenced by the ability to identify the threat and understand the adversary's capabilities and intentions in the information environment.

A-11. It is essential that the elements of the command's IO capabilities and vulnerabilities are integrated into the command intelligence plan. This leads to COAs that synchronize the elements of IO and IO-related activities into the commander's warfighting plans. Integration of the full-spectrum aspects of IO ensures that the relative importance of information is recognized in the development of a synchronized OPLAN. Further, integration of IO into the planning process provides a methodology for analyzing the threat from a knowledge base that enhances protection of friendly systems and assets while exposing windows of opportunity for attack or exploitation. IPB is the best process we have for understanding the battlefield and the options it presents to friendly and threat forces.

A-12. Provide Intelligence support to IO includes, but is not limited to, the following:

- **Provide Intelligence Support to Offensive IO.** The Intelligence BOS supports offensive IO by providing information to identify critical enemy C2 nodes. Intelligence also helps to identify enemy systems and procedures that may be vulnerable for offensive IO. Additionally, intelligence plays a key role in evaluating and assessing the effectiveness of offensive IO.
  - *Provide Intelligence Support to PSYOP.* This task identifies the cultural, social, economic, and political environment of the AO. It identifies target groups and subgroups and their location, conditions, vulnerabilities, susceptibilities, cultures, attitudes, and behaviors. PSYOP influence foreign target audiences in the AO to support achieving the commander's goals in the AO.
    - Identify the cultural, social, economic, and political environment of the AOI; for example, adversary mechanisms for political control, adversary communication and broadcast systems used to elicit support from the populace, and current and past adversary propaganda activities and their effectiveness.
    - Identify target groups and subgroups and their location, conditions, vulnerabilities, susceptibilities, cultures, attitudes, and behaviors. This includes determining the audience demographics of popular radio and television programs and periodicals; groups influenced by media personalities and political cartoons.
    - Identify impact of planned PSYOP on individuals outside the targeted group (for example, multinational partners, neighboring populations).
  - *Provide Intelligence Support to Military Deception.* This task identifies the capabilities and limitations of the adversary's

intelligence-gathering systems and identifies adversary biases and perceptions.

  - Profiles of key adversary leaders.
  - Cultural, religious, social, and political characteristics of the country and region.
  - Sources of military, economic, or political support.
  - Adversary decisionmaking processes, patterns, and biases.
  - Adversary perceptions of the military situation in the AO.
  - Capabilities and limitations of adversary CI and security services.

- *Provide Intelligence Support to Electronic Attack.* This task supports electronic attack employing jamming, electromagnetic energy, or directed energy against personnel, facilities, or equipment. It identifies critical adversary information systems and C2 nodes. It includes determining and presenting the adversary's electronic OB and their information system infrastructure. The enemy's C2 system vulnerabilities and the means they use to protect their C2 systems is part of the electronic OB.

- **Provide Intelligence Support to Defensive IO.** The Intelligence BOS supports defensive IO by providing information to identify threat IO capabilities and tactics. Intelligence provides information relating to CND, physical security, OPSEC, counterdeception, and counterpropaganda. The Intelligence BOS supports defensive IO by providing information to identify threat IO capabilities and tactics. Intelligence provides information relating to CND. Provide Intelligence Support to OPSEC—identify capabilities and limitations of the adversary's intelligence system to include adversary intelligence objectives and the means, methods, and facilities used by the threat to collect, process, and analyze information—supports the identification of indicators that could be interpreted or pieced together to penetrate EEFI in time to be useful to adversaries.
- **Provide Intelligence Support to Activities Related to IO.** The Intelligence BOS when operating outside US territories supports activities related to IO under some circumstances.
  - *Provide Intelligence Support to CMO.* This task allows military intelligence organizations to collect and provide information and intelligence products concerning foreign cultural, social, economic, and political elements within an IO in support of CMO. Identify cultural, social, economic, and political environment of the AOI, including—
    - Population demographics.
    - Civilian populace attitudes, alliances, and behavior.
    - Availability of basic necessities (food, clothing, water, shelter, medical care) and the ability of the populace to care for itself.
    - Access to medical care.
    - Locations and potential routes, destinations, and assembly areas or sites of displaced persons.
    - Local government type, status, organization, and capabilities.

  - Availability of local material and personnel to support military operations.
  - NGOs in the AOI, their agenda, resources, and capabilities.
- *Provide Intelligence Support to Public Affairs.* This task identifies the multinational and foreign public physical and social environment, as well as world, HN national, and HN local public opinion, in addition to the propaganda and misinformation capabilities, activities, targets, themes, and dissemination means of the adversary. Identify world, national, and local public opinion (location, biases or predispositions, and agenda of national and international media representatives in the AOI, and trends reflected by the national and international media).

Appendix B

Linguist Support

ROLE OF LINGUISTS

B-1. Military operations are highly dependent on foreign language support. The requirement to communicate with and serve on multinational staffs, communicate with local populations, and exploit enemy forces necessitates the use of linguists. The growing focus on multinational operations increases the competition for limited linguist resources that are vital for mission success. This appendix establishes the framework and process to access, prioritize, and employ the Army's limited organic linguist resources.

LINGUISTIC SUPPORT CATEGORIES

B-2. Foreign language support requirements of US Armed Forces typically fall into one of four broad categories:

- Intelligence and Information Gathering. This category includes the traditional SIGINT and HUMINT disciplines, as well as foreign language support to FP and exploitation of open-source information.
- CMO. This category encompasses all functions relating to military interaction with the civilian population. Foreign language support is critical to CMO in areas such as government liaison, legal agreements, medical support and operations, law enforcement, engineering projects, public safety, security and population control, CA, and PSYOP.
- Logistics. This category consists of foreign language support to sustainment or transportation functions. These include logistical contracting, port, railhead, airhead, or transshipment operations and convoy operations.
- Multinational Operations and Liaison. This category includes the coordination of military operations and liaison with multinational partners, previously unaffiliated nations, and at times adversary or former adversary nations. Multinational operations are becoming more common and increasingly important.

DETERMINING LINGUIST REQUIREMENTS

B-3. To identify linguist requirements, the staff conducts mission analysis and identifies specified or implied tasks requiring foreign language support. Other critical factors are the organization or echelon of command and the location of the mission. The staff uses these criteria to determine the allocation of linguists, such as one linguist team per echelon of command, one linguist per piece of equipment, or one linguist team per location where the function is to be performed. The staff then applies task organization and scheme of maneuver to determine the number of linguists needed for an operation.

B-4. The staff must analyze each linguist assignment to determine the minimum level of foreign language proficiency needed. While interpretation for a peace negotiation requires not only outstanding linguistic capability but also cultural acumen, the translation of routine documents (with the aid of a dictionary) requires a much different skill set. Poor identification of linguist proficiency requirements can tie up the best linguists in less effective roles, creating linguist shortfalls in other areas.

B-5. The relative importance of each of the four linguist support categories is mission dependent. For example, during a NEO civil and military coordination would probably not be as critical as intelligence and information gathering. However, the situation is reversed for a humanitarian assistance mission in which CMOs have a significant impact on mission success. Identifying these "dynamics" helps the commander and staff prioritize linguist requirements.

B-6. Determining linguist requirements for any operation can be difficult because each operation is unique. However, commanders and staffs with a basic knowledge of organic Army linguistic assets, foreign language resource alternatives, and MI skills can successfully assess, prioritize, and employ linguists in support of their military operations.

PLANNING AND MANAGING LINGUIST SUPPORT

B-7. Commanders must consider the linguist requirements as part of their MDMP for every CONPLAN and OPLAN assigned to their commands. Prior staff planning and identification of linguist requirements should prompt commanders to initiate linguist support requests and identify command relationships prior to actual operations. If the mission analysis reveals requirements for linguistic support, the commander must identify what foreign languages are needed, the foreign language proficiency levels needed for each assignment, and the best source of linguists. In addition, if the mission includes intelligence and information collection, the commander must identify MI collection skills required. During mission analysis, the commander should consider linguist requirements for every CONPLAN and OPLAN assigned to his command.

LINGUIST CATEGORIES

B-8. The commander and staff must identify linguist requirements by category:

- Category I – Have native proficiency in the target language (level 4-5) and an advanced working proficiency (Interagency Language Round Table [ILRT] level 2+) in English. They may be locally hired or from a region outside the AO. They do not require a security clearance. They must be screened by the Army CI support team.
- Category II – Are US citizens screened by Army CI personnel and are granted access to SECRET by the designated US government personnel security authority. Have native proficiency in the target language (level 4-5) and an advanced working proficiency (ILRT 2+) in English.
- Category III – Are US citizens screened by Army CI personnel and are granted either TS/SCI clearance or an interim TS/SCI clearance by the

designated US government personnel security authority. Meet a minimum requirement of ILRT level 3. They are capable of understanding the essentials of all speech in a standard dialect. They must be able to follow accurately the essentials of conversation, make and answer phone calls, understand radio broadcasts and news stories, and oral reports (both of a technical and non-technical nature).

PRIMARY STAFF RESPONSIBILITIES

B-9. Primary staff at each echelon has responsibilities for evaluating requirements and managing linguist support. The responsibilities include but are not limited to those discussed below. In addition, each staff section is responsible for determining its linguist support required to meet its operational missions.

Assistant Chief of Staff, G1 (S1):

- Identify linguist requirements needed to support G1/S1 functions in all contingency areas. G1/S1 requirements for linguist support include but are not limited to the following:
  - Coordinate with local authorities on matters of civilian hire, finance, and recordkeeping.
  - Contract for local hire personnel.
  - Coordinate for local morale support and community activities.
  - Coordinate with local authorities for postal operations.
  - Support for administration, counseling, personal affairs, and leave for LN and third-country national (TCN) personnel.
  - Coordinate for local medical support.
  - Liaison with multinational counterparts.
- Linguist staffing and linguist replacement management.
- Identify foreign language skill identifiers for all assigned, attached, or OPCON Army linguists.
- Identify all Army foreign language skilled soldiers not identified on electronic Military Personnel Office System (eMILPO) and Defense Integrated Management Human Resource System (DIMHRS). The Standard Installation Division Personnel System (SIDPERS) was replaced by eMILPO.
- Deploy and provide administrative support of DA and DOD civilian linguists.
- Hire, contract for, and provide administrative support of LN linguists.
- Procure Army foreign language support personnel for screening local labor resources.

Assistant Chief of Staff, G2 (S2):

- Identify linguist requirements needed to support G2/S2 functions in all contingency areas. G2/S2 requirements for linguist support include but are not limited to—
  - Evaluate and/or use local maps and terrain products in operations.
  - Process for MI purposes material taken from EPWs or civilian internees.

  - At lower echelons, conduct tactical questioning of refugees, detainees, and EPWs.
  - Assess local open-source information for intelligence value.
  - Coordinate intelligence and liaison with multinational and HN counterpart.
- Determine, during the initial IPB, all foreign languages (spoken and written) and dialects needed for mission accomplishment.
- Collect, process, produce, and disseminate information derived from linguist sources.
- Provide intelligence training for MI linguists employed in AOs.
- Coordinate for security investigations, as necessary, for local hire linguists.
- Provide support to CI screening of contracted linguists and LN labor force.

Assistant Chief of Staff, G3 (S3):

- Identify linguist requirements needed to support G3/S3 functions in all contingency areas. G3/S3 requirements for linguist support include but are not limited to—
  - Operational coordination and liaison with multinational and HN counterparts.
  - Translate OPORDs and OPLANs for use by multinational counterparts.
- Consolidate unit linguistic requirements and establish priorities.
- Develop linguist deployment and employment plans.
- Develop plans to train linguists and to use linguists for training the force in AO's foreign language survival skills. In addition to global language skills, linguists must have training in specific vocabulary used in the AO; for example, terms used for military, paramilitary, civilian or terrorist organizations, and ethnic groups within the area, nomenclatures of equipment used, and other military or technical vocabulary. Training in the specific dialect used in the AO would also be beneficial.
- Assign, attach, and detach linguists and linguist teams.
- Integrate additional or replacement linguists through operational channels.
- Recommend modernization and development of linguist systems and methods.
- Coordinate mobilization and demobilization of RC linguist support.
- Plan linguist usage for deception operations.
- Plan linguist support for movement of EPWs, detainees, and refugees.
- Coordinate evaluation of linguist support by all staff elements.

Assistant Chief of Staff, G4 (S4):

- Identify linguist requirements needed to support G4/S4 functions in all contingency areas. G4/S4 linguist requirements for linguist support include but are not limited to—

- Procure local supply, maintenance, transportation, and services.
- Coordinate logistics at air and seaports of debarkation.
- Contract with local governments, agencies, and individuals for sites and storage.

- Provide logistical, supply, maintenance, and transportation support to attached linguists.

Assistant Chief of Staff, G5 (S5):

- Identify linguist requirements needed to support G5/S5 functions in all contingency areas. G5/S5 linguist requirements for linguist support include but are not limited to—
  - Determine civilian impact on military operations.
  - Minimize civilian interference with combat operations.
  - Inform civilians of curfews, movement restrictions, and relocations.
  - Provide assistance to liaison with HN and multinational agencies, dignitaries, and authorities.
  - Promote positive community programs to win over support.
  - Determine if multinational operations PSYOP efforts are mutually planned and synchronized.
  - Interpret support to assist resolution of civilian claims against the US Government.
  - Solicit linguistic and cultural knowledge support to protect culturally significant sites.
  - Use linguistic and cultural support to identify cultural and religious customs.
- Assist the G1 in the contracting of local hire linguists.
- Identify foreign language requirements for CMOs.

Assistant Chief of Staff, G6 (S6):

- Identify linguist requirements needed to support G6/S6 functions in all contingency areas. G6/S6 linguist requirements for linguist support include but are not limited to—
  - Coordinate suitable commercial information systems and services.
  - Coordinate with multinational forces on command frequency lists.
  - Coordinate signal support interfaces with HN and multinational forces.
- Manage RF assignments for supporting SIGINT linguist elements.
- Support linguist operations with internal document reproduction, distribution, and message services.
- Integrate automation management systems of linguist units.

SPECIAL STAFF OFFICER RESPONSIBILITIES

B-10. If no special staff officer is assigned the duties below, the corresponding coordinating staff officer should assume those responsibilities. Linguist requirements for special staff officers include but are not limited to the following staff officers.

Liaison Officer:

- Should speak the required foreign language. If not, he requires a translator or interpreter for all aspects of his duties.
- Request interpreters to assist when representing the multinational operations.
- Translate orders, maps, traces, overlays, and documents into multinational foreign languages.

Civilian Personnel Officer:

- Recruit, interview for suitability, and hire civilian local labor force if required.
- Negotiate host country on labor agreements.

Dental Surgeon:

- Administer dental care to support humanitarian mission requirements.
- Rehabilitate, construct, and gain usage of existing dental facilities as required.

Finance Officer:

- Support the procurement process of local goods and services not readily available through normal logistical channels.
- Ensure limited non-US and US pay functions to foreign national, HN, civilian internees, and EPWs are provided.
- Ensure all necessary banking functions are performed in theater.

Surgeon:

- Support medical humanitarian assistance and disaster relief operations.
- Provide medical care of EPWs and civilians within the command's AO.
- Coordinate medical laboratory access in AO.
- Determine the nature of local health threats to the force through populace interviews.
- Determine the identity of local or captured medical supplies.

Veterinary Officer:

- Determine source and suitability of local foods.
- Assist the local population with veterinary service needs.

Chemical Officer:

- Identify enemy force chemical weapons and equipment.
- Communicate NBC risks to supported populations.

Engineer Coordinator:

- Procure proper local materials to support engineering missions.
- Communicate engineering project requirements to contracted local work force.

- Communicate engineering project impact on local landowners and other affected parties.
- Determine, in coordination with G2/S2, suitability of local topographic maps and terrain products.
- Assess environmental concerns of HN and local populations in combined operations.

Provost Marshal:

- Support dislocated and civilian straggler control activities.
- Support internment and resettlement operations, to include displaced civilians.
- Support weapons buy-back programs, as required, and work closely with civil-military liaisons for payments to local officials.
- Support counter-drug and customs activities.
- When authorized, help foreign civil authorities maintain control.
- Conduct liaison with local LEAs.

PSYOP Officer:

- Produce approved PSYOP propaganda and counter-propaganda media.
- Evaluate PSYOP impact on target audience.

Air Defense Coordinator:

- Identify enemy air defense artillery (ADA) weapons and radars.
- Communicate air defense warnings to supported populations.
- Communicate air defense project requirements to contracted local work force.

Safety Officer:

- Provide safety training to local labor force.
- Communicate warnings of dangerous military operations and other hazards to local populace.

Transportation Officer:

- Coordinate commercial and local transportation needs.
- Coordinate movement scheduling and routes with multinational forces and/or HN.

PERSONAL STAFF OFFICER RESPONSIBILITIES

B-11. Personal staff officers are under immediate control of the commander and have direct access to the commander. Most personal staff officers also perform special staff officer duties, working with a coordinating staff officer. These assignments are on a case-by-case basis, depending on the commander's guidance and the nature of the mission; they are very common in stability operations and support operations. Linguist requirements for special staff officers include but are not limited to the following staff officers.

Chaplain:

- Coordinate religious support with multinational partners.
- Determine the impact of local population religious group faiths and practices on military operations.
- Provide religious support to the community to include hospital patients, EPWs, refugees, and civilian detainees.
- Conduct liaison with local population religious leaders in close coordination with the G5.

Public Affairs Officer:

- Act as the commander's spokesman for all communication with external media.
- Assess the accuracy of foreign media interpretation of Public Affairs Office (PAO) releases.
- Assess and recommend news, entertainment, and other information (assisting G5) for contracted services foreign nationals.

Staff Judge Advocate:

- Translate and interpret foreign legal codes, SOFAs, and international laws.
- Determine local environmental laws and treaties through translation services.
- Assess the treatment of EPWs and civilian internees.
- Translate documents to support G4 in local contracts.

SOURCES OF LINGUISTS

B-12. There are various sources that a commander can use to obtain the linguists necessary to support operations. It is vital to know the advantages and disadvantages of each type of linguist and to carefully match the available linguists to the various aspects of the operation.

ARMY LANGUAGE-QUALIFIED MOSs

B-13. The AC MI language-dependent military occupational specialities (MOSs) are 98G with a skill qualification identifier (SQI) of L/352G (Cryptologic Communications Interceptor/ Locator), and their related WO fields. Some soldiers in MOS 96B (All-Source Intelligence Analyst), MOS 97B (CI Agent), MOS 97E/351E (HUMINT Collector) and MOS 98C (SIGINT Analyst), and their related WO fields are trained in foreign languages. Using soldiers in the MOSs mentioned above has many advantages. They are already trained in the military system, are not subject to deployment restrictions (a limiting factor with civilian linguists), have a security clearance and, as US personnel, support the command's interests. The major disadvantage to utilizing these individuals for general foreign language support is that in doing so, they are removed from their primary MI functions. They should be used only in linguistic duties that include intelligence potential. For example, a HUMINT collector (97E) provides linguist support to a medical assistance team as a method to provide access to the local population to determine their attitudes toward US Forces.

B-14. Non-MI Army language qualified MOSs include some enlisted and WOs in career management fields 18 (Special Forces), 37 (PSYOP), 180A (Special Forces); and commissioned officers with a branch code 18 (Special Forces); and functional areas 39 (PSYOP and CA) and 48 (Foreign Area Officer). Particular attention must be paid to the recorded language proficiency and test date of these individuals since the standards vary by field. The same advantages and disadvantages apply as with the AC MI linguists.

B-15. RC language-dependent MOSs include those listed above in the AC. RC linguists have the same set of advantages and disadvantages as listed above for AC language-dependent MOSs. The RC also includes linguists in MOS 97L (translator/interpreter). The 97Ls are specifically trained to be a translator and interpreter. They have the same advantages as the AC linguists. An added advantage is that since their sole job is translation and interpretation, they do not have to be removed from another job in order to be used as a linguist. Their major disadvantage is that they have no additional skill that gives them dual functionality.

Army Linguists Not DOD Trained

B-16. The Army also includes numerous soldiers of all grades who are proficient in a foreign language and are receiving Foreign Language Proficiency Pay (FLPP) but whose primary duties do not require foreign language proficiency. They may have attended a civilian school to learn a foreign language, or they may have acquired proficiency through their heritage. They have the advantage of being trained soldiers and are therefore readily deployable to all areas of the battlefield.

B-17. These soldiers may have the specific vocabulary and military skill knowledge for certain linguist support missions. For example, a supply sergeant who speaks the local language would be an invaluable asset to the G4. There are disadvantages in that they already have another job and units are reluctant to give up personnel especially if they are in key positions. Their capabilities are difficult to assess. Since they are not required to take the Defense Language Proficiency Test (DLPT) if they are not receiving FLPP, it is often difficult for the G1/S1 to identify them as a linguist or for a non-linguist to judge the level of their foreign language capability.

Other Service Linguists

B-18. Other service linguists have the advantage of deployability, loyalty, and clearance, but must often learn the Army system and specific Army vocabulary. They are also difficult to obtain since their parent service probably also lacks a sufficient number of trained linguists. Other service linguists, however, will be valuable in joint operation centers and joint activities. When serving the JTF headquarters, Army commanders and staffs must be aware of the linguists in the other services in order to plan for the participation and optimize their employment.

US Contract Linguists

B-19. US civilians can be contracted to provide linguist support. They have an advantage over LN hires in that their loyalty to the US is more readily evaluated, and it is easier for them to be granted the necessary security clearance. However, there are usually severe limitations on the deployment and use of civilians. A careful assessment of their language ability is important because, in many cases, they use "old fashioned" terms, or interject US idioms. If the linguists are recent émigrés, the use of the language in their country of origin could be dangerous to them, or their loyalty may reside with their own country when at odds with US interests.

Multinational Linguists

B-20. Multinational linguists have their own set of advantages and disadvantages. These linguists may be unfamiliar with the US military system unless they have previously participated in a multinational operation with US forces. They may have a security clearance, but clearances are not necessarily equal or reciprocal, automatically guaranteeing access to classified or sensitive information between nations. They support the command's interest but may have differing priorities or responsibilities within their assigned AOs. These linguists also are already fulfilling specific duties for their own nation, which may also have a shortage of linguists. The major disadvantage to acquiring and maintaining multinational linguist support is that they are outside the C2 (via military authority or military contract) of the US forces. These linguists will be valuable in multinational operations centers and activities.

Local National Contract Linguists

B-21. LN hires will provide the bulk of your linguist support. They are usually less expensive to hire than US civilians and will know the local dialect, idioms, and culture. The expertise of these linguists in particular areas or subject matters can be an asset. However, there are several potential problems with using LN hires, to include limited English skills and loyalty considerations. Therefore, a screening interview or test is necessary to determine their proficiency in English. These individuals must also be carefully selected and screened by CI personnel (with US linguist support) initially and periodically throughout their employment. Their loyalty is always questionable. Local prejudices may influence them, and they may place their own interests above those of the US.

EVALUATING LINGUIST PROFICIENCY

B-22. Commanders and staffs must understand the Army linguist proficiency evaluation system in order to effectively plan for and employ linguists. Evaluation and reevaluation of linguist proficiency is covered in detail in AR 611-6, Section III. Language testing is required for all Army personnel in a language-dependent MOS, who have received foreign language training at government expense, who are receiving FLPP, or who are in a language-required position regardless of MOS. Other Army personnel who have knowledge of a foreign language are encouraged to take the proficiency test and may work as linguists.

B-23. The Army uses the DLPT to determine foreign language proficiency levels. DLPTs are listed by foreign language in DA Pam 611-16. In foreign languages where no printed or recorded test exists, oral interview tests are arranged. The DLPT is an indication of foreign language capability, but it is not the definitive evaluation of an individual's ability to perform linguist support.

B-24. AR 611-6, Appendix D, Sections 1 through 4, describes the proficiency levels for the skills of speaking, listening, reading, and writing a foreign language based on the interagency roundtable descriptions. The plus-level designators, shown as a "+" symbol, are used to designate when a linguist is above a base level, but not yet to the capability of the next level. For example, 2+ would indicate a better than limited working proficiency in the foreign language. The six "base levels" of proficiency, as established by DLPT and/or oral exam, are—

- Level 0 (No proficiency). The soldier has no functional foreign language ability. Level 0+. The minimum standard for Special Forces personnel indicates a memorized proficiency only.
- Level 1 (Elementary proficiency). The soldier has limited control of the foreign language skill area to meet limited practical needs and elementary foreign language requirements.
- Level 2 (Limited working proficiency). The linguist is sufficiently skilled to be able to satisfy routine foreign language demands and limited work requirements.
- Level 3 (General professional proficiency). The linguist is capable of performing most general, technical, formal, and informal foreign language tasks on a practical, social, and professional level.
- Level 4 (Advanced professional proficiency). The linguist is capable of performing advanced professional foreign language tasks fluently and accurately on all levels.
- Level 5 (Functionally native proficiency). The linguist is functionally equivalent to an articulate and well-educated native in all foreign language skills; and reflects the cultural standards of the country where the foreign language is natively spoken.

B-25. The above proficiency base levels designate proficiency in any of the four language skills: listening, reading, speaking, and writing. The most evaluated skills on the DLPT are reading and listening. These tests are not for use in evaluating linguists above the 3 proficiency level. Most Army linguist DLPT scores show only two skill levels: listening and reading (for example, 2+/3, or 3/1+).

SUSTAINING MILITARY LINGUIST PROFICIENCY

B-26. Language proficiency diminishes with lack of use and absence of exposure to the foreign language. To ensure combat readiness, commanders should require all military linguists receive periodic language training. In-country language immersion training, in-garrison contracted language instructors, on-line foreign newspapers, and foreign radio broadcasts are all examples of language training resources. Funding for language training is available through MACOM language training program funds.

Glossary

The glossary lists acronyms and terms with Army or joint definitions, and other selected terms. Where Army and joint definitions are different, (Army) follows the term. Terms for which FM 2-0 is the proponent FM (authority) are marked with an asterisk (*) and followed by the number of the paragraph (¶) where they are discussed. For other terms, the number of the proponent FM follows the definition. JP 1-02 and FM 1-02 are posted in the Joint Electronic Library, which is available online and on CD-ROM.

- Use this URL to access JP 1-02 online: http://www.dtic.mil/doctrine/jel/doddict/.
- Use this URL to access FM 1-02 online: http://www.dtic.mil/doctrine/jel/service_pubs/101_5_1.pdf.
- Follow this path to access JP 1-02 on the Joint Electronic Library CD-ROM: Mainmenu>Joint Electronic Library>DOD Dictionary.
- Follow this path to access FM 1-02 on the Joint Electronic Library CD-ROM: Mainmenu>Joint Electronic Library>Service Publications>Multiservice Pubs> FM 101-5-1.

A2C2	Army Airspace Command and Control
AA	avenue of approach
AAMDC	Area Air and Missile Defense Command
ABCS	Army Battle Command System
AC	Active Component
ACE	analysis and control element
ACINT	acoustic intelligence (water)
ACOUSTINT	acoustic intelligence (atmosphere)
ACR	armored cavalry regiment
ACT	analysis and control team
ADA	air defense artillery
ADCON	administrative control
ADDO	assistant deputy director for operations
ADO	air defense officer
ADP	automated data processing
AFACSI	Air Force Assistant Chief of Staff for Intelligence
AFIWC	Air Force Information Warfare Center
AFMIC	Armed Forces Medical Intelligence Center

AFOSI	Air Force Office of Special Investigation
AGM	attack guidance matrix
AIA	Air Force Intelligence Agency
AITR	aided target recognition
ALOC	administrative and logistics operations center
AMC	US Army Materiel Command
AMD	Air Missile Defense
AMHS	Automated Message Handling System
AMIB	Allied Military Intelligence Battalion
***analysis**	Determination of the significance of the information, relative to information and intelligence already known, and drawing deductions about the probable meaning of the evaluated information. (¶1-58)
AO	area of operations
AOC	area of concentration
AOI	area of interest
AOIR	area of intelligence responsibility
AOR	area of responsibility
AR	Army Regulation
ARAT	Army Reprogramming Analysis Team
ARCENT	US Army Central Command
ARFOR	Army forces
ARISC	Army Reserve Intelligence Support Center
ARL	airborne reconnaissance low
ARNG	US Army National Guard
ARSOF	Army Special Operations Forces
ASCC	Army Service Component Commander
ASPO	Army Space Program Office
AT	antiterrorism
ATCAE	Army Technical Control Analysis Element
ATCCS	Army Tactical Command and Control System
ATR	automatic target recognition
ART	Army tactical task
BDA	battle damage assessment
BDU	battle-dress uniform

BFACS	Battlefield Functional Area Control System
C/J2	combined J2
C2	command and control
CA	civil affairs
CASO	Collection Activities and Source Operations
CAT	crisis action team
CBRN	chemical, biological, radiological, and nuclear
CBRNE	chemical, biological, radiological, nuclear, and high-yield explosive
CCD	coherent change detection
CCIR	commander's critical information requirement
CD-ROM	compact disc-read only memory
CE	counterespionage
CECOM	Communications-Electronics Command
CED	captured enemy document
CEE	captured enemy equipment
CEM	captured enemy materiel
CFSO	counterintelligence force protection source operations
CGS	common ground station
CI	counterintelligence
CIA	Central Intelligence Agency
CIAC	counterintelligence analysis cell
CICA	Counterintelligence Coordinating Authority
CITO	Central Imagery Tasking Office
CJCS	Chairman, Joint Chiefs of Staff
CMEC	Captured Materiel Exploitation Center
CMF	career management field
CMISE	Corps Military Intelligence Support Element
CMO	civil-military operations
CNA	computer network attack
CND	computer network defense
CNE	computer network exploitation
CNN	Cable News Network
CNO	computer network operations

COA	course of action
COCOM	combatant command (command authority)
COG	center of gravity
COMINT	communications intelligence
COMSEC	communications security
CONOPS	contingency operations
CONPLAN	contingency plan
CONUS	continental United States
COP	common operational picture
COSCOM	Corps Support Command
***counterintelligence**	**1.** Information gathered and activities conducted to protect against espionage, other intelligence activities, sabotage, or assassinations conducted by or on behalf of foreign governments or elements thereof, foreign organizations, or foreign persons, or international terrorist activities. (JP 1-02) **2.** (Army) Counterintelligence counters or neutralizes intelligence collection efforts through collection, CI investigations, operations, analysis and production, and functional and technical services. CI includes all actions taken to detect, identify, exploit, and neutralize the multidiscipline intelligence activities of friends, competitors, opponents, adversaries, and enemies; and is the key intelligence community contributor to protect US interests and equities. (FM 2-0) (¶1-111)
CP	command post
CRN	command radio net
CS	combat support
CSG	cryptologic support group
CSS	combat service support
CSSCS	Combat Service Support Control System
CT	Counterterrorism
***cueing**	The use of one or more sensor systems to provide data that directs collection by other systems. (¶2-53)
CWT	combat weather team
DA	Department of the Army
DAC	Department of the Army Civilian
DAPDM	Department of the Army Production and Dissemination Management
DCI	Director of Central Intelligence
DCS, G2	Deputy Chief of Staff, Intelligence

***debriefing** The systematic questioning of individuals to procure information to answer specific collection requirements by direct and indirect questioning techniques. (¶6-4)

DF direction finding

DHS Defense HUMINT Service

DIA Defense Intelligence Agency

DIAC Defense Intelligence Analysis Center

DIMHRS Defense Integrated Management Human Resource System

DIO Defensive Information Operations

DISCOM Division Support Command

DISUM daily intelligence summary

DLPT Defense Language Proficiency Test

DMS Director of Military Support

DOCEX document exploitation

***document exploitation** The systematic extraction of information from all media formats in response to collection requirements. (¶6-4)

DOD Department of Defense

DOJ Department of Justice

DOT Department of Transportation

DP decision point

DPM dissemination program manager

DS direct support

DSM decision support matrix

DST decision support template

EA electronic attack

EAC echelons above corps

ECOA enemy course of action

EEFI essential elements of friendly information

ELINT electronic intelligence

EM electromagnetic

eMILPO electronic Military Personnel Office System

EMS electromagnetic spectrum

E-O electro-optical

EOB electronic order of battle

EOD explosive ordinance disposal

EP	electronic protection
EPW	enemy prisoner of war
ES	electronic warfare support (Army)
***evaluating**	**1.** Comparing relevant information (RI) on the situation or operation against criteria to determine success or progress. (FM 6-0) **2.** In intelligence usage, appraisal of an item of information in terms of credibility, reliability, pertinence, and accuracy. (FM 2-0) (¶1-26 and 1-98)
EW	electronic warfare
EWS	electronic warfare support (joint)
FBI	Federal Bureau of Investigation
FDA	functional damage assessment
FDU	Force Design Update
FEMA	Federal Emergency Management Agency
FFIR	friendly force information requirement
FISINT	foreign instrumentation signals intelligence
FISS	Foreign Intelligence Security Service
FLPP	Foreign Language Proficiency Pay
FM	Field Manual
FMA	foreign materiel acquisition
FME	foreign materiel exploitation
FMEP	Foreign Materiel Exploitation Program
FMP	Foreign Materiel Program
FMT	Foreign Materiel for Training
FP	force protection
FRAGO	fragmentary order
FS	fire support
FSAC	Fire Support Armaments Center
FSE	fire support element
FSO	fire support officer
G1	Assistant Chief of Staff, Personnel
G2	Assistant Chief of Staff, Intelligence
G3	Assistant Chief of Staff, Operations
G4	Assistant Chief of Staff, Logistics
G5	Assistant Chief of Staff, Civil/Military Affairs

G6	Assistant Chief of Staff, Command, Control, Communications, and Computer Operations
G7	Assistant Chief of Staff, Information Operations
GEOINT	geospatial intelligence
GMI	general military intelligence
GPS	Global Positioning System
GS	general support
GS-R	general support-reinforcing
HN	host nation
HOC	HUMINT operations cell
HPT	high-payoff target
HQ	headquarters
HQDA	Headquarters, Department of the Army
HSI	hyperspectral imaging
HUMINT	human intelligence
***human intelligence**	Collection by a trained HUMINT collector of foreign information from people and multimedia to identify elements, intentions, composition, strength, dispositions, tactics, equipment, personnel, and capabilities. It uses human sources and a variety of collection methods, both passively and actively, to gather information to satisfy the commander's intelligence requirements and cross-cue other intelligence disciplines. (¶1-106 and ¶6-1)
HVT	high-value target
I/R	internment and resettlement (operations)
I&W	indications and warnings
ICC	Intelligence Coordination Center (US Coast Guard)
ICF	intelligence contingency fund
ICL	intelligence coordination line
IDC	Information Dominance Center
IIR	intelligence information report
IM	information management
IMA	individual mobilization augmentee
IMEP	International Materiel Evaluation Program
IMETS	Integrated Meteorological System
IMINT	imagery intelligence

***indications and warning** (joint) Those intelligence activities intended to detect and report time-sensitive intelligence information on foreign developments that could involve a threat to the United States or allied and/or coalition military, political, or economic interests or to US citizens abroad. It includes forewarning of enemy actions or intentions; the imminence of hostilities; insurgency; nuclear or non-nuclear attack on the United States, its overseas forces, or allied and/or coalition nations; hostile reactions to US reconnaissance activities; terrorists attacks; and other similar events. (¶2-19)

***indicator** (Army) Positive or negative evidence of threat activity or any characteristic of the AO which points toward threat vulnerabilities or the adoption or rejection by the threat of a particular capability, or which may influence the commander's selection of a COA. Indicators may result from previous actions or from threat failure to take action. (FM 2-0) (¶5-14)

INFOSEC information security

***infrastructure** **1.** (joint) All building and permanent installations necessary for support, redeployment, and military forces operations (e.g., barracks, headquarters, airfields, communications facilities, stores, port installations, and maintenance stations). (JP 4-01.8) **2.** (Army) In intelligence usage, the basic underlying framework or feature of a thing; in economics, basic resources, communications, industries, and so forth, upon which others depend; in insurgency, the organization (usually hidden) of insurgent leadership. (FM 2-0) (¶5-8)

INR Bureau of Intelligence and Research, Department of State

INS Immigration and Naturalization Service

INSCOM US Army Intelligence and Security Command

intelligence **1.** (joint) The product resulting from the collection, processing, integration, analysis, evaluation, and interpretation of available information concerning foreign countries or areas. **2.** Information and knowledge about an adversary obtained through observation, investigation, analysis, or understanding. (¶1-3)

***intelligence coordination line** A line that designates the boundary between AOIRs. The G2/S2 establishes ICLs to facilitate coordination between higher, lateral, and subordinate units; coordinates with the G3/S3 to direct subordinates to track enemy units and HPTs in their areas; and hands over intelligence responsibility for areas of the battlefield. The establishment of ICLs ensures that there are no gaps in the collection effort; that all echelons are aware of the location, mission, and capabilities of other assets; facilitates asset cueing, and provides timely exchange of information between assets. The G2/S2 keeps abreast of collection activities in progress (all echelons) and battlefield developments through the ICLs. (¶3-13)

intelligence discipline (joint) A well-defined area of intelligence collection, processing, exploitation, and reporting using a specific category of technical or human resources. There are seven major disciplines: human intelligence, imagery intelligence, measurement and signature intelligence, signals intelligence (communications intelligence, electronic intelligence, and foreign instrumentation signals intelligence), all-source intelligence, technical intelligence, and counterintelligence. (¶1-104)

***intelligence process** A theoretical model used to describe intelligence operations. The intelligence process is not a framework for actual operations because the functions and common tasks occur in parallel as opposed to sequentially. (¶1-91)

***intelligence products** Intelligence products are generally placed in one of six categories: I&W, current, general military, target, S&T, and CI. The categories are distinguished from each other primarily by the purpose for which the intelligence was produced. The categories often overlap, and the same intelligence data can be used in each of the categories. (¶1-5)

INTREP intelligence report

INTSUM intelligence summary

***intelligence requirements** (Army) Those requirements generated from the staff's IRs regarding the enemy and environment that are not a part of the CCIR (PIR and FFIR). Intelligence requirements require collection and can pro-vide answers in order to identify indicators of enemy actions or intent, which reduce the uncertainties associated with an operation. Significant changes (i.e., branches and sequels) with an operation usually lead to changes in intelligence requirements. (¶2-7)

***intelligence running estimate** The intelligence running estimate is a continuous flow and presentation of relevant information and predictive intelligence that, when combined with the other staff running estimates, enable the decisionmaker's visualization and situational understanding of the AOI in order to achieve information superiority. The intelligence running estimate requires constant verification to support situational understanding of the current situation as well as predictive assessments for future operations. (¶5-30)

***intelligence synchronization** The task that ensures ISR operations are linked to the commander's requirements and respond in time to influence decisions and operations. The intelligence officer, with staff participation, synchronizes the entire collection effort to include all assets the commander controls, assets of lateral units and higher echelons units and organizations, and intelligence reach to answer the commander's CCIR (PIR and FFIR). (¶1-24)

***intelligence synchronization plan**	The plan the intelligence officer uses, with staff input, to synchronize the entire collection effort to include all assets the commander controls, assets of lateral units, and higher echelon units and organizations, and intelligence reach to answer the commander's CCIR (PIR and FFIR). (¶1-38)
intelligence warfighting function	The related tasks and systems that facilitate understanding of the operational environment
IO	information operations
IPB	intelligence preparation of the battlefield
IPIR	Initial Phase Imagery Report
IPTF	international police task force
IR	information requirement
IRINT	infrared intelligence
IRR	Individual Ready Reserve
ISM	intelligence synchronization matrix
ISR	intelligence, surveillance, and reconnaissance
J2	Joint Staff Directorate, Intelligence
J2A	Joint Staff Directorate, Intelligence (Administrative)
J2J	Joint Staff Directorate, Intelligence (Joint)
J2M	Joint Staff Directorate, Intelligence (Management)
J2O	Joint Staff Directorate, Intelligence (Operations)
J2P	Joint Staff Directorate, Intelligence (Assessments, Doctrine)
J2T	Joint Staff Directorate, Intelligence (Targeting)
J2X	Joint Staff Directorate, Intelligence (CI & HUMINT)
J3	Joint Staff Directorate, Operations
J5	Joint Staff Directorate, Civil Affairs
JAC	Joint Analysis Center
JCMEC	Joint Captured Materiel Exploitation Center
JCO	Joint Commission Observers
JCS	Joint Chiefs of Staff
JDEC	Joint Document Exploitation Center
JDISS	Joint Deployable Intelligence Support System
JFC	joint force commander
JIAWG	joint interagency working group
JIC	Joint Intelligence Center

JIF	Joint Interrogation Facility
JISE	joint intelligence support element
JOA	joint operations area
JP	Joint Publication
JSTARS	Joint Surveillance Target Attack Radar System
JTF	joint task force
JTMD	joint table of mobilization and distribution
JWICS	Joint Worldwide Intelligence Communications System
LAN	local area network
LASINT	laser intelligence
LCC	land component commander
LEA	law enforcement agency
LIDAR	laser radar
LN	local national
LNO	liaison officer
LOC	line of communication
LOS	line of sight
LPT	logistics preparation of the theater
LTIOV	latest time information is of value
MACOM	major Army command
MASINT	measurement and signature intelligence
***materiel**	(Army) In intelligence usage, the all-encompassing term for the weapons systems, equipment, apparatus, documents, and supplies of a foreign military force or nonmilitary organization. (¶2-24)
MCIA	Marine Corps Intelligence Agency
MCOO	modified combined obstacle overlay
MCS	Maneuver Control System
MDMP	military decisionmaking process
MEA	munitions effects assessment
METL	mission-essential task list
METT-TC	mission, enemy, terrain and weather, troops and support available, time available, civil considerations
MI	military intelligence
MLT	machine language translation

MNFC	multinational force commander
MO	method of operation
MOPP	mission-oriented protective posture
MOS	military occupational specialty
MP	military police
MSC	major subordinate command
MSE	mobile subscriber equipment
MSI	multi-spectral imagery
MSIC	Missile and Space Intelligence Center
MTI	moving target indicator
MTOE	modified table of organization and equipment
MTW	major theater war
NAI	named area of interest
NAIC	National Air Intelligence Center
NATO	North Atlantic Treaty Organization
NBC	nuclear, biological, and chemical
NCA	National Command Authorities
NCO	noncommissioned officer
NCR	national cryptologic representative
NCTI	non-cooperative target identification
NEO	noncombatant evacuation operation
NFH	National Defense Headquarters, Canada
NGA	National Geospatial-Intelligence Agency
NGO	non-governmental organization
NIC	National Intelligence Center
NIIRS	National Imagery Interpretability Rating Scale
NIMA	National Imagery and Mapping Agency
NIST	National Intelligence Support Team
NLT	not later than
NMCC	National Military Command
NMIC	National Maritime Intelligence Center
NMJIC	National Military Joint Intelligence Center
NRO	National Reconnaissance Office
NRT	near-real time

NSA	National Security Agency
NSC	National Security Council
NUCINT	Nuclear Intelligence
NVD	night vision device
OB	order of battle
OE	operational environment
OMA	Office of Military Affairs
OMB	Office of Management and Budget
OMT	operational management team
ONDCP	Office of National Drug Control Policy
ONI	Office of Naval Intelligence
ONIR	overhead non-imaging infrared
OPCON	operational control
OPLAN	operation plan
OPORD	operation order
OPSEC	operations security
OSC	operations support cell
OSCE	Organization for Security and Cooperation in Europe
OSD	Office of the Secretary of Defense
OSINT	open-source intelligence
PAO	Public Affairs Office
PDA	physical damage assessment
PIAP	Police Information Assessment Program
PIO	police intelligence operations
PIR	priority intelligence requirement
PME	peacetime military engagement
POC	point of contact
POD	port of debarkation
POE	port of embarkation
***predictive intelligence**	Intelligence analysis conclusions, assessments or products that attempt to define, describe, present, or portray the future situation or conditions of the enemy, +terrain, weather and civil considerations. (¶1-3)
***priority intelligence requirements**	(joint) Those intelligence requirements for which a commander has an anticipated and stated priority in his task of planning and decision-making. (FM 2-0) (¶1-32)

***process/ processing**	A function of the intelligence process that involves converting collected data, which is not already in a comprehensible form when it is reported, into a form that is understandable and suitable for analysis and production of intelligence. Examples of processing include developing film, enhancing imagery, translating a document from a foreign language, converting electronic data into a standardized report that can be analyzed by a system operator, and correlating dissimilar or jumbled information by assembling like elements before the information is forwarded for analysis. (FM 2-0) (¶1-96)
***production**	In intelligence usage, conversion of information into intelligence through the integration, analysis, evaluation, and interpretation of all source data and the preparation of intelligence products in support of known or anticipated user requirements. (¶4-20)
PSYOP	psychological operations
QRC	quick reaction capability
QRF	quick reaction force
R&D	research and development
RADINT	radar intelligence
RC	reserve component
RDEC	Research, Development, and Engineering Center
***reconnaissance, surveillance, and target acquisition/intelligence surveillance and reconnaisssance**	A full spectrum combined arms mission that integrates ground and air capabilities to provide effective, dynamic, timely, accurate and assured combat information and multi-discipline actionable intelligence for lethal and non-lethal effects/decisions in direct support of the ground tactical commander.
REMBASS	Remotely Monitored Battlefield Sensor System
***requirements management**	The intelligence task that develops a prioritized list of what information needs to be collected and produced into intelligence, dynamically updates and adjusts those requirements in response to mission adjustments/changes, and places a latest time intelligence is of value to ensure intelligence and information is reported to meet operational requirements. (¶2-83)
RF	radio frequency
RFI	request for information
RFW	radio frequency weapon
RI	relevant information
RM	requirements management
ROE	rules of engagement
ROK	Republic of Korea
ROZ	restricted operating zone

RSO&I	reception, staging, onward movement, and integration
RSOC	Regional Security Operations Center
RSTA	reconnaissance, surveillance, and target acquisition
RSP	render safe procedure
RTCAE	Regional Technical Control and Analysis Element
running estimate	A staff estimate, continuously updated, based on new information as the operation proceeds. (FM 6-0) (¶5-28)
S&T	scientific and technical
S&TI	scientific and technical intelligence
S1	Staff Officer, Personnel
S2	Staff Officer, Intelligence
S2X	Staff Officer, Intelligence (CI & HUMINT)
S3	Staff Officer, Operations
S4	Staff Officer, Logistics
S5	Staff Officer, Civil Affairs
S6	Staff Officer, Command, Control, Communications, and Computer Operations
SAP	special access program
SAR	synthetic aperture radar
SATCOM	satellite communications
SBCCOM	Chemical and Biological Defense Command
SCI	sensitive compartmented information
SCIF	sensitive compartmented information facility
SIDPERS	Standard Installation Division Personnel System
SIGINT	signals intelligence
SIGO	signal officer
SII	statement of intelligence interest
SIO	senior intelligence officer
SIPRNET	Secret Internet Protocol Router Network
SIR	specific information requirement
SITMAP	situation map

***situation template**	(Army) A depiction of a potential adversary course of action as part of a particular adversary operation. Situation templates are developed on the adversary's current situation (for example, training and experience levels, logistic status, losses, and disposition), the environment, and adversary doctrine or patterns of operations. The commander dictates the level to depict the adversary based on the factors of METT-TC (at minimum two levels of command below the friendly force) as a part of his guidance for mission analysis. (FM 2-0) (¶1-58)
SJA	staff judge advocate
SMU	special mission unit
SOCOM	satellite communications on-the-move
SOF	special operations forces
SOFA	status of forces agreement
SOP	standing operating procedure
SOTA	SIGINT operational tasking authority
SR	special reconnaissance
SRTD	signals research and target development
SSO	special security office
STANAG	standardization agreement (NATO)
STICEUR	Scientific and Technology Intelligence Center Europe
SUPR	supplemental imagery report
SWO	staff weather officer
TA	target acquisition
TACON	tactical control
***tactical questioning**	The expedient initial questioning for information of immediate tactical value. (FM 2-0) (¶6-6)
TAI	target area of interest
TCAE	technical control and analysis element
TCN	third-country national
***technical control**	The technical function to ensure adherence to existing policies or regulations and provide technical guidance for MI activities, particularly HUMINT, SIGINT, and CI operations. (FM 2-0) (¶3-23)
TECHINT	technical intelligence
TECOM	Test and Evaluation Command
TENCAP	Tactical Exploitation of National Capabilities
TEU	Technical Escort Unit

THREATCON	threat condition
TIB	Theater Intelligence Brigade
TIG	Theater Intelligence Group
TIM	toxic industrial material
TLP	troop-leading procedure
TOC	tactical operations center
TPFDD	Time Phased Force Deployment Data
TRADOC	US Army Training and Doctrine Command
TSA	target system assessment
TSCM	Technical Surveillance Countermeasures
TSE	theater support element
TTP	tactics, techniques, and procedures
TVA	threat vulnerability assessment
UAV	unmanned aerial vehicle
UN	United Nations
US	United States
USAF	United States Air Force
USAR	United States Army Reserve
USC	United States Code
USCENTCOM	United States Central Command
USCG	United States Coast Guard
USCS	United States Cryptologic System
USEUCOM	United States European Command
USJFCOM	United States Joint Forces Command
USMC	United States Marine Corps
USNORTHCOM	United States Northern Command
USPACECOM	United States Space Command
USPACOM	United States Pacific Command
USSID	United States Signal Intelligence Directive
USSOCOM	United States Special Operations Command
USSOUTHCOM	United States Southern Command
USSPACECOM	United States Space Command
USSTRATCOM	United States Strategic Command
USTRANSCOM	United States Transportation Command

VJ2	Vice J2
VTC	video teleconference
WAN	wide area network
WARNO	warning order
WMD	weapons of mass destruction
WO	warrant officer

Bibliography

The bibliography lists field manuals published with new numbers followed by old number.

DOCUMENTS NEEDED

These documents must be available to the intended users of this publication.

JP 1-02. *Department of Defense Dictionary of Military and Associated Terms.* June 1999. [Online] Available http://www.dtic.mil/doctrine/jel/doddict/

FM 1-02/MCRP 5-2A. *Operational Terms and Symbols* (Drag Draft). 21 February 2003.

READINGS RECOMMENDED

These sources contain relevant supplemental information.

ARMY PUBLICATIONS

Most Army doctrinal publications are available online: http://155.217.58.58/atdls.htm

AR 381-10. *US Army Intelligence Activities.* 1 July 1984.

AR 381-12. *Subversion and Espionage Directed Against the U.S. Army (SAEDA).* 15 January 1993.

AR 381-20. *The Army Counterintelligence Program.* 15 November 1993.

AR 611-6. *Army Linguist Management.* 16 February 1996.

AR 715-9. *Contractors Accompanying the Force.* 29 October 1999.

FM 1 (100-1). *The Army.* 14 June 2001.

FM 2-19.402. *STRYKER Brigade Combat Team Intelligence Operations.* 1 March 2003.

FM 2-19.602. *Surveillance Troop.* 1 March 2003.

FM 2-33.5/ST. *Intelligence Reach Operations.* 1 June 2001.

FM 3-0. *Operations.* 14 June 2001.

FM 3-05.102. *Army Special Operations Forces Intelligence.* 31 August 2001 (formerly FM 34-36).

FM 3-05.30. *Psychological Operations.* 19 June 2000.

FM 3-06.11. *Combined Arms Operations in Urban Terrain.* 28 February 2002.

FM 3-13. *Information Operations.* 28 November 2003 (formerly FM 100-6).

FM 3-19. *NBC Reconnaissance.* 19 November 1993.

FM 3-19.1. *Military Police Operations.* 22 March 2001.

FM 3-19.4. *Military Police Leaders' Handbook.* 4 March 2002.

FM 3-19.40. *Military Police Internment/Resettlement Operations.* 30 September 1987.

FM 3-20.98. *Reconnaissance Platoon.* May 2001.

FM 34-2. *Collection Management and Synchronization Planning.* 8 March 1994.

FM 34-52. (FM 2-22-3) *Human Intelligence.* 28 September 1992.

FM 3-90. *Tactics.* 4 July 2001.

FM 3-90.3. *The Mounted Brigade Combat Team.* 1 November 2001.

FM 3-100.4. *Environmental Considerations in Military Operations.* 15 June 2000, Change 1, 11 May 2001.

FM 3-100.21. *Contractors on the Battlefield.* 26 March 2000.

FM 5-170. *Engineer Reconnaissance.* 5 May 1998.

FM 6-0. *Mission Command: Command and Control of Army Forces.* 11 August 2003.

FM 6-20-10. *Tactics. Techniques, and Procedures for the Targeting Process.* 8 May 1996.

FM 6-121. *Tactics, Techniques, and Procedures for Field Artillery Target Acquisition.* 25 September 1990.

FM 7-0. *Training the Force.* 22 October 2002.

FM 7-10. *The Infantry Rifle Company.* 14 December 1990.

FM 7-15. *Army Universal Task List.* 31 August 2003.

FM 7-92. *The Infantry Reconnaissance Platoon and Squad (Airborne, Air Assault, Light Infantry).* 23 December 1992.

FM 7-100. *Opposing Force Doctrinal Framework and Strategy.* 1 May 2003.

FM 17-97. *Cavalry Troop.* 3 October 1995.

FM 17-98. *Scout Platoon.* 10 April 1999.

FM 19-1. *Military Police Support for the Airland Battle.* 23 May 1988.

FM 27-10. *The Law of Land Warfare.* 18 July 1956.

FM 31-20-5. *Special Reconnaissance Tactics, Techniques, and Procedures for Special Forces.* 23 March 1993.

FM 34-2-1. *Tactics, Techniques, and Procedures for Reconnaissance and Surveillance and Intelligence to Counterreconnaissance.* 19 June 1991.

FM 34-3. *Intelligence Analysis.* 15 March 1990.

FM 34-8-2. *Intelligence Officer's Handbook.* 1 May 1998.

FM 34-10. *Division Intelligence and Electronic Warfare Operations.* 25 November 1986.

FM 34-25. *Corps Intelligence and Electronic Warfare Operations.* 30 September 1987.

FM 34-37. *Echelons Above Corps (EAC) Intelligence and Electronic Warfare (IEW) Operations.* 15 January 1991.

FM 34-54. *Technical Intelligence.* 30 January 1998.

FM 34-80. *Brigade and Battalion Intelligence and Electronic Warfare Operations.* 15 April 1986.

FM 34-130. *Intelligence Preparation of the Battlefield.* 8 July 1994.

FM 4-0. *Combat Service Support.* 29 August 2003.

FM 41-10. *Civil Affairs Operations.* 14 February 2000.

FM 100-25. *Doctrine for Army Special Operations Forces.* August 1999.

FM 101-5. *Staff Organization and Operations.* 31 May 1997.

FM 101-5-1. *Operational Terms and Graphics.* 30 September 1997.

FM 101-5-2. *US Army Report and Message Formats.* 29 June 1999.

ST 2-22.7. *Tactical Human Intelligence and Counterintelligence Operations.* April 2002.

ST 2-33.5. *Intelligence Reach Operations.* 1 June 2001.

ST 2-50. *Intelligence and Electronic Warfare Assets.* June 2002.

ST 2-50.4. *Combat Commanders' Handbook on Intelligence.* September 2001 (formerly FM 34-81, 28 September 1992).

JOINT PUBLICATIONS

Most joint publications are available online: http://www.dtic.mil/doctrine/jel/

JP 0-2. *Unified Action Armed Forces (UNAAF).* 10 July 2001.

JP 1. *Joint Warfare for the US Armed Forces.* 14 November 2000.

JP 1-06. *Joint Tactics, Techniques, and Procedures for Financial Management During Joint Operations.* 22 December 1999.

JP 2-0. *Joint Doctrine for Intelligence to Operations.* January 2000.

JP 2-01. *Joint Intelligence Support to Military Operations.* 20 November 1996.

JP 2-01.1. *Joint Tactics, Techniques, and Procedures for Intelligence Support to Targeting.* 9 January 2003.

JP 2-01.3. *Joint Tactics, Techniques, and Procedures for Joint Intelligence Preparation of the Battlespace,* 24 May 2000.

JP 2-02. *National Intelligence to Joint Operations.* 28 September 1998.

JP 2-03. *Joint Tactics, Techniques, and Procedures for Geospatial Information and Services Support to Joint Operations.* 31 March 1999.

JP 3-0. *Doctrine and Joint Operations.* 1 February 1995.

JP 3-07. *Joint Doctrine for Military Operations Other Than War.* 16 June 1995.

JP 3-07.2. *JTTP for Antiterrorism.* 17 March 1998.

JP 3-08. *Interagency Coordination During Joint Operations,* Volume II. 9 October 1996.

JP 3-09. *Doctrine for Joint Fire Support.* 12 May 1998.

JP 3-09.1. *Joint Tactics, Techniques and Procedures for Laser Designation Operations.* 28 May 1999.

JP 3-11. *Joint Doctrine for Operations in Nuclear, Biological, and Chemical (NBC) Environments.* 11 July 2000.

JP 3-13. *Joint Doctrine for Information Operations.* 9 October 1998.

JP 3-18. *Joint Doctrine for Forcible Entry Operations.* 16 July 2001.

JP 3-35. *Joint Deployment and Redeployment Operations.* 7 September 1999.

JP 3-51. *Joint Doctrine for Electronic Warfare.* 7 April 2000.

JP 3-55. *Doctrine for Reconnaissance, Surveillance, and Target Acquisition Support for Joint Operations (RSTA).* 14 April 1993.

JP 3-57. *Joint Doctrine for Civil-Military Operations.* 8 February 2001.

JP 3-60. *Joint Doctrine for Targeting.* 17 January 2002.

JP 4-0. *Doctrine for Logistic Support of Joint Operations.* 6 April 2000.

JP 5-0. *Doctrine for Planning Joint Operations.* 13 April 1995.

JP 5-00.2. *Joint Task Force Planning Guidance and Procedures.* 13 January 1999.

PUBLIC LAWS AND OTHER PUBLICATIONS

The United States Code is available online: http://uscode.house.gov/usc.htm

EO 12333. *United States Intelligence Activities.* 4 December 1981.

CJCS Instruction 1301.01, Policy and Procedures to Assign Individuals to Meet Combatant Command Mission-Related Temporary Duty Requirements. 1 July 2001.

CJCSM 3500.04C. *Universal Joint Task List.* 1 July 2002.

Military Standard 2525B, *DOD Interface Standard, Common Warfighting Symbology.* 30 January 1999.

PUBLICATIONS UNDER DEVELOPMENT

FM 2-01. *Intelligence Synchronization.* When published tentatively in July 2004, FM 2-01 will supersede FM 34-2, 8 March 1994, and FM 34-2-1, 19 June 1991.

FM 2-01.2. *Counterintelligence Analysis.* When published tentatively in April 2005, FM 2-01.2 will supersede FM 34-60, 3 October 1995.

FM 2-01.3. *Intelligence Preparation of the Battlefield.* When published tentatively December 2004, FM 2-01.3 will supersede FM 34-130, 8 July 1994.

Index

Entries are by paragraph number.

D-E-F

G-H

I

J-L

Field Manual
No. 3-19.15

Headquarters
Department of the Army
Washington, DC, 18 April 2005

Contents

Distribution Restriction: Approved for public release; distribution is unlimited.

*This publication supersedes FM 19-15, 25 November 1985.

Page

Preface

Field Manual (FM) 3-19.15 addresses continental United States (CONUS) and outside continental United States (OCONUS) civil disturbance operations. Today, United States (US) forces are deployed on peacekeeping, peace enforcement, and humanitarian assistance operations worldwide. During these operations, US forces are often faced with unruly and violent crowds intent on disrupting peace and the ability of US forces to maintain peace. Worldwide instability coupled with increasing US military participation in peacekeeping and related operations requires that US forces have access to the most current doctrine and tactics, techniques, and procedures (TTP) necessary to quell riots and restore public order.

In addition to covering civil unrest doctrine for OCONUS operations, FM 3-19.15 addresses domestic unrest and the military role in providing assistance to civil authorities requesting it for civil disturbance operations. It provides the commander and his staff guidance for preparing and planning for such operations. The principles of civil disturbance operations, planning and training for such operations, and the TTP employed to control civil disturbances and neutralize special threats are discussed in this manual. It also addresses special planning and preparation that are needed to quell riots in confinement facilities are also discussed.

In the past, commanders were limited to the type of force they could apply to quell a riot. Riot batons, riot control agents, or lethal force were often used. Today, there is a wide array of nonlethal weapons (NLW) available to the commander that extends his use of force along the force continuum. This manual addresses the use of nonlethal (NL) and lethal forces when quelling a riot.

Appendix A complies with current Army directives, which state that the metric system will be incorporated into all new publications.

The proponent for this publication is HQ, TRADOC. Send comments and recommendations on Department of the Army (DA) Form 2028 (*Recommended Changes to Publications and Blank Forms*) directly to Commandant, US Army Military Police School (USAMPS), ATTN: ATSJ-DD, 401 MANSCEN Loop, Fort Leonard Wood, MO 65473-8929.

Unless this publication states otherwise, masculine nouns or pronouns do not refer exclusively to men.

Chapter 1

Operational Threats of the Civil Disturbance Environment

As the Cold War ended with the fall of the Berlin Wall and the collapse of Soviet communism in 1989, the Army was no longer concerned with containing the aggressive Soviet Union. The US Army began to transform because the battlefield on which it would fight had changed. Operations Just Cause, Joint Endeavor, Desert Shield, and Desert Storm have all been recognized as containing nontraditional battlefields. Following these operations, the Army was called upon to conduct peacekeeping, peace enforcement, and humanitarian assistance operations. These operations required US forces to protect refugees and humanitarian workers and keep hostile factions separate to enforce peace.

GENERAL CAUSES FOR CIVIL UNREST

1-1. In these modern times, demonstrations, civil unrest, public disorder, and riots happen for a number of reasons. Some of these reasons are economic hardships, social injustices, ethnic differences (leading to oppression), objections to world organizations or certain governments, political grievances, and terrorist acts. An event can be triggered by a single cause or a combination of causes. For example, operations in the Balkans involving civil unrest and riots were the result of ethnic hatred, a lack of civil authority, food shortages, a revolution, and religious-based fighting factions.

1-2. Demonstrations may range from simple, nonviolent protests that address specific issues, to events that turn into full-scale riots. Gatherings in protest are recognized rights of any person or group, regardless of where US forces may be operating. This fundamental right is protected under the *Constitution of the US*. During peacekeeping or peace enforcement operations, US forces should never violate basic civil or human rights. Most protesters are law-abiding citizens who intend that their protests be nonviolent, but some protest planners insist that the event involve some kind of violence. Often, in the media, protesters can gain sympathy for their cause by prompting authorities to take physical action against them. Violence is often the result of demonstrators beginning to conduct unlawful or criminal acts and authorities (who are responsible for the safety and welfare of all) enforcing the laws of the municipality, state, or nation. The depth of violence is determined by the willingness of demonstrators to display and voice their opinions in support of their cause.

1-3. Commanders must be aware of the possibility that some individuals or groups within an organized demonstration may have the intent to cause disruption, incite violence, destroy property, and provoke the authorities. The tactical situation and actions of the crowd should dictate control and enforcement options. Agitators and criminal infiltrators within the crowd can lead to the eruption of violence. Inciting a crowd to violence or a greater intensity of violence through the use of brutish enforcement tactics should be avoided. Publicity can be detrimental to authorities and beneficial to crowds because it can further their causes.

1-4. Community unrest results in urban conflicts that arise from highly emotional social and economic issues. Economically deprived inner city residents may feel that they are treated unjustly or ignored by people in power and authority. Tensions can build quickly in a community over a variety of issues, such as hunger, poor employment opportunities, inadequate community services, poor housing, and labor issues. Tension in these areas creates the potential for violence. When tensions are high, it takes a small (seemingly minor) incident, rumor, or act of injustice to ignite groups within a crowd to riot and act violently. This is particularly true if community relations with authorities are part of the problem.

1-5. Significant ethnic differences in a community can create an atmosphere of distrust, even hatred. Unrest among ethnic groups competing for jobs, living areas, and sparse essentials can cause an eruption of civil disorder and/or riots. As emotions run high, violence becomes likely.

1-6. Terrorist organizations may infiltrate groups within a demonstrating crowd. These terrorist groups may intend to embarrass their government or other governments. Terrorist infiltrators can be used to provoke crowds as a diversion, as part of a demonstration, or as cover for terrorist acts.

CROWD DEVELOPMENT

1-7. Crowds are a gathering of a multitude of individuals and small groups that have temporarily assembled in the same place. These small groups are usually comprised of friends, family members, or acquaintances that represent a group belief or cause. Individuals assume a sense of anonymity—they are viewed as just another face in the crowd. People in small groups are known only to companions in their group and to others in the gathering that have come from the same neighborhood or community. Commanders must consider how the individuals assembled and how they are interacting during the gathering process.

GATHERINGS

1-8. The assembly process of a gathering refers to the movement of people from different locations to a common location within a given period. This largely determines who participates. Creating a gathering from a crowd is a process with a beginning, middle, and end. As shown in *Figure 1-1* a crowd has an assembly process that leads to the gathering of the crowd, which is always followed by a dispersal process.

1-9. Gatherings are often assisted by the activities of individuals or groups with a specific agenda, such as yelling catchy slogans and cheers that everyone can easily pick up and join in on. Some groups are so well organized that they can prestage leaders to infiltrate a gathering. This creates unity, even inciting newcomers to join their cause. The first phase of the gathering is the assembly process. It can occur in one of two ways—impromptu or organized.

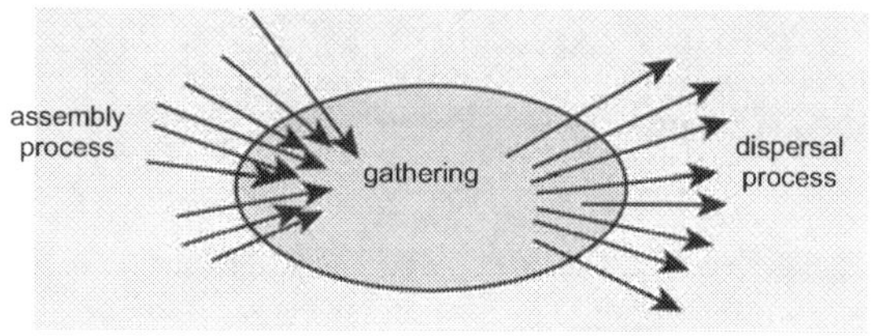

Figure 1-1. Crowd Building

IMPROMPTU GATHERINGS

1-10. Impromptu gatherings usually develop informally and are mostly done by word of mouth (one person to another or one group to another). Participants spread intelligence by telling one another when, where, and what is happening and inviting them to participate. An example of an impromptu gathering would be a gathering at a secured food distribution point after receiving information (by word of mouth) that a large truck carrying much-needed supplies and food is about to arrive. In this case, hunger would be the driving force causing the migration of people to the food distribution facility.

ORGANIZED GATHERINGS

1-11. Like an impromptu gathering, an organized gathering can also involve individuals and groups passing on information to one another. Passing intelligence of police activity and occurrences and on when and where events will take place helps organizers to prestage participants. Rarely is one represented group responsible for pulling together a gathering. Organized gatherings rely heavily on established groups that attract people to gather. Recent examples of these well-organized groups are anarchists, anti-globalization groups, and anti-free-enterprise groups. Groups representing extreme religious faiths and ethnic organizations have been common too. Some gatherings involve groups from rival, or even warring, factions.

1-12. Organized gathering processes rely more on centralized planning and organization. One or more of the groups offer organizers lists of individual names and groups, which they contact as potential participants. Modern technologies are available to these organizers through telephone banks, mailing lists, or e-mail addresses. In preparation for a long-term event, some group organizers have the means to provide logistical support of transportation, food, and water to participants.

> During the preelection period, Bosnian-Serbs (in a bid not to lose any of the territory they had won during the war) began to organize this fiercely contested city. Serbian mobs attacked US forces with bricks, rocks, metal bars, and firebombs. Organization for this altercation started days before, as many witnessed the prestaging of the items to be used in the attack. Wheelbarrows, 55-gallon metal drums, and other large containers were filled with the ammunition that would later be used to attack peacekeepers.
>
> Brcko, Bosnia
> September 1996

CROWD BUILDING

1-13. The second phase of the gathering process is the building of the crowd. Old concepts of crowds based on stereotypes are no longer true. Studies show that—

- Crowds are made of many entities, and all participants are not the same.
- Crowds are not made of isolated individuals but of a minority of individuals and a majority of small groups who may or may not be acquainted with each other.
- Groups and individuals in the crowd are not unanimous in their motivation.
- Groups and individuals in the crowd are usually not anonymous to each other, especially different groups gathered at the same place.
- Individuals within a crowd are not given to unique emotional displays; although known within a group, most individuals would not want to draw attention to themselves.
- Groups within a crowd will often act in unison in an attempt to achieve their agendas.
- Crowds are generally not unique or distinguished by violence or unlawful conduct.

DISPERSAL PROCESS OF A GATHERING

1-14. The last and final phase of any gathering of a crowd is the dispersal process. It involves the movement of people from the common location where they assembled to one or more alternate locations. The dispersal process ends the gathering of a crowd or at least begins its decline. The dispersal can occur on a routine, emergency, or coerced basis.

ROUTINE DISPERSAL

1-15. The routine dispersal may be specified in advance. It can also be included in the assembly instructions given by the organizers of an event.

EMERGENCY DISPERSAL

1-16. An emergency dispersal occurs when people evacuate an area in result of an unexpected crisis, such as a fire, explosion, bomb threat, or terrorist act. Individuals in such an emergency quickly recover from the initial shock. They keep their wits about them and improvise ways of extricating themselves and their companions from the dangerous situation. However, tragedies have occurred in some emergency dispersal situations. Therefore, it is important that forces on the scene are careful not to misinterpret what is occurring.

COERCION DISPERSAL

1-17. Coercion dispersal is caused by the use of force at some level. This is not necessarily the best way to force the dispersal of a crowd. The negotiated management of crowds is the preferred method and has proven to be highly successful in getting crowd organizers to police themselves, especially if the demonstration and/or protest leaders are available and willing to participate.

CROWD DYNAMICS

1-18. Understanding crowds and how individuals and groups form crowds through the gathering process is important because these issues apply to the dynamics of a crowd. Under most circumstances, gathered crowds are orderly and present little or no problems for authorities. Crowds consist of people who, although very motivated and passionate, are also subject to their own need for creature comforts. Rainy, cold, and nasty weather has a way of disheartening all but the few highly motivated and disciplined individuals. If problems exist, they usually fall into the following three categories:

- **Public disorder.** Public disorder is a basic breach of civic order. Individuals or small groups assembling have a tendency to disrupt the normal flow of things around them.
- **Public disturbance.** Public disturbance is designed to cause turmoil on top of the disruption. Individuals and groups assembling into a crowd begin chanting, yelling, singing, and voicing individual or collective opinions.
- **Riot.** A riot is a disturbance that turns violent. Assembled crowds become a mob that violently expresses itself by destroying property, assaulting others, and creating an extremely volatile environment.

1-19. Being part of a crowd of people has certain effects on different people. Each individual in a crowd is susceptible to behaving in a way that is contrary to their normal behavior. Some reasons for these behaviors are as follows:

- Crowds provide individuals with a sense of anonymity. With so many others, an individual realizes that he is just another face in the crowd, giving a sense of invulnerability.
- Crowd and individual behaviors are impersonal by nature. The "them-against-us" attitude affords those within the crowd the ability to freely (without hesitation or reservation) be verbally abusive, throw objects, or attack anyone who gets in the way.

- Crowds provide individuals with the idea that their moral responsibilities have shifted from themselves as a person to the crowd as a whole. Large numbers of people discourage individual behavior, and the urge to imitate others within the crowd is strong. Individuals look to others around them for cues of what to do next, disregarding their own background and training. Often, it is only the strong, well-disciplined person who can resist the prevailing behavior of a crowd.
- Crowd behavior influences the actions of both the disorderly individuals of the crowd and the authorities tasked to control them.

1-20. Individuals within a crowd are at times driven by deep-felt emotions. Emotional contagion is the most dramatic psychological factor of crowd dynamics. It provides the crowd with a temporary bond of psychological unity. Lasting long enough, this unity can push a simple organized crowd into a mob. Normal law and authority are rejected en masse under these conditions, increasing the potential for violence and panic to erupt.

1-21. Panic can erupt quickly, especially when crowds turn into mobs. Individuals within the mob can easily sense that their safety and well-being are at risk, putting them in a "fight-or-flight" mentality. Adding to the panic and confusion is the use of riot control agents (RCAs) by authorities in an attempt to gain control. Individuals in a mob, during the heat of confrontation, may attempt to leave and find that there are no escape routes and that roads are blocked. This can often lead to violent, physical attacks.

CROWD TYPES

1-22. Gaining an understanding of crowds and the dynamics that cause individuals or groups to join together in a crowd should be every leader's responsibility. Active, expressive, acquisitive, and hostile are not acceptable terms used to describe a crowd. They are the motives inferred from the actions the crowd takes. No gathering of a crowd is persistently or exclusively active, expressive, acquisitive, or hostile. For the sake of description, crowds can be identified as casual, sighting, agitated, or mob-like.

CASUAL CROWDS

1-23. Casual crowds are identified as individuals or small groups with nothing in common to bind them together. If they have an agenda, it is their own. They arrive separately and leave separately. Casual crowds are made up of individuals or small groups occupying the same common place, such as a shopping mall where these individuals or small groups meet.

SIGHTING CROWDS

1-24. Sighting crowds are similar to casual crowds with one additional element—an event. People migrate as a crowd to sporting events, are attracted to fires and accidents, and attend music concerts. Individuals or small groups gather at these events for the same purpose. It is the event and/or one's curiosity that compels a crowd to come together.

AGITATED CROWDS

1-25. Agitated crowds add responses that are based on the elements (people, space, and event). Individuals with strong emotional feelings within a crowd can quickly spread and infect the rest of the crowd. As more people within the crowd become emotionally involved, a sense of unity can develop, causing changes in the overall demeanor of the crowd. Yelling, screaming, crying, and profane name-calling are all associated with an agitated crowd.

MOB-LIKE CROWDS

1-26. Mobs have all the elements found in the first three types of crowds, with the addition of aggressive, physical, and sometimes violent actions. Under these conditions, individuals within a crowd will often say and do things they usually would not. Extreme acts of violence and property damage are often a part of mob activities. Mobs consist of (or involve) the elements of people and groups being mixed together and becoming fluid.

CROWD TACTICS

1-27. During a public disorder or disturbance, individuals and small groups within a crowd use any number of tactics to resist authority and disrupt and add turmoil in order to achieve their goals. These tactics can be unplanned or planned and violent or nonviolent. The more organized and purposeful a crowd becomes, the more likely a tactic will be used.

NONVIOLENT BEHAVIOR

1-28. As explained earlier in this chapter, most gatherings of individuals and small groups into a crowd do not involves violent behavior. A public disorder or disturbance usually involves some harmless name-calling, demonstrations to express views, corporate yelling and chanting, and even singing and dancing. In some instances, there may be property damage and the erection of makeshift barricades. All four types of crowds can be nonviolent, but mobs and agitated crowds have the greatest tendency to turn violent. Nonviolent actions of a crowd are disruptive because they are in direct conflict with what authorities want them to do, such as refusing to leave when directed, locking arms, and sitting in front of or around areas and buildings that the authorities are attempting to clear.

VIOLENT BEHAVIOR

1-29. A crowd that becomes a mob can be very violent and destructive. Although some nonviolent activities occur, violent crowds strike out physically at bystanders or others in the crowd, destroying both private and government property, setting fires, and employing bombs (in extreme cases). The only limitations for violent crowds are their own imaginations, the training of their leaders, and the materials readily available.

1-30. Violent crowds may erect barricades and physical barriers to impede the movement of authorities and prevent them from entering certain areas and/or buildings and to hide their activities. Common materials used for barricades include vehicles, felled trees, furniture, felled light poles, fencing materials, or anything else on hand and readily available.

1-31. Setting fires and preventing firefighters access to the structure by blocking it is another violent crowd tactic. Often, fires are set to create confusion in authorities or as a diversion to another activity.

RIOTS

1-32. A riot is one or more groups or individuals who are part of a larger crowd that involves threats of violence against persons or property. In some cases, a crowd will continue to gather until it evolves into a riot.

1-33. Riots vary considerably in both targets and players. A communal riot, for instance, deals with deep-seated ethnic, religious, and language differences. Commodity riots involve an attack on property by acts of vandalism, looting, and arson. Protest riots, such as the riot around the World Trade Organization (WTO) Assembly in Seattle, Washington (30 November 1999), illustrated individuals and groups aggressively and sometimes violently acting out or voicing their opposition to the assembly. The Democratic Convention in Chicago, Illinois, in 1968 reflects a riot that directly targeted police and authority in general. Celebration riots occur across the US as a result of home team victories in sporting events, among other reasons. Celebrating crowds look to make the moment more memorable through raucous acts that demonstrate their joy or happiness, for example, the riots that took place in Chicago in 1992 as a result of the Chicago Bulls winning the National Basketball Association (NBA) Championship.

1-34. Commanders and leaders must be aware that highly organized groups of protesters and/or demonstrators have developed tactics to disrupt the control force. These crowd tactics were published in handbooks for communist organizers during the Cold War, and today the Internet contains sites that are devoted to sharing these tactics with the general public and various protest groups. Although these tactics are somewhat outdated, commanders and leaders must be aware of them and their potential consequences. Many of these groups, such as anarchists, often mock the tactics employed against large crowds; for example, they may identify the police as "Darth Vader Cops" when the control force response is in full riot control gear. Commanders must be aware that well-organized crowds may attempt to engage the control force, surround it, and overpower it by sheer numbers. To avoid this, commanders must do a detailed terrain analysis (intelligence preparation of the battlefield [IPB]) of the area to include all approach and exit routes in the area. The control force flanks have to be protected to avoid being enveloped by the mob. To avoid being enveloped, the on-site commander should keep the crowd at a comfortable distance from the control force formation, which is usually the lethal zone of his NL munitions. Establishing predetermined rally points for the control force is critical in the event that this type of tactic is used by the crowd.

TACTICS USED TO DEFEAT AUTHORITIES

1-35. Organized mobs will try to defeat the control force by employing several different types of tactics. These tactics include the following:

- Constructing barricades.
- Using Molotov cocktails, smoke grenades, rocks, and slingshots.
- Feinting and flanking actions.

BARRICADES

1-36. By constructing barricades, the rioters are trying to protect themselves from assault by the control force. If the control force does advance toward the barricade, rioters hope that the barricade will split the control force and leave it vulnerable to actions by the mob.

MOLOTOV COCKTAILS, SMOKE GRENADES, ROCKS, AND SLINGSHOTS

1-37. Angry mobs will often attempt to disrupt the control force by throwing rocks and other projectiles. As the situation deteriorates, the mob may escalate the violence by using a battery of slingshots that will pellet the control force with a barrage of projectiles ranging from small rocks to marbles. They may also use smoke grenades (homemade or store-bought) to mask their movements. Finally, Molotov cocktails are used against personnel, employed vehicles, and portions of the control force. The mob will attempt to gain a position above an armored vehicle to enable them to drop a Molotov cocktail into an open hatch.

FEINTING AND FLANKING ACTIONS

1-38. Organized mobs may attempt to disrupt the movement of the control force by feinting an assault. When the control force moves to blunt the assault, the mob will assault the exposed flanks in an attempt to split the control force and envelop a portion of it.

WEAPONS

1-39. Mobs will often use various types of weapons against authorities to achieve their agendas and goals. It is extremely important that leaders train soldiers to recognize possible threats of the various weapons used by crowds or mobs, enabling the soldiers to react with the appropriate minimum force.

VERBAL ABUSE

1-40. In almost every instance of a civil disturbance or riot, verbal abuse will be an aggressive tool. Obscene language, racial remarks, taunts, ridicules, and jeers should always be expected. It is apparent that the purpose for using verbal abuse is to anger, demoralize, and provoke a physical response. Undisciplined, untrained soldiers who face such an attack could cause the situation to escalate. Just one provoked action of a soldier could be interpreted as an act of brutality by the media.

PERCEIVED INNOCENT PEOPLE OR THE WEAK

1-41. Women, children, and the elderly have often been used as pawns in civil disturbance operations. On many occasions, the innocent and the weak are moved to the front of the crowd and used as a barrier. With the innocent and the weak directly facing authorities, aggressive and violence-prone individuals or groups behind them attempt to provoke authorities to react. Often, this tactic is to exploit the situation in the eyes of the media for propaganda value. Despite the perception of sometimes being weaker, women, children, and the elderly can be just as committed to the insurgency as those behind them.

BARRICADES

1-42. Barricades are made of anything large or heavy enough to impede or prevent the movement of authorities. Trees, vehicles, fires, and furniture are all examples of items used to erect a barricade.

THROWN OBJECTS

1-43. In many cases, a crowd will throw anything in their reach that can be picked or picked up. Rotten fruit and/or vegetables, rocks, bricks, bottles, sticks, and pieces of lumber are most often used because they are readily available. More sophisticated types of thrown objects are improvised bombs, such as petrol bombs or tennis balls with nails sticking out of them. Never discount that these individuals and groups could acquire hand-thrown explosives, such as sticks of dynamite or grenades.

VEHICLES AND OTHER OBJECTS

1-44. Vehicles and other objects can be placed on higher ground (atop a hill or in a building) to disrupt civil disturbance operations. For example, a vehicle or tire can be set on fire and rolled down a hill, carts or barrels filled with hazardous or flammable materials can be pushed down a hill, or burning furniture and petrol bombs can be thrown from two-story or higher buildings or toward a formation of soldiers.

FIRE

1-45. Rioters have set buildings, vehicles, and other structures on fire to block the advances of authorities and to create confusion and diversions. If property damage is the goal of a crowd, fire is an effective tool. A sniper, for example, may set a building on fire to divert attention or provide maximum target possibilities. Other tactics used, particularly in the Balkans, were to flood the lower floors or basements of structures with flammable liquids or gas and then ignite it in an opportune moment.

FIREARMS

1-46. Sniping or massing fire from within the crowd, in buildings, or other covered positions can (at times) be effective tools for terrorists using a crowd to cover their acts. Firearms are anything that fires a projectile toward a target of opportunity.

EXPLOSIVES

1-47. A bomber's imagination is the only limiting factor in the use of explosives. Explosives can be used as simple diversions to block the advances of formations or provide an escape for rioters. The worst use of explosives is to purposely kill, injure, or demoralize authorities that are protecting and assuring the safety of everyone.

Chapter 2

Control Force Operations

Information is the key to developing civil disturbance plans. One must know who the demonstrators are; when, where, and why they are demonstrating; what their capabilities are; and what their possible course of action (COA) is. A commander's need for current, valid information cannot be overemphasized. The commander must learn as much as he can about the participants (their motivations, strategies, tactics, targets, and dedication). The more knowledge a commander has about the participants, the better equipped he is to counter their actions. He needs sound information to decide how to best use his available resources.

CIVIL DISTURBANCE PLANNING CONSIDERATIONS

2-1. The side that possesses better information and uses that information to gain understanding has a major advantage over its opponent. Forces that have this advantage can use it to bring about changes in attitudes, decisions, and actions in assembled demonstrators. In planning for crowd control or civil disturbance operations, planners must decide what data is needed to develop threat assessments. This decision is based on a multitude of information sources.

STANDARDIZED POSTURES IN COMPANY TACTICAL STANDING OPERATING PROCEDURES

2-2. Within squads, platoons, and companies, equipment may be increased or decreased, as needed. Some examples are as follows:

- Choose the M9 pistol for extraction and apprehension teams. The use of a long weapon (for example, an M16 with an M203, an M4 with an M203, or a 12-gauge shotgun) with NL munitions capability is also recommended, especially for overwatch personnel.
- Add nonstandard weapons such as shotguns for greater NL capabilities.

NOTE: The shotgun is used to protect the M203 gunner as he reloads.

- Add NL munitions to existing organic weapons systems, such as the M203.
- Arm soldiers in the front line of the formation with their standard weapon. If the weapon is a long weapon, it should be carried across the back from left to right or vice versa, with the butt up and muzzle down. Ensure that the weapon is cleared and that the magazine is in the appropriate ammunition pouch.

- Balance the mix of weapons and munitions according to the mission, enemy, terrain, troops, time available, and civilian considerations (METT-TC).
- Establish command relations and determine who has or gives the authority to fire NL munitions.
- Maximize the distance and barriers between the crowd and control formations. Use NL munitions to create a standoff distance.
- Maintain a lethal overwatch of the control force at all times.
- Maintain a reserve force at all times to reinforce the control force, as needed.
- Create NLW range cards for static positions.
- Consider environmental conditions and their effect on the performance of NL munitions. Proper storage, periodic inspections, and the rotation of NL munitions are critical to maintaining the effectiveness and viability of NL munitions.

WEAPON AND AMMUNITION CONFIGURATIONS

2-3. Individuals designated as NL shooters must have the means to transition to lethal rounds, if required. Lethal rounds are carried separately from NL rounds so that the shooter will not confuse them in the heat of confrontation. However, in the heat of confrontation where a lethal option is necessary, NL rounds can be used lethally by adjusting the point of aim to a vulnerable part of the body and closing in on the distance to the target.

2-4. Squad leaders should designate NL shooters within their squad. Ideally, the squad should not change its organization to accommodate NLW.

2-5. Commanders should not dispatch NL patrols. They should plan a combat or security patrol with NL capabilities. Soldiers are never sent in harm's way without lethal protection. NLW are only considered additional tools for the mission and not a mission in itself. There is no such thing as a NL mission.

2-6. Soldiers manning a static position should have NL capabilities depending on METT-TC. Shotguns and the M203 work well at static positions.

2-7. Recovery, apprehension, and/or extraction teams should be established before deployment. Team members should be equipped with personal protective equipment (PPE), an M9, NL munitions and weapons, and Flex-Cufs®. They should have some training in open-hand control, pain compliance, and handcuffs and/or Flex-Cufs.

2-8. Crowd control formations (see *Chapter 6*) should be well trained and well rehearsed. Rapid, coordinated movements of a well-trained and well-rehearsed control force can often be a strong enough deterrent. A lethal overwatch marksman always covers the control force.

2-9. During a NL engagement, the use of designated marksmen (DM) provides confidence and safety to those facing a riot. The DM in an overwatch position scans the crowd to identify threats and designate personnel for recovery and lethal rounds firing (if a lethal threat is presented). They are ideally suited for flank security and countersniper operations.

NONLEHTAL PLANNING

2-10. The fundamental principles of mission planning are not changed by the incorporation of NLW into a plan. NLW are intended to be operational-enhancement tools. The first concern of a commander is the successful completion of the mission. NL munitions provide a commander with a wider range of response options, but these munitions are not a replacement for lethal capabilities. Commanders must always be capable of answering a lethal attack with a lethal response.

2-11. When a commander commits his soldiers and equipment to a crowd control situation, he commits his forces with the additional tools of NL capabilities. This is an addition to the force continuum that the force commander now has available to him. Commanders that properly employ NL munitions and weapons have a tactical advantage over those who rely completely on lethal means.

2-12. When developing a tactical standing operating procedure (TSOP) (see *Chapter 6*), units should consider the following:

- Crowd control formations.
- Extraction teams.
- Apprehension teams.
- Lethal overwatch marksmen and/or observer teams.
- Reserve and/or security forces.

MILITARY WORKING DOG TEAMS

2-13. Military working dog (MWD) teams may be employed with a control force formation as a method of increasing crowd apprehension about approaching or engaging the formation. The teams should be in the rear of the formation in plain sight of the crowd, but in front of the command element and the M33A1 squad RCA disperser. The MWD teams work back and forth behind the formation as an intimidation measure. The presence of the MWD may produce a profound psychological effect on the crowd. These teams may also be used to help control individuals who have been captured by the recovery and apprehension teams.

CAUTION

Do not unleash an MWD on a crowd.

COMBAT CAMERAMEN

2-14. Video and still cameramen should make a photographic record of the individuals in the crowd who are leaders and instigators. Events must be documented to hold personnel, factions, gangs, or groups accountable for acts that violate law, destroy property, or cause physical harm. Electronically recording events aids in the prosecution of such cases and eliminates the sense of anonymity that people in large crowds often feel.

INFORMATION OPERATIONS

2-15. Information superiority helps forces to anticipate problems and requirements. It allows commanders to control situations earlier and with less force, creating the conditions necessary to achieve the optimal end state. Public affairs, psychological operations (PSYOP), and civil military operations are activities that will allow the commander to control situations earlier and with less force.

2-16. Information is available from a multitude of sources. A diversity of sources is the best approach because it prevents biased perspectives. Primary sources are as follows:

- Open sources.
  - Libraries.
  - Newspapers and news periodicals.
  - Radio and television.
  - Internet.
- Law enforcement sources.
  - Local law enforcement agencies.
  - National law enforcement agencies.
- Military sources.
  - Department of Defense (DOD) intelligence community (most restrictive source).
  - Local military intelligence (MI) field offices.

2-17. Police intelligence operations (PIO) is one of the five military police functions. This function is a process of actively and passively collecting information that is of a police, criminal, or combat nature. As military police perform the other four functions of maneuver and mobility support (MS), area security (AS), law and order (L&O), and internment and resettlement (I/R), they are gathering information that supports, enhances, and contributes to the commander's protection program, situational awareness, and battlefield visualization by portraying relevant threat information that may affect operational and tactical environments.

2-18. IPB is a continuous process for analyzing the threat and the environment of a specific geographic area. During the IPB process, the Intelligence Officer (US Army) (S2) or Assistant Chief of Staff, G2 (Intelligence) (G2) uses all available databases, intelligence sources and products, and related MI discipline to analyze the threat and the environment. The PIO function supports this process by providing the S2 with collected police, criminal, and combat information that can directly and significantly contribute to the success of the MI effort. In addition to combat information, the PIO function provides additional information on possible criminal threats and COAs. This is intended to support the S2 IPB process and can be used by the commander to upgrade force protection. (See *FM 19-10.*)

THREAT ANALYSIS

2-19. Crowd control requires its own thought process. Emphasis should be on prevention rather than confrontation. In combat, military forces are taught to fight and eliminate threats. In crowd control, military forces must deal with noncombatants that have internationally recognized rights. These rights must be respected while maintaining public order. This is an issue that law enforcement agencies (LEAs) have been struggling with for years.

2-20. Dealing with crowd control incidents is a matter of using some basic guidelines. It also includes asking the right questions in a logical manner so that key issues are not omitted. The commander must use these guidelines in context with METT-TC and the location of the incident (CONUS or OCONUS).

PREINCIDENT PLANNING

2-21. This phase of planning begins before the incident and is initiated at the operational level with guidance from the strategic level. The planning includes guidance on crowd control and addresses responsibilities, training, organization, operating procedures, use of force (CONUS), and/or rules of engagement (OCONUS). The most difficult and productive decisions are those made in the preincident planning process.

Avoid Confrontation

2-22. Crowd situations are highly unpredictable, but one thing seems certain—confrontation will likely cause crowd resistance. When pushed, people tend to resist opposition to the realization of their purposes.

Focus on Prevention

2-23. Planning should key in on the prevention of unfavorable outcomes. Experience has shown some LEAs attempting to help crowds accomplish their goals within the law that have been beneficial and even led to conceding some violations for the purpose of avoiding confrontation. However, LEAs maintain a law enforcement presence, which signifies social restraint. LEAs also decide when and where they will not compromise and the amount of force to use.

Define Goals

2-24. When defining a goal, deciding what must be accomplished is the first step. Defining the goal is fundamental. However, actually working toward and accomplishing the goal is easier said than done, as the process can drive the situation. Commanders and leaders must be aware of this. The military force must focus on what they are trying to accomplish.

BACKGROUND INFORMATION AND INTELLIGENCE

2-25. Seek to know as much as possible about social protest groups (within the limits of the law, see *Appendix B*) before an incident. This will provide insight into the organization and its functions and provide a warning as to what to expect. When you know as much about these organizations as possible, then it is possible to anticipate their next move. See *Appendix C*.

2-26. Continually assess what is to be accomplished against what others are trying to accomplish. Some groups may not have goals that conflict with those of US forces, but their actions may. Other groups may have goals and activities that do conflict. In either case, the assessment of group goals compared to the goals of US forces helps to understand and avoid potential conflicts.

CIVIL DISTURBANCE OPERATIONS PREPARATION

2-27. Counterdemonstration workgroups are a mixture of traditional and nontraditional staff proponents brought together to coordinate the resources to be employed by task force (TF) units before a potential civil disturbance. Meetings of the counterdemonstration workgroup are held on a regular basis (generally weekly) and are chaired by the Assistant Chief of Staff G3 (Operations and Plans) (G3). The following staff proponents may be involved, and others may be involved as needed:

- Public affairs office (PAO).
- Joint military commission (JMC), if organized.
- Provost marshal office (PMO).
- PSYOP.
- G2, Assistant Chief of Staff G5 (Civil Affairs) (G5), Assistant Chief of Staff G6 (Signal) (G6).
- Civil affairs (CA).
- Army airspace command and control (A2C2).
- Chaplain.
- Engineer.
- Surgeon.
- Fire support element (FSE).

RELATIONSHIPS DEVELOPMENT

2-28. Working relationships between commanders and protest group leaders are increasingly seen as the best means for preventing bad outcomes in crowd situations. This is called the negotiated management model of crowd control. LEAs in large metropolitan cities in the US and Europe practice it.

2-29. Open dialog helps develop working relationships between commanders and protest group leaders, providing an opportunity to communicate clearly. Talking allows group leaders to tell authorities and tactical commanders what they want to accomplish. It also allows authorities and commanders to tell group leaders what they are prepared to do and how they might respond to certain crowd behaviors. Such communications can do much to resolve issues and prevent violence.

2-30. Commanders may not be able to talk to all leaders before a crowd assembles. Some groups do not have recognized leaders and are ad hoc organizations. Other groups may have several leaders, but only some of those leaders will negotiate. In such cases, meetings with some leaders may tell you who the other leaders are and provide critical information. Negotiations may also encourage more moderate leaders to do things that will support the commander.

2-31. Initiate communication with group leaders to work out issues before an event. Commanders should make a concentrated effort to maintain a working relationship with the leaders of protest groups.

2-32. Communication also means persuasion. Commanders should make a concerted effort to win over demonstration leaders. Commanders may tell group leaders that they want to help them complete their mission in a safe manner. They should encourage demonstrators to protest in an acceptable way and, if possible, offer favors to get them to do just that.

NOTE: Commanders must only offer what they are willing and able to deliver.

2-33. Several rules apply in negotiations, and credibility is key. Only communicate necessary information and those actions that authorities and commanders intend to do. Negotiations are made from a position of strength, and by negotiating an agreement may be reached. However, contingencies should be developed in the event that the agreement is violated.

2-34. Working relationships between commanders and protest group leaders often result in protest groups policing themselves. This is one of the basic premises of negotiated management. Protesters are more likely to listen to their own leaders, as opposed to listening to US forces. Protest organizers are likely to buy into a cooperative effort and agree to ground rules set during preprotest negotiations. Many protest leaders seek to stay on the right side of the law.

ACCEPTABILITY

2-35. Deployed US forces will find themselves engaged in crowd control operations under difficult circumstances. The host nation (HN) may have groups that do not accept the presence of US forces, which will make imposing order and the protection of citizens difficult at best. US forces will be under intense media and political scrutiny (an environment much like what LEAs operate in on a day-to-day basis).

2-36. Winning in this environment is not like winning in combat. US forces may appear to be invincible and formidable, but they risk being portrayed as oppressors. Thus, US forces can lose by appearing to win. Groups that perceive themselves as oppressed will readily seek victimhood in an effort to gain the support of public opinion. Winning in this environment is about seizing and holding the moral high ground. US forces must maintain the authority and legitimacy of what they are doing.

2-37. Projecting a favorable image will require outreach to local leaders and citizens. It will also require developing a relationship with the media. For example, the Los Angeles Sheriffs Department often invites the media to accompany their tactical commander during crowd control situations. This shows that they have nothing to hide. It also provides an opportunity for individuals to see the commander's side of an event.

2-38. In this environment, commanders must consider how actions will play among several audiences—local, allied, US, and international. The media effect is inescapable. In some cases, it will be difficult to accommodate all of these audiences. However, the most important audience will be the US public, which is key to the continued support of US forces.

CROWD ASSESSMENT QUESTIONS

2-39. Assessing crowds requires its own set of questions. These questions should be answered before a crowd assembles or as quickly as possible if a gathering occurs without notification. In some cases, all the questions may not be answered. These questions represent a logical way to think about crowds. The questions are as follows:

- **Who are they? What is the overarching identity of the crowd? Are they strikers, ethnic factions, or social protesters? Do they identify themselves as strikers, ethnic groups, religious factions, or protesters against some perceived social injustice?** Understanding who they are will indicate what they may do. It may be possible to determine identities and goals from advance assembling instructions, leaflets distributed to bystanders, placards and banners, and chants and songs.
- **What are their goals?** What the group wants to accomplish by assembling could determine the extent to which they can be accommodated once they have assembled. They may only seek recognition for their cause (being seen and heard). If so, this goal is usually easy to accommodate. However, some groups may have more demanding goals, for example the demonstrators in Seattle who sought to stop the WTO. Goals that cannot be accommodated make confrontation very likely.
- **What is the composition of the crowd and are there any known factions?** Seattle demonstrations against the WTO were comprised of groups that were protesting environmental issues, wages, and child labor laws. Differing goals and the resulting friction were evident between such organizations as the Ruckus Society and the American Federation of Labor-Congress of Industrial Organizations (AFL-CIO). Factions within a crowd represent threats and opportunities.
- **What are they capable of doing?** Protest groups often claim that they will assemble large numbers of people to produce some disruptive action. However, there is often a difference between the claim and reality. An organization may claim that it can mass a 100,000 people, but in reality can only get 40,000. Organizers exaggerate for the following reasons: they want to boost the morale of their own people and they want the media to report that they have strength in numbers. Studying the past activities of a group may provide indications of what they are capable of doing in the future.
- **What are their traditional behaviors or cultural repertoires?** What people do during protests is not universal. It varies with the group and the culture. Social protest organizations and striking unions will carry placards and banners. Other groups will protest in a more quiet way, like the 5,000 women in Sarajevo whose standard Sunday behavior was to sit and block traffic. They were protesting the loss of male relatives in the Bosnian war. Understanding the goals of

the protesters can be helpful in deciding how to respond to their behaviors.

- **When and where will they assemble?** Every protest organizer has a time and place for assembling and perhaps a destination for the crowd to move toward. If the organizer attempts to mobilize large numbers of participants, the time and place for assembling and dispersing must be made known in the instructions. This information may be stated in the mobilization instructions or disseminated by an informal network, such as word of mouth. Such impromptu networks in densely populated areas can enable rapid assembling.
- **Where will they go?** Many crowds have destinations. Organizations may march a specified distance to ensure that their cause gets sufficient attention. Commanders need to know the route to minimize disruptions to the rest of the community. In Los Angeles, LEAs attempt to reroute traffic and prevent congestion caused by protest marches. It is also necessary to provide security along the route to prevent counterdemonstrators from confronting the marchers. This will prevent an even greater problem for the community.
- **What are the possible targets of violence?** Riots, in particular, may focus on target facilities. In the 1992 Los Angeles riots, gun stores were major targets. In developing countries, targets may be more basic. In Somalia, throngs of people stormed food supply facilities. In Haiti, riots broke out over trash dumps where US forces had discarded the remnants of meals, ready to eat (MREs).
- **What is the worst-case scenario?** The worst-case scenario must be recognized for the sake of avoidance. This may be when a peaceful crowd degenerates into a violent riot.
- **When and where will they disperse?** Crowds have a life cycle that includes how they disperse (see *Chapter 1*). Commanders must consider this. It is essential that there be one or more avenues that individuals can use to disperse. These dispersal routes should be clearly marked, visible, and open-ended. There may be a need for US forces to assist the crowd with dispersal. For example, in Los Angeles a 70,000-man march went from east to central Los Angeles. Once at their destination, they engaged in peaceful activities and then dispersed. The LEAs provided transportation back to the assembly area (AA); otherwise, the protesters would have returned to the AA in a disorganized and unsupervised manner.
- **Are there plans for subsequent gatherings?** A crowd may disperse for a short time so participants can take care of personal needs. This could be days or only a few hours.

LEGAL CONSIDERATIONS

2-40. The legal authority in which a peace operation is conducted defines the parameters of the operation. All commanders, leaders, and soldiers must know the legal authority that regulates their operations.

MANDATES

2-41. There are two types of mandates—United Nations (UN) and non-UN. Mandates are the authority under which an operation is conducted. UN mandates are created from resolutions by the UN Security Council (UNSC) or the UN General Assembly (UNGA) that authorize and define an operation. Non-UN mandates are created from treaties, accords, resolutions, or agreements evolving from other international or regional organizations.

STATUS OF FORCES AGREEMENT AND STATUS OF MISSION AGREEMENT

2-42. A status of forces agreement (SOFA) and a status of mission agreement (SOMA) are key documents that define the legal authority and responsibilities of a force and force personnel participating in an operation. A SOFA or SOMA may be a treaty or memorandum of understanding. It is an agreement negotiated between the UN and the host countries, which details the rights, privileges, immunities, and nature of services to be provided to the force and its personnel and their responsibilities and obligations.

INTERNATIONAL LAW

2-43. Legally, international law affects most stability operations and support operations (SOSO). These laws consist mainly of international agreements, treaties, and customary international law, which include the agreements and laws known as the law of war. International agreements prescribe the rights, duties, powers, and privileges of nations relative to particular undertakings. International agreements will affect US participation in SOSO in the—

- Right of US forces to enter a foreign country.
- Status of US forces while in the foreign country.
- Construction and operation of US bases of operations.
- Aircraft overflights and landing rights.
- Processing of claims for any damage to persons and property of the foreign country.

2-44. These agreements or customary international laws govern all aspects of operations carried out. In the absence of a viable HN government or proper international agreement, many aspects of the operations are controlled by domestic (US) or customary international law.

UNITED STATES LAW

2-45. Operations in peace and conflict must comply with US law, whether as a statute, executive order, regulation, or other directive from a branch or agency of the federal government. The *Uniform Code of Military Justice (UCMJ)* applies to questions of military justice. Various statutes, *Executive Order 12333; Section 3, Code of Federal Regulations (CFR);* and DOD and service regulations govern all intelligence-gathering activities. The staff judge advocate (SJA) must actively advise and participate in all activities from initial planning to redeployment.

HOST NATION LAW

2-46. If a viable HN government exists, whether at the national or the local level, all laws of the HN will apply to US forces in that country unless an international agreement provides otherwise. The types of laws that may inhibit US operations are in the fields of immigration, labor, currency exchange, the procurement of goods and services, customs and taxes, and criminal and civil liabilities. A SOFA usually covers these areas before an operation begins. Therefore, planners must understand the law to assess it and determine if it will adversely affect the operation. Assistance may be available from the local US embassy or the command judge advocate; if not, the command may have to rely on other sources for guidance. If local law hinders the operation, the commander must inform the US embassy and request that it negotiate a solution.

CIVIL AUTHORITIES

2-47. There will be times that US forces will deploy to nations where the infrastructure of civil authorities has survived the conflict and is functional. Local authorities and police are responsible for ensuring the safety of their citizens, including the UN High Commissioner for Refugees (UNHCR). US forces assist in this by providing general military security and by facilitating negotiations. In countries where the civil authority infrastructure has collapsed, the UN mandate may require the coalition of forces to help ensure the safety of citizens until their own civil authorities can be rebuilt.

RULES OF ENGAGEMENT

2-48. The rules of engagement (ROE) are the commander's rules for the use of force (RUF). Commanders are responsible for interpreting, drafting, disseminating, and training on the ROE. Ensure that the SJA assists throughout the ROE development process. This will ensure that the ROE will not improperly constrain actions, but are still consistent with domestic and international law and policies and orders of the chain of command.

2-49. Clearly stated ROE are published before Army forces are committed. The highest military authority, with input from subordinate commanders, will continually evaluate the ROE and modify them as appropriate. When the use of NLW has been authorized in the ROE, the decision to use NLW should be delegated to the lowest possible level, preferably the platoon or squad level. This requires that all personnel, not just leaders, have a clear understanding of the ROE and the commander's intent (see *FM 3-22.40*).

2-50. In peace operations, the use of force is restrained while the diplomatic solution is sought through negotiation or mediation. Accordingly, ROE are more restrictive in peace operations, making them more politically sensitive. In today's world, peace operations are usually conducted by a coalition of forces operating under the purview of the UN charter and customary international law. Therefore, the UN may mandate certain restraints. Using overwhelming force could compromise diplomatic efforts to reach a peaceful settlement. Even a single, relatively small engagement could jeopardize the legitimacy of the peacekeeping force and upset negotiations or mediations. Commanders must beware that in any confrontation ROE decisions made by soldiers can have strategic political implications on current and future operations. Therefore, it is vital that leaders and soldiers hold a common understanding of when, how, and to what degree force is to be used. This requires training.

MULTINATIONAL OPERATIONS

2-51. Multinational operations include military forces from different nations. These operations are usually interagency operations. Conducting operations with foreign military partners, like operations with civilian partners, is uncommon to many soldiers, so a clear understanding of this different environment is necessary.

2-52. The majority of US peace operations will be part of a UN peace operation. Their multinational character merits particular attention because national interests and organizational influence may compete with doctrine and efficiency. Commanders can expect contributing nations to adhere to national policies and priorities, which at times complicate the multinational effort.

2-53. When working within a multinational force and with allies, do the following:

- Train coalition partners and allies on tactics and techniques.
- Remember that teamwork and trust are essential.
- Achieve unity of effort.
- Consider equipment and communications capability and interface.
- Strive to eliminate sources of confusion and misunderstanding.
- Remember that military doctrine varies from country to country.
- Remember that national culture influences the way units operate.
- Remember that language barriers represent a significant challenge.

SCALABLE EFFECTS

2-54. Scalable effects concepts are a measured approach in response to a crowd gathering. By recognizing a use-of-force policy, soldiers must be taught and understand that they use the minimum force necessary. Without the appearance of a graduated response, the gathering crowd may consider actions as excessive, causing a possible escalation of hostilities or violence. Do the following to aid in the scalable effects process:

- Try to persuade the crowd to quietly disperse by talking with leaders.
- Use translators as necessary.
- Let the first approach be by local authorities (the mayor or police).
- Pass out handbills requesting that the crowd return home.
- Use video and still cameras to photograph individuals and events for later use at trials.
- Give warnings before moving to the next level of force.

2-55. Ideally, the force should be positioned out of sight of the crowd. Have the troops move into position with few shouted commands. For maximum effect on the crowd, have them form their formation decisively and professionally. Do the following to aid in the formation process:

- Display force in a graduated manner (for example, a UH 60, hardened vehicles, and soldiers with charged weapons).
- Exploit the psychological effects of show of force.
- Demonstrate force (do not use on unarmed civilians).
- Escalate the PSYOP message.
- Highlight the target pointer.
- Demonstrate sniper precision strike capability.
- Employ riot control measures.
- Use NL munitions.
- Use RCAs (if approved).
- Move through the crowd using riot control formations and movement techniques.
- Use the graduated response matrix (GRM)

2-56. With the increased participation of US forces in contingency operations around the world, it accentuates the need to establish procedures for applying graduated military responses to situations that threaten these missions. Numerous GRMs and similar products exist throughout the military. These products graphically portray available responses in a graduated manner. The intent of these products is to give the on-scene commanders a list of options with which to control or defuse a situation before it gets out of hand. Most threats can be eliminated without loss of life or collateral damage by effectively applying the resources available.

PLAN A GRADUATED RESPONSE MATRIX

2-57. The planning for and development of a GRM begins with the mission analysis portion of the military decision-making process (MDMP). Some missions require US forces to enforce treaties or accords to protect the lives of civilians in uncertain or hostile environments (such as noncombatant evacuation operations [NEO]) or to provide large-scale humanitarian assistance. These missions require some sort of graduated response criteria to maintain order and prevent uncertain environments from becoming hostile. The development of a GRM requires the following seven steps:

- Identify the need for a GRM.
- Establish a team to develop the GRM.
- Develop targets.
- Coordinate staff functions.
- War-game.
- Receive command approval.
- Operate deliberately.

Identify the Need for a Graduated Response Matrix

2-58. Not all missions require a GRM. The decision to use a GRM requires careful consideration. Once it is agreed that a GRM is necessary, it requires guidance from the commander regarding the response options available.

Determining the appropriate responses is based on the facts, assumptions, and limitations identified during mission analysis.

2-59. Planners (staff) must agree on the intent of the GRM. The GRM can be used as a training and rehearsal tool. It provides leaders with the most likely vignettes that can be incorporated into a COA analysis, predeployment training, and rehearsals. The GRM can also be used as a handy reference during situations that require graduated responses.

Establish a Team to Develop the Graduated Response Matrix

2-60. Establish a GRM development team with a broad range of skills, knowledge, and professionalism. The following list of individuals makes up a typical GRM team:

- Provost marshal or senior military police advisor.
- FSE.
- Brigade legal officer.
- PSYOP representative.
- Land information warfare or information operations officer.

2-61. Since the GRM is designed to give commanders graduated options for dealing with both hostile and nonhostile threats to the mission, this team composition allows for target selection, application of the ROE, and attack methods using both NL and lethal means.

Develop Targets

2-62. The FSE, in conjunction with the S2 section, develops targets for both lethal and NL attacks. In the case of stability operations, these targets are usually not the conventional specific point or piece of equipment on the ground. They are more situational than specific. The GRM identifies situations or acts that subordinate elements could face during the mission. The sample GRM in *Figure 2-1* shows three possible situations or acts that on-scene commanders could expect to encounter. From the targeting standpoint, these are groups of more specific targets.

2-63. During mission analysis, the fire support officer (FSO) identifies both lethal and NL assets available to the unit. A tactical PSYOP team (TPT) attached to the unit is an example of a NL attack asset that should not be overlooked. The following are examples of what the FSO should look for:

- RCAs.
- TPTs.
- Electronic warfare assets.
- CA teams.
- Information operations (IO) teams.
- Artillery smoke projectiles.
- Aircraft (AH-64s, OH-58Ds, and AC-130s).
- Mortars.

2-64. The lethal assets described could be used in a NL show of force or demonstration to defuse a situation before it requires lethal force. The critical element of this mission analysis is not to focus solely on lethal assets. In

<table>
<tr><th>Crowd Action</th><th colspan="4">Nonlethal Response</th></tr>
<tr><td></td><td>Command Presence/PSYOP</td><td>Show of Force</td><td>Demonstration of Force</td><td>Riot Control Means (if approved)</td></tr>
<tr><td>Unarmed Civilians</td><td rowspan="3">- Ensure that the on-site commander of the TPT directs the broadcast of the dispersal proclamation and/ or passes handbills.
- Escalate the tone of the dispersal proclamation from information to a warning of force.</td><td rowspan="3">- Display force along with escalating the dispersal proclamation.
- Display force in a graduated manner, such as a helicopter hovering over a crowd or soldiers with charged weapons.
- Exploit the psychological effect of a show of force.</td><td>- Do not demonstrate force toward unarmed civilians.</td><td rowspan="2">- Employ RCAs at the point of penetration.
- Use PSYOP to exploit the psychological effect.
- Move through the crowd using riot control formations and movement techniques.</td></tr>
<tr><td>Armed Civilians (knives, clubs)</td><td rowspan="2">- Display force along with escalating the dispersal proclamation.
- Highlight the target pointer.
- Demonstrate sniper precision strike capabilities.</td></tr>
<tr><td>Armed Crowds/ Military (firearms)</td><td>- Do not use RCAs—they may escalate the situation.</td></tr>
<tr><td colspan="5">Hostile intent/hostile act occurs by armed threat.</td></tr>
<tr><th>Crowd Action</th><th colspan="4">Lethal Response</th></tr>
<tr><td></td><td>Sniper Response</td><td>Small Arms Direct Fire</td><td>Aerial Fires</td><td>Indirect Fires</td></tr>
<tr><td>Unarmed Civilians</td><td rowspan="3">- Ensure that target leaders or troublemakers are targeted.
- Use the minimum response necessary.
- Exploit the psychological effect of an attack.</td><td rowspan="3">- Determine that sniper attack is insufficient.
- Consider demonstrating capabilities.
- Exploit the psychological effect of a lethal response.
- Escalate gradually, starting with a small caliber, single round and work up to a large caliber, automatic.</td><td>- Determine that small arms direct fire is ineffective.
- Use the minimum response necessary.
- Use a minimal precision strike initially; use subsequent fires based on the situation.
- Exploit the psychological impact of each strike.</td><td>- Determine that air assets are unavailable or ineffective.
- Use the minimum response necessary.
- Ensure that the response is directed by the on-site commander.</td></tr>
<tr><td>Armed Civilians (knives, clubs)</td><td colspan="2" rowspan="2">- CAS/indirect fires must be authorized by the MACOM commander.
- Consider requesting permission for use when—
• All lesser means have been ineffective.
• There are physical eyes on target.
• Proximity to civilians has been considered.
• Risk to friendly forces/evacuees outweighs the risk of collateral damage.</td></tr>
<tr><td>Armed Civilians/ Military (firearms)</td></tr>
</table>

Figure 2-1. Sample GRM Card

stability operations, the Army wants to prevent acts of hostility first and then be prepared, if necessary, to apply lethal force.

2-65. Graduated responses can range from command presence through the show of force, a demonstration, the use of RCAs, and other techniques (such as the application of lethal force using snipers, small arms, AC-130s, and indirect fires).

Coordinate Staff Functions

2-66. To coordinate staff functions the rest of the GRM team assembles to complete the escalation sequence for each response. PYSOP and legal representatives are critical attendees during the escalation sequencing process. In the area of psychological operations, the TPT must exploit the effects of all responses.

2-67. The legal officer evaluates each escalation of force option and graduated response to ensure that it is consistent with the ROE. The GRM is designed to recommend applications of force consistent with the ROE, yet not limit the leader or individual soldier's right of self-defense. A GRM should clearly show that if a hostile act occurs, lethal options would be first and foremost.

2-68. In the case of lethal responses, the commander's guidance must again be applied. For example, lethal responses are allowed only in self-defense.

War-Game

2-69. Once the types of escalations for each potential graduated response are determined and annotated, the GRM must then be war-gamed. The staff must walk through each act or situation from the on-scene commander's standpoint.

Receive Command Approval

2-70. Once the GRM has been war-gamed, it must be submitted to the commander for approval. This is the final check to ensure that the GRM team has applied the commander's guidance correctly and met his intent.

Operate Deliberately

2-71. Operating deliberately affords commanders ample opportunity to plan and prepare for possible civil disorder situations. Through the effective gathering of information and a working cooperation with local government and police officials, commanders can often be made aware of dates, times, locations, and what groups may assemble before the operation. The purpose is to gather as much information as possible who is involved, where they are assembling, what incident promoted the activity, and what seems to be the prevailing attitude of the assembling crowd.

2-72. Civil disturbance operations are dynamic, ever-changing environments requiring effective communications both up and down the chains of command. They require advance preparation and planning using the established troop-leading procedures (TLP) outlined in *FM 3-19.4*. The eight-step process of TLP aids commanders and their subordinate leaders in planning and preparing for an operation. Most steps can be accomplished concurrently, but rarely is there enough time to go through each step in detail. Despite this, leaders must follow these steps to ensure that nothing is left out of the planning process. As outlined in *FM 3-19.4*, the following are the eight steps of TLP:

- Receive and analyze the mission.
- Issue a warning order.
- Make a tentative plan.
- Initiate movement.
- Conduct a reconnaissance.
- Complete the plan.
- Issue the order.
- Supervise, refine, and rehearse.

CONDUCT HASTY OPERATIONS

2-73. Conducting hasty operations seems to be the norm rather than the exception. Unlike deliberate operations, hasty civil disorder operations are reactionary in nature with little or no time for planning. In these situations, the event of a crowd gathering is already underway. There is very little, if any, advanced warning of the unfolding situation, and commanders are usually put in the position of sending their soldiers into an already volatile, and perhaps hostile, environment.

2-74. Commanders should be in immediate and constant communication with local civil and police authorities. The purpose is to gather as much information as possible about who is involved, where they are assembling, what incident promoted the activity, and what seems to be the prevailing attitude of the assembling crowd.

2-75. It is essential that commanders operating in environments where civil disturbance operations are likely ensure that their soldiers receive training at all levels. Commanders and subordinate leaders need to instinctively rely on the same TLP that they would in a deliberate operation. See *FM 3-19.4.*

CONDUCT TACTICAL OPTIONS

2-76. In the peacekeeping and peace enforcement environment, the use of conventional firearms or the threat of their use may not be the solution to a situation where US forces must separate two belligerent, hostile ethnic groups or prevent a similar group from entering an area that is off limits to them.

2-77. Crowd control options are often combined. Commanders choose their options based on an evaluation of the particular crowd. Commanders select any combination of control techniques and force options they think will influence the particular situation (METT-TC). Commanders must always try to choose the response that can be expected to reduce the intensity of the situation.

Monitoring

2-78. Monitoring an assembled crowd consists of gathering necessary intelligence and watching them to determine progress and development. It is this gathered information that helps commanders select the appropriate actions. Gathered information can often be used to help defuse a situation through persuasion.

2-79. Monitoring is continuous. Without timely information, responses could cause the situation to escalate beyond what it should have or cause a response to be inadequate based on the situation.

2-80. Commanders may task teams with specific missions to monitor crowd activity and note any new developments. Monitoring done by these observation teams gives the commander up-to-the-minute information so he can gauge crowd activity and intent in relation to the overall situation.

2-81. Observation teams monitor crowd activities to gather information. They observe and report on crowd size, location, and mood and on the developing situation. An observation team may consist of a marksman, a radio operator, and an observer equipped with binoculars. They may be posted strategically on rooftops and other high terrain that overlooks the crowd. Sometimes observers use helicopter-mounted observation devices. This also affords security for the operational forces. Commanders must know where their observation teams are located so they are not mistaken for teams of snipers.

2-82. The timely flow of information may allow the commander to influence the outcome of the situation with simple negotiations. Monitoring is appropriate when more decisive action is not feasible because of crowd size or when the intensity of a situation might escalate. It is particularly useful in large, nonviolent demonstrations. Monitoring can serve as an interim measure until more control forces arrive. It includes establishing communication with crowd leaders to convey official interest and intent to the crowd. Monitoring also includes efforts to gain the cooperation of crowd leaders.

2-83. Communication with crowd leaders and participants can help a commander control a situation without more severe measures. If communications exist with crowd leaders, the authorities may be able to divert the leaders or the crowd from their stated or apparent goal. Pressure can be put on leaders to channel the crowd into an area that minimizes disruption to the community and aids crowd control operations. March routes and demonstration areas can be limited to those that will help contain the crowd and reduce their potential for disrupting the community. Pressure can be positive (offering concessions) or negative (with deterrents).

2-84. If the commander can gain the cooperation of crowd leaders, it can decrease the potential for disorder, and if crowd leaders seek cooperation from authorities, officials should try to be accommodating. Crowd leaders can be placed in liaison positions between the crowd and the control force. Crowd leaders can be made responsible for managing the crowd by policing their own activities.

2-85. Taking still pictures or videotaping the faces of individuals within a crowd may prevent or reduce unlawful and violent acts. This could lessen their sense of anonymity. If needed, photographs or videotapes can be used as evidence for prosecution. To be effective, crowd members must see their presence being recorded. The photographer or cameraman should be in uniform to let the crowd know who he or she is. Recorders must be close enough to the crowd to be seen easily, but not close enough to be endangered.

Blocking

2-86. It is not uncommon for protests and demonstrations against the US or its forces to occur outside the main entrances to military installations or US Embassies. Because of the sensitive nature of equipment and information that can be found at these locations, blocking unauthorized entry is critical.

2-87. Blocking is the physical denial of advance upon a facility or an area by a crowd. Commanders may have to task their forces to block a crowd. Blocking physically denies crowd advance. Crowd control formations, particularly line formations, along with barricades can be used to block advancing crowds. Barricades of vehicles, concertina wire, and water-filled barrels are used to block or channel the movement of the crowd.

Dispersing

2-88. Dispersing is taking deliberate actions to fragment an assembled crowd in order to prevent the destruction of property or prevent injury. It is extremely effective against smaller crowds in congested urban environments. However, this may increase and spread lawless activity rather than reduce it. Thus, the commander's forces must control dispersal routes and the areas in which the dispersal will occur. Forces must protect the facilities that could be considered likely targets of opportunity for dispersing individuals or groups. Dispersal of the crowd may require apprehension of small groups still active in the area. Proclamations, shows of force, crowd control formations, and RCAs can disperse crowds.

2-89. Avoid (if possible) the dispersion of crowds into wide-open areas because it gives the crowd the opportunity to grow rapidly in size. The direction of crowd movement is difficult to anticipate and channel.

2-90. Once the crowd has started to disperse, it may occasionally be necessary for the forces to advance, keeping the crowd on the move and in the right direction. A crowd on the move should not be hurried to avoid panic. At no time should the crowd be cornered in a position where there is the perception of no escape. This invokes the "fight-or-flight" syndrome, possibly escalating violent activity.

2-91. Issuing a proclamation (see *Figure 2-2,* page 2-20) can help disperse a crowd. Proclamations officially establish the illegal nature of crowd actions and put the populace on notice that the situation demands special measures. Proclamations prepare the people for the presence of military authority and it tends to inspire respect. Proclamations support law-abiding elements and psychologically bolster military forces trying to restore order. They also demonstrate the gravity of the situation to all concerned and are excellent ways to make a commander's intentions known to a crowd. It is also a good way to reduce crowd size before direct action is taken. Commanders can make a verbal proclamation similar to the following statement: "*In the name of the President of the United States, I command that you disperse and retire peaceably to your homes.*" Such a proclamation may even make direct action unnecessary.

I. DISPERSAL

Unlawful Gathering

"Attention! Attention! This area must be cleared at once! Further unlawful behavior will not be tolerated. Clear this area at once or the necessary force to do so will be used."*

Demonstration

"Disperse and retire peaceably! Disperse and retire peaceably! Attention all demonstrators!"

"The demonstration in which you are participating ends at _______. The permit that was agreed to by the leaders of the demonstration expires at that time. All demonstrators must depart from the _______ by ______. All persons who wish to leave voluntarily may board the buses. These buses will go to the ______. Those who wish to take buses should move to _______. Those demonstrators who do not leave voluntarily by _______ will be arrested and taken to a federal detention center. All demonstrators are urged to abide by the permit."*

Warning of Looting

"Return to your homes! Someone may be looting them at this moment! During a disturbance, criminal activity is at its peak. Your family or your property may be in danger."*

II. EMPLOYMENT OF TROOPS

Employment

"Attention! Attention! Troops are present in this area. They are preparing to advance. Order must and will be maintained. Disperse peaceably and leave this area. To avoid possible injury, leave at once."

"Disperse now and avoid possible injury! Disperse now and avoid possible injury!"*

(Repeat until troops are committed.)

When Troops Are Committed

"Troops are advancing now. They will not stop until this crowd is dispersed and order is restored. To avoid injuries, leave the area at once. Return to your homes as peaceful citizens. Troops have their orders, and they will not stop until the crowd is dispersed. Do not get hurt. Leave this area."*

III. PRESENCE OF CHILDREN

(Used in conjunction with other announcements.)

"Attention! Attention! Do not attempt to cause further disorder. Disperse now in an orderly manner and avoid possible injury to children. Return at once to your homes."*

IV. ADDITIONAL INSTRUCTIONS

*Indicate the method, streets, and direction that the crowd should use when dispersing.

Figure 2-2. Sample Proclamation

2-92. In making a proclamation, a commander may consider imposing a time limit. However, the situation may change, and not imposing a time limit leaves the commander free to choose other actions when he wants. A proclamation must be specific in its instruction.

2-93. If a time limit is stated, it must be a reasonable length of time for the crowd to comply with the instructions. When drafting a proclamation, the commander must consult closely with the SJA. He must use the simplest language possible to maximize the effects of the proclamation. If proclamations must be translated to a local language, the translation must be made with great care.

2-94. The commander at the scene may direct that a proclamation be issued over public address (PA) systems. The force of the words used in the proclamation must be gauged to the composition of the crowd. If the crowd consists of normally law-abiding citizens who are presently assembled to show disagreement with an existing situation, the proclamation requires less force. On the other hand, if the crowd consists of militant rioters, the proclamation requires more force. The text may take a number of forms, depending on the situation.

2-95. A show of force is often a useful measure for dispersing a crowd. When troops arrive, the psychological impact of their arrival can be used. Soldiers can dismount from the helicopters, buses, or trucks in plain sight of the crowd, but they must be far enough away to prevent a provoked attack of thrown objects. The first echelon to dismount from the vehicle secures the surrounding area.

2-96. When small groups are scattered throughout a large disturbance area, a show of force can be made by marching troops conducting motor marches, conducting saturation patrolling, and setting up static posts. Sometimes marching well-equipped, highly-disciplined soldiers in view of a crowd may be all that is needed to convince them to disperse and retire peaceably. On the other hand, a show of force may attract people to an event. It may also provoke a nonviolent crowd into a violent confrontation. Intelligence can help best in preparing a show-of-force response.

Containing

2-97. Containment is the process of limiting a crowd to the area they are presently occupying. It is a suitable option when the crowd must be prevented from spreading to surrounding areas and communities. Additionally, it prevents those outside the already assembled crowd from joining the gathering. Containment is useful when apprehensions become necessary for preventing those in the crowd from escaping. Crowd control formations, perimeter patrols, and barriers are effective methods to accomplish containment. In all instances, caution must be used to avoid the "fight-or-flight" syndrome common to people feeling trapped with no escape.

2-98. Armored vehicles, armored security vehicles (ASVs), and up-armored high-mobility multipurpose wheeled vehicles (HMMWVs) are adaptable to roadblock operations as they can serve as barriers. They also provide added protection for the soldiers inside and those outside manning the roadblocks because they provide an easily accessible barrier for them to crouch behind.

2-99. Military vehicles traveling at close intervals in a column formation next to a crowd are largely a psychological barrier (see *Chapter 6*). They can be used to contain a large, fast-moving crowd. The moving cordon creates a temporary obstacle between the crowd and the line beyond which they will not be allowed to cross. A well-trained driver in a mobile cordon can do a better job than dismounted soldiers. In order to execute safe, quick reversals of direction for mobile cordons on narrow roads, the soldiers executing this formation must be highly-trained vehicle operators.

2-100. By-the-number commands issued over vehicle radios are the most common method for coordinating a cordon movement. Vehicles equipped with PA systems can also prove effective to issue both commands to troops and directives to the crowd. The cordon speed is no more than 5 miles per hour with an interval of at least 20 feet. Blue and red lights, sirens, blinkers, and horns may or may not be used. Armored vehicles can also serve as mobile command posts. When used properly, vehicles provide security, communications, and mobility.

DECIDE TO CONTAIN OR DISPERSE

2-101. Current crowd control doctrine places emphasis on crowd dispersal. Forced dispersal may result in a crowd breaking up into multiple groups that scatter over a large area. This may pose even greater public order problems and may pose a continued threat to control forces. A crowd is often better controlled by means of containment (confining its activities to a given area). A crowd has a limited duration, and its numbers are likely to diminish as individual needs take precedence over those of the crowd.

2-102. During the G8 summit protest in Ottawa, Canada, in June 2002, several thousand people showed up to demonstrate against the evils of world corporate leaders. Canadian Police determined that their best response to this event was a "soft response." Throughout the two-day event, hundreds of officers in standard patrol uniforms operated in four-officer teams, prestaged throughout the city. As the demonstrators began to assemble and eventually took to marching, officers provided escorts both in the front and rear of the enormous crowd along with officers marching to the flanks of the demonstration. This "soft approach" frustrated demonstrators, especially those who had it within their agenda to cause property damage and personal injury. Even more frustrating to the demonstrators was the heavy downpour of rain they endured on both days. Their organizations quickly disbanded out of concern for their own creature comforts. In the end, 10,000 people participated in the protest, with only three arrests made in two days.

USE PATROLS

2-103. Alert, aggressive patrolling of the disturbance area deters the gathering of crowds. The use of saturation patrolling is most effective for this purpose. It allows information to be collected, and it creates the psychological impression of the control force being everywhere at once.

2-104. Standard military transport vehicles provide mobility and communications for area coverage. Soldiers must be deployed with enough vehicles to provide the flexibility to handle all situations in the disturbance area. The vehicles can be modified with sandbags, wire screens, or similar materials to protect against sniper fire and thrown objects. If sniper fire is expected or has been encountered, then up-armored HMMWVs or light armored vehicles should be used for patrolling.

2-105. Armored vehicles patrolling an area of violence provide an added psychological effect and allow troops to maneuver close to snipers to make an apprehension. They may also be used as rescue vehicles to extract hostages or people surrounded by a hostile crowd.

2-106. Varying patrol routes and times for mounted and dismounted patrols keep lawbreakers from being able to pick a safe place or time to act. Patrols are particularly useful in preventing overt arson and violence. Patrol members can also spot and promptly report fires. Whenever possible, military patrols are integrated with civil police patrols. Joint patrols conserve military forces and aid civilian military communications. They also help troops become familiar with an area quickly. In addition, the civil police are available to make apprehensions, if needed. Patrol members must practice proper standards of conduct and fair treatment of civilians at all times. They are performing an important community relation and control function.

2-107. Motorized patrols, because of their speed and mobility, provide timely reconnaissance and broad area coverage. Motorized patrols are in radio contact with TF headquarters (HQ). They can make periodic contact with foot patrols and stationary posts, while patrolling their own areas. Moreover, motorized patrols can respond quickly to calls for help from other patrols and guard posts. Motorized patrols should have at least three vehicles with three men in each. Strength in numbers is necessary for protection. When planning patrol routes, avoid areas where the patrols can be isolated or placed in jeopardy. In addition, motorized patrols are equipped with fire extinguishers to put out small fires, thus reducing the burden on the fire department.

2-108. Besides foot and motorized patrols, air patrols provide a third dimension in directing the overall control effort in the disturbance area. They can perform reconnaissance and surveillance and provide near real time formation over the disturbance area. They are an excellent means of providing timely information to the commander concerning the demonstration area and its surrounding perimeter. They can monitor the actions of rioters, the extent of damage, the status of access routes, locations and conditions of road barriers, and other important conditions.

USE DELAYING TACTICS

2-109. AOs that units are assigned to patrol are often larger than the unit has resources to manage effectively. In addition, units must manage taskings, contingency plans, and other commitments, which will quickly drain available manpower. This problem will manifest itself when factions threaten to riot or gather in large numbers to demonstrate. In order for the commander to stall for time while he shifts unit assets, he may want to delay the arrival of buses and vehicles full of demonstrators. This technique is based on the assumption that demonstrators are moving to the demonstration site by vehicle. Key and secondary routes into the demonstration site must be analyzed by the S2 when he does the IPB.

NOTE: There should be an overwatch when this technique is attempted.

2-110. As demonstrator buses and/or vehicles move toward the demonstration site, soldiers in two HMMWVs block the road just ahead of the approaching vehicles. Once the demonstrators dismount their vehicles and begin to close in on the soldiers, the soldiers leapfrog (see *Figure 2-3*) 1 kilometer farther down the road to repeat the process again. This is repeated as often as necessary. This process, if repeated enough times, should discourage the demonstrators enough that they give up and go home. It should also provide enough time to shift assets to the demonstration site.

2-111. Demonstrators are just as capable of blocking secondary and key routes leading into a demonstration or riot site to prevent relief forces from assisting with riot control as US forces are capable of preventing demonstrators access to the site. One must remember that demonstrators are often well organized and have handheld radios and other items. A great example of this was the 1999 demonstration against the WTO in Seattle, Washington. Police had erected temporary portable fences in the downtown areas to keep demonstrators from disrupting the delegates. Determined to be disruptive, the demonstrators relocated the temporary fences by not allowing police to enter the area. Police were forced to back patrol cars over their own fences to gain access.

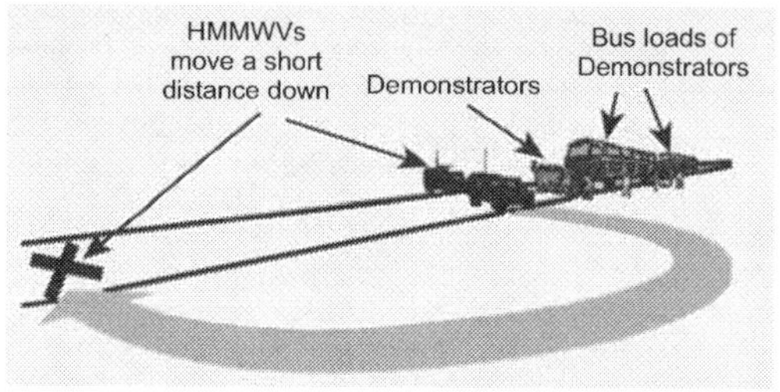

Figure 2-3. Delaying Tactics

Chapter 3

Apprehension, Search, and Detention

During the course of a civil disturbance, some members of the crowd may take part in unlawful activities, such as looting and assault. This will require the civil authorities or attached elements to apprehend, search, and detain people who are participating in the demonstration. Circumstances of the unfolding situation may require US military forces to be called upon to search, take custody of, and detain people who are participating in some form of protest and violating the law. Search, apprehension, and detention operations are conducted to halt these violations and to deter future violations.

LEGAL CONSIDERATIONS

3-1. All apprehensions are made by the civil police force, unless it is not possible for them to do so. Individuals must be known to have committed an illegal act, or there must be probable cause to believe that an individual committed such an act to be apprehended. If it becomes necessary for a military control force to apprehend or temporarily detain such violators, control force members often do so with the approval of the civil authorities. This ensures that all searches, apprehensions, and detentions (if necessary) are conducted within the parameters of the law.

3-2. Commanders operating overseas, similar to Bosnia and Kosovo, may find themselves without an established government or LEAs to work with. Commanders on the ground are then thrust into positions of authority, making them responsible for the actions, conduct, searches, apprehensions, and detentions performed by their troops.

3-3. If US military forces are called upon by the local authorities, certain policies have to be observed during search, apprehension, and detention operations. First, treat all people firmly but with reasonable courtesy and dignity. Remember that participating in a legal demonstration to express views is a right of all people, not just Americans in the US. The action, attitude, and behavior of US soldiers performing these operations are very important.

3-4. Our forces should never be seen as a military policing force on our own soil or as an occupying force in another country. Treating people with contempt, hostility, or excessive force increases the likelihood of resistance and violence. Searching people, placing them under apprehension, and detaining them without probable and just cause or without concern for their constitutional rights create problems and hinder due process. Numerous cases exist where authorities violated an individual's rights, resulting in the prosecution of police authorities or civil suits being waged against the officers and their leaders. This includes the military personnel assisting those officials.

3-5. It is very important that civil law enforcement (if available) be present and supervise all activities. Military personnel conducting a search or making an apprehension must carry out all the procedures carefully within the parameters of their training and the authority given to them. For each search and apprehension performed, the apprehension element must document the specific conduct of the person that violated the law.

3-6. Apprehending officers and supporting US servicemen must provide careful descriptive data for each subject. This description must be sufficient so that, at a later time, the suspected person can be clearly identified as the subject in question. Names and addresses of witnesses to the violation must be obtained. This information can be recorded on *Department of Defense (DD) Form 2708 (Receipt for Inmate or Detained Person).* Apprehension teams must retain and tag each item of physical evidence, such as weapons or stolen goods, that supports the apprehension on *DA Form 4137 (Evidence/Property Custody Document).* The team must give the owner a receipt for all property collected as evidence and retained.

TYPES OF SUBJECTS

3-7. There are two types of subjects (cooperative and uncooperative); they are as follows:

- **Cooperative.** Cooperative subjects actively and willingly comply with instructions. Cooperative subjects are usually compliant and offer no resistance to verbal directions or physical actions. However, if they provide any resistance to authority, it is usually passive. Some examples of this resistance are as follows:
  - Yelling.
  - Screaming.
  - Tightening the body.
  - Pulling away.
  - Letting the body go limp.
- **Uncooperative.** Uncooperative subjects will display preattack cues. Uncooperative subjects will actively resist being apprehended. Often, this resistance comes in the form of physical actions, such as—
  - Assaulting verbally.
  - Refusing to be handcuffed.
  - Refusing to be searched.
  - Attempting to push away, forcing a pursuit.
  - Wrestling with the apprehension team.
  - Striking out at the apprehension team with fists, hands and elbows, and feet.
  - Striking out at the apprehension team with weapons, such as knives, clubs, bats, bricks, rocks, or firearms.

APPREHENSION

3-8. US military forces will be in a supporting role to the local civilian authorities or acting as the control force in civil disturbance operations. When making apprehensions, it is best to use apprehension teams. These teams provide quick, organized responses to developing situations. Teams can be organized at the team, squad, or platoon level depending on the number of apprehensions expected. The intelligence information gathered helps to prepare for events where apprehensions may be necessary.

3-9. Each team (if possible) consists of a civilian police or apprehension officer, security element, and recorder. Team apprehension officers make the actual apprehension with the help of the security elements. If there is not a civilian police officer available to act as the apprehension officer, then it may be an officer or noncommissioned officer (NCO).

3-10. Ensuring that the apprehension process is properly conducted and documented, determine who is to be apprehended. Then inform the subject of why he is under apprehension and supervise his handling.

3-11. Security elements help the apprehension officer handle subjects. Security elements move, restrain, and search subjects under the supervision of the apprehension officer. The primary responsibility of the security elements is to watch the crowd and act as a blocking force, keeping the crowd from interfering with the apprehension. Recorders document the event, preferably by video, and complete *DD Form 2708*. If possible, recorders also photograph the apprehension officer with each subject. This aids in the identification process because it links the apprehension officer to the subject and provides the apprehension officer with documentation of the apprehension for use in court.

3-12. Electronically capturing the events as they unfold is essential in all civil disturbance operations. If personnel and equipment are available, videotape the events as they happen or, at a minimum, take still photographs to provide pictorial documentation that may be used later in court. Recording the scene before, during, and after the apprehension provides necessary evidence and can also help eliminate additional hostile and illegal activities. Having control force members take videos or pictures can help control crowd reaction. It reduces the feelings of anonymity that a crowd often enjoys and makes crowd members less prone to unruly or illegal acts.

3-13. Verbal commands given to the subject should be in a voice and manner appropriate for the purpose and the situation. Commands should be simple, concise, and in an understood language. Verbal commands should be convincing and convey an expectation that the instructions will be obeyed. With desperate, frightened people, sympathy combined with an attitude of firm assurance on the soldier's part may avoid additional violent behavior or trouble. An attitude of forceful authority may be needed with an uncooperative person. Commands given by soldiers must be done in a manner and context so that the subject has no confusion as to what the soldier is ordering them to do.

3-14. A subject will need to be transported from the area. In CONUS operations, civil authorities will usually be responsible for providing the transportation of a subject. Exceptions to this may be if the demonstration occurs on an installation. OCONUS operations may require US forces to provide vehicles, such as trucks, buses, or sedans, that will be used to transport a subject. Whenever possible, vehicles used to transport a subject should be modified with barriers to separate them from the drivers. If a barrier cannot be provided, a guard is to be placed in the vehicle. Passenger compartments are checked for items that could be used as weapons. When guards are required, the guard always sits with his weapon away from the prisoner. Prisoners must be placed where they can be best controlled, but they are not fastened to the vehicle with hand irons or Flex-Cufs. When seat belts are available they must be fastened. See *Soldier Training Publications (STPs) 19-95C1-SM and 19-95C24-SM-TG.*

3-15. Female escorts should always be used to transport a female subject. If female escorts are not available, measures must be taken to avoid false charges of molestation or sexual misconduct. Net control stations should be notified of the departure and arrival times and vehicle mileage readings (before and after). At least two male escorts must be present at all times when a female escort is not available. These procedures also apply when the male and female roles are reversed.

FLEX-CUFS

3-16. The two main positions used to place Flex-Cufs on a subject are the standing-supported and the prone.

3-17. When using the standing-supported position, the search man—

- Escorts the subject to a wall or other sturdy vertical surface (see *Figure 3-1*).

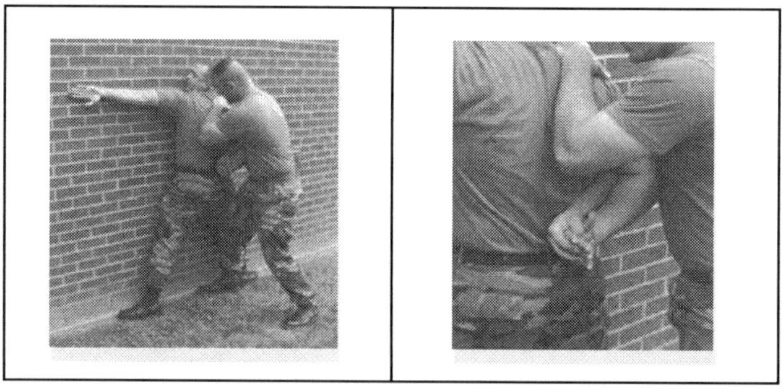

Figure 3-1. Standing-Supported Position

- Instructs the subject (verbally) to do the following:
  - Turn his head away.
  - Place his free arm (palm out, thumb down, and elbow locked) against the surface of the wall.

  - Spread his feet until instructed to stop.
  - Place his heels against the wall.
- Uses the leg closest to the controlled arm to pin the subject's leg (above the knee) to the wall.
- Reinforces verbal commands with pain compliance techniques if the subject fails to follow instructions.
- Maintains a double 90° hold on the controlled arm with his inside hand. With his free hand, the search man grasps the middle portion of the Flex-Cuf, keeping one cuff up and one cuff down.
- Places the first cuff (bottom cuff) on the subject's controlled arm, maintaining wrist manipulation (see *Figure 3-2*).
- Reaches through the free cuff with his outside hand and instructs the subject to place his free hand in the center of his back. With his outside hand, the search man takes the subject's free hand and completes the restraint by securing the cuff. He ensures that there is at least one-finger distance between the cuff and the subject's wrist (see *Figure 3-3,* page 3-6).

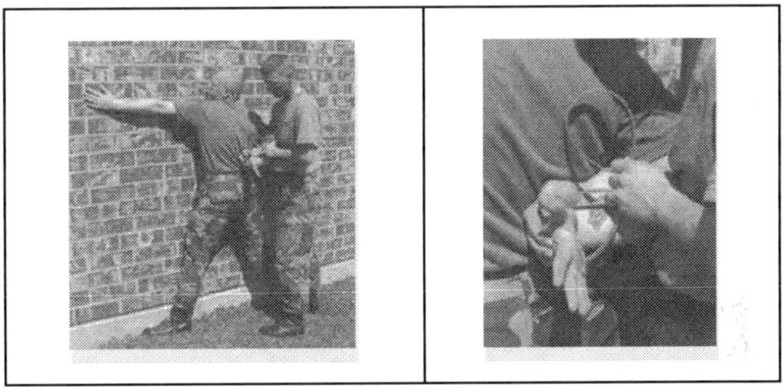

Figure 3-2. Cuffing the First Hand From the Standing Supported Position

3-18. When using the prone position, the search man—

- Grasps the Flex-Cuf with his outside hand, maintains tension on the subject's arm, and places the first cuff on the subject's controlled wrist (see *Figure 3-4,* page 3-6). Next, the search man secures the cuff, keeping at least a one-finger distance between the cuff and the wrist.
- Reaches through the free cuff with his outside hand and instructs the subject to place his free hand in the center of his back (see *Figure 3-5,* page 3-7).
- Reinforces verbal commands with pain compliance techniques if the subject fails to follow instructions. The search man bends the subject's controlled arm (using the natural break of the elbow as a hinge) toward the hand in the middle of the subject's back. With the outside hand, the search man takes the subject's free hand and completes the restraint by securing the cuff.

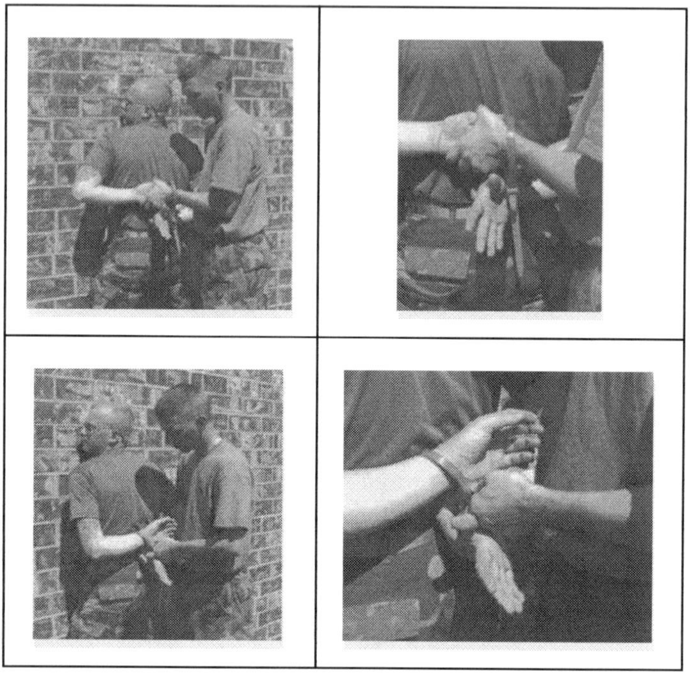

Figure 3-3. Cuffing the Second Hand From the Standing Supported Position

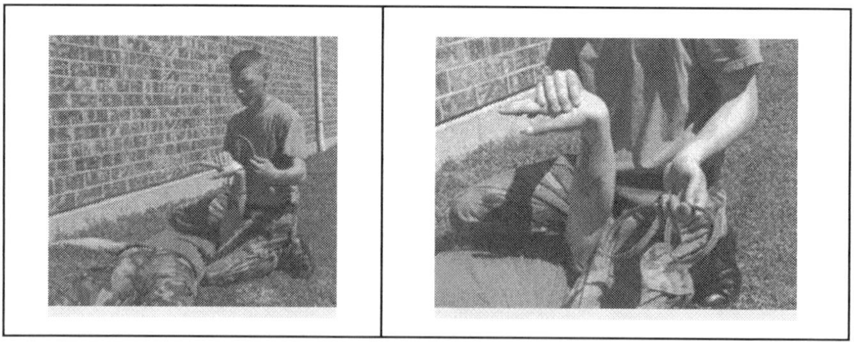

Figure 3-4. Cuffing the First Hand From the Prone Position

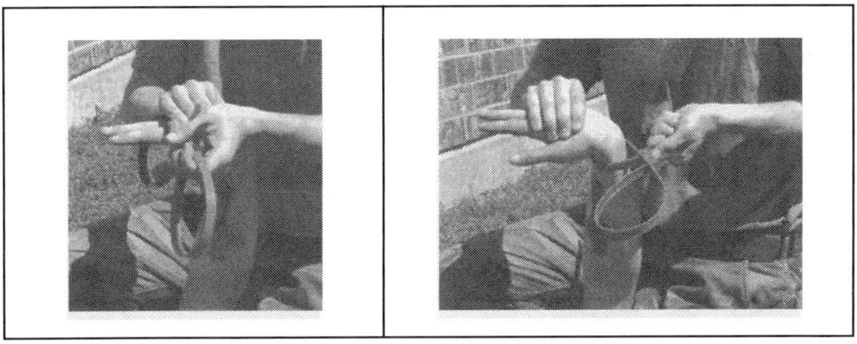

Figure 3-4. Cuffing the First Hand From the Prone Position (continued)

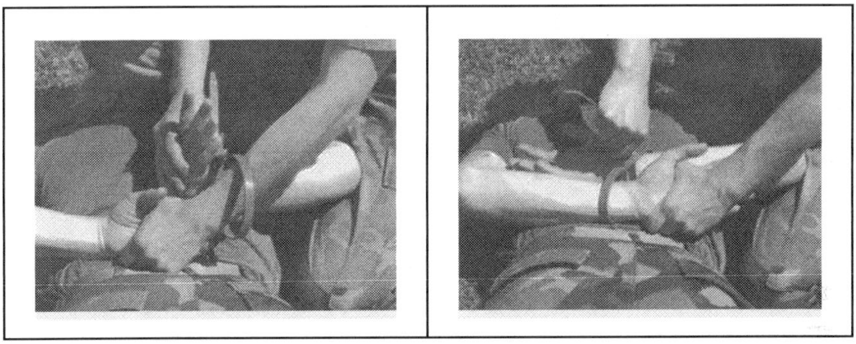

Figure 3-5. Cuffing the Second Hand From the Prone Position

SEARCHES

3-19. Initial contact with the subject is the most dangerous time during any apprehension; therefore, initial contact must be a swift and decisive team effort. Do not give the subject time to think or react. Distract his attention when possible. Apprehension teams are comprised of a search man, threat cover man, and security man (if necessary). Their responsibilities are as follows:

- The search man—
  - Issues verbal commands.
  - Conducts searches.
  - Handles radio communication.
  - Holsters his weapon before approaching a subject and informs the threat cover man that he is holstering his weapon.
  - Performs control procedures and handcuffs the subject.

  - Moves the subject to a safe area.
  - Removes weapons (if any) and sensitive evidence from the subject.
  - Disengages after the subject is restrained.
  - Controls and escorts the subject to a detention area.
- The threat cover man—
  - Provides cover for the search man.
  - Discourages escape and resistance.
  - Discourages hostile interference.
  - Alerts the search man of threats.
  - Resists distractions and maintains target acquisition or scanning.
  - Assesses threats constantly and prepares for the unexpected.
- The security man (if necessary)—
  - Provides cover for the search man.
  - Discourages hostile interference.
  - Alerts the search man of threats.
  - Resists distractions and maintains target acquisition or scanning.
  - Assesses threats constantly and prepares for the unexpected.

Search Principles

3-20. A subject should always be put at a disadvantage and searched immediately, but circumstances may dictate postponing the search until a safer place can be found. At a minimum, a quick pat down can be done en route to a safer location. The longer the apprehension takes, the greater the risk of drawing a hostile crowd and violent acts toward the control force.

3-21. Crowds often sympathize with a subject and try to help him or escalate the activities of the crowd toward violence. A subject's actions may also incite the crowd to violence. The longer a subject is the focal point of an apprehension, the easier it is to stir a crowd to hostile or violent acts. Apprehension teams should choose to move the subject out of the crowd's view. For the safety of the team, always apply Flex-Cufs or handcuffs and escort the subject to an area where a search can be done safely.

3-22. Search team members never operate alone; searches are conducted by a minimum of two team members. Initial steps must be taken to put the subject at a physical disadvantage. Before conducting a search (regardless of rank, sex, position, age, or physical condition), a team member applies the handcuffs or Flex-Cufs. Verbal commands must be short, clear, and distinct.

3-23. A search is conducted on everyone that has been apprehended to ensure that the team and those being apprehended are safe. The search man must abide by the following procedures when conducting a search:

- Search in a clear, secure area (when possible).
- Be ready to counter any resistance or an assault by the subject. Consider the subject armed until proven (by search) otherwise.
- Employ the appropriate level of force immediately, based on the actions of the subject.

- Do not step over the subject when he is at a disadvantage on the ground; go around him and remain out of sight.
- Do a 360° search of the subject's waistline for weapons. Search obvious and inconspicuous areas. Inconspicuous areas such as the groin, armpits, chest, and small of the back must be searched on both men and women; however, do not linger in these areas.

NOTE: Conduct same-gender searches when possible. If mixed gender searches are necessary for speed or security, conduct them in a respectful manner and avoid any action that could be interpreted as sexual in nature. To prevent allegations of sexual misconduct, the apprehension officer shows careful control of the soldiers who perform mixed-gender searches.

- Maintain security.
- Brief subjects quickly on what to do and who to follow.
- Remember that when conducting a search, a subject's body can be divided for a more efficient and thorough search. For example, search from head to toe or from side to side.

NOTE: The search man announces the items found and secures weapons and contraband.

- Military forces (when supporting civil authorities) may perform searches of people and property that are not usually subject to military law during a civil disturbance.
- Remember that it is always better for civilian police to search civilians and civilian property (when possible).
- Remember that if military forces are called upon to support police it is usually a last resort and there are not enough civilian officers available to accomplish the mission.
- Military personnel may search people incident to an apprehension.

3-24. The prescribed methods of searching someone are the frisk, the standing, and the prone. The method of search used largely depends on the situation and the mission. Searches are made not only of the subject and his personal effects such as purses, bags, and wallets, but also of the immediate surrounding area. This prevents the subject being searched from grabbing a weapon or destroying evidence. It may also be necessary for military forces to search private property, including vehicles, if there is—

- Reasonable belief that a person has committed or is committing a violent crime and is hiding in a building or a vehicle.
- Reasonable belief that a vehicle contains weapons or instruments of violence.
- Probable cause for searching a building or a vehicle.
- Probable cause to believe that (unless immediate action is taken) evidence of a crime will be destroyed before a warrant can be obtained.

NOTE: Search teams should always consist of two personnel—one to conduct the search, and the other to provide security and overwatch for the soldier conducting the search. Before being searched, a subject should always be restrained and put at a disadvantage by

securing his hands with handcuffs or Flex-Cufs. Ask the subject if there is anything in his pockets that could hurt the searcher, such as knives or needles.

Frisk Search

3-25. The frisk search (also known as the stop and frisk, pat down, and terry stop) is a search incident to a lawful stop. Subjects are not under apprehension at this point, but if it were an apprehension situation, handcuffs or Flex-Cufs would be applied first. Frisk searches are used when there is reasonable belief that a subject may be armed with anything that could be considered a safety threat.

3-26. Legal and regulatory circumstances for conducting a frisk search are covered in the *Manual for Courts-Martial (MCM)*. When a lawful stop is performed, the person stopped may be frisked for weapons when that person is reasonably believed to be armed and dangerous.

3-27. The primary purpose of a frisk search is safety. Frisk-search techniques are to be used only on a compliant subject. Frisk searches only cover the outer garments. Do not search the pockets of the individual unless a weapon is suspected. A frisk search begins when the search man informs the subject of the intent to search.

- During the preparation phase, the search man—
  - Tells the subject that he must perform a search for weapons for the subject's safety and his own.
  - Tells the subject to turn and face away from him and instructs the subject to keep his hands out to his sides where they are visible.

NOTE: The cover man providing security moves around the subject and positions himself out of reach and in front of the subject.

  - Instructs the subject to spread his feet until instructed to stop (more than shoulder width) and point his toes outward compromising his balance (see *Figure 3-6*).

NOTE: Do not kick at the subject's feet to spread them apart; it is unnecessary and unprofessional.

  - Tells the subject to bring both of his hands back to the small of his back with palms out, thumbs up, and the back of his hands together.
  - Instructs the subject not to move (see *Figure 3-7*).
- During the performance phase the search man—
  - Approaches the subject from behind (if possible) at a 45° angle, off center. The search man approaches with caution to ensure that his feet are to the outside of the subject's stance. See *Figure 3-8*, page 3-12.
  - Stays just inside arms reach and grips the subject's fingers with his support hand. The search man firmly grips at least one finger on each of the subject's hands (both thumbs, both index fingers, or an index and middle finger). An overhand or an underhand grip can be used effectively.

Figure 3-6. Approaching the Subject to Perform a Frisk Search

Figure 3-7. Giving Instructions to Subject Before Frisk Search

NOTE: The primary purpose of gripping the fingers is to control the subject's hands while performing the frisk search (see *Figure 3-9*, page 3-12). If the subject is going to resist or become aggressive, he will attempt to free his hands first. If the subject attempts to pull his fingers from the search man's grip, the search man must respond instantly by increasing his grip. If the grip on the fingers is lost, the search man should immediately push away from the subject and create a safe distance. This will allow the search team to respond with the necessary force to regain control of the situation.

Figure 3-8. Performing a Frisk Search

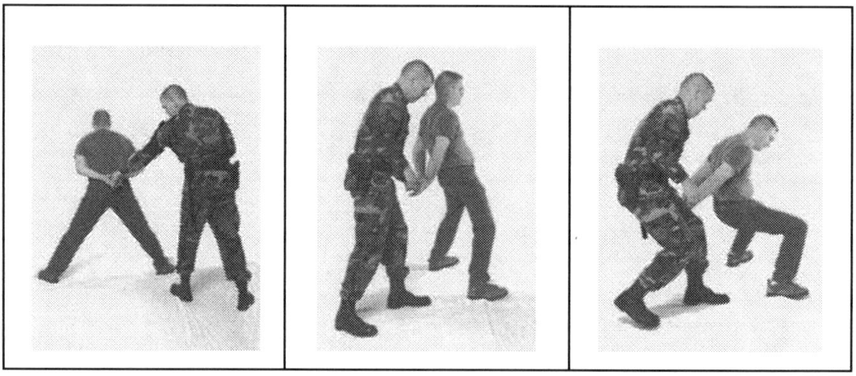

Figure 3-9. Gripping the Subject's Fingers During a Frisk Search

- Ensures that the subject always looks away from him.
- Places his strong hand on the outside of the subject's elbow (similar to the handcuffed escort position).

NOTE: The search man searches one side (the front, back, or side) at a time from head to foot.

- Divides the subject (visually) straight down his middle and does not cross the line while conducting the search.

NOTE: The search man is just feeling or searching for possible weapons, so an open hand should be used.

- Works at arm's reach. It is the sense of touch not the sense of sight that is being used. While performing the search, search men should stay behind and to the side of the subject. This provides the search man with a tactical advantage and prevention from being struck by the subject's elbows or knees if he should begin to resist.

3-28. When searching below the subject's knees, the search man takes a half step back and to the outside. He squats down, but does not drop his knees to the ground.

3-29. If no weapons are found and the subject is to be detained, the hands are secured with Flex-Cufs and the subject is escorted to the holding area. If the subject is to be released, collect information (name and address) and escort him away from the crowd.

Standing Search

3-30. Safety is the primary concern when dealing with a subject under apprehension. Therefore, the search man must protect himself by restraining subjects under apprehension before conducting the standing search. All subjects under apprehension have the propensity to turn violent or hostile. Putting a subject at a physical disadvantage provides safety for the subject and the search man. It limits the subject's ability to hide, destroy, or remove any evidence he may have on him.

3-31. Apply Flex-Cufs to the subject quickly during the apprehension. It is impossible to predict what a subject will do when put under apprehension. A subject may initially be cooperative and compliant, but as the subject becomes aware of his situation he may be overcome with emotion and begin to struggle or pull away.

3-32. There is no legal or regulatory prohibition against opposite-gender searches. Common sense prevails and whenever possible, searches should be done by soldiers of the same gender to avoid obvious questions and complaints. Once a subject is put at a physical disadvantage by cuffing and is more easily controlled, it affords the necessary time to get someone of the same gender there to conduct the search.

- During the preparation phase, the search man—

NOTE: Standing searches always begin with the subject in the handcuffed escort position (see *Figure 3-10,* page 3-14).

  - Takes his outside hand (farthest from subject) and places it on the outside of the subject's elbow while his inside hand (closest to subject) has a reverse-handshake grip of the subject's hand. At this point, both hands are on the same arm (see *Figure 3-10,* page 3-14).
  - Ensures that his position is on the same side of the subject's body being controlled (behind and to the side). When changing sides, change hand position.
  - Instructs the subject to spread his feet until instructed to stop (usually more than shoulder width) and point his toes outward, compromising the subject's balance.

Figure 3-10. Handcuffed Escort Position

NOTE: Do not kick at a subject's feet to spread them; it is unnecessary and unprofessional. If the subject becomes noncompliant or aggressive, pain compliance is usually the best tool to use, but only use enough force to gain compliance.

- Uses his outside hand to conduct the search. His inside hand is always used to maintain control during the search.
- During the performance phase, the search man—

NOTE: Only one side (the front, back, and side) of the subject is searched at a time from head to toe (see *Figure 3-11*).

- Divides the subject (visually) (straight down the middle using the spine as the centerline). He does not cross this line while conducting the search. Searches must be systematic, from head to foot. If there is an interruption, it must be started all over again.
- Searches with an open hand, using a rub down or squeezing method. Do not only pat down like in the frisk search.
- Works at arms reach. It is the sense of touch, not the sense of sight that is being used. To prevent the possibility of being struck by an elbow or knee, the search man stays behind and to the side of the subject.
- Takes a half step back and to the outside when searching below the subject's knees. The search man squats down, but does not drop his knees to the ground. Maintaining positive control is essential throughout the search, so never the reverse handshake is never released.
- Change sides when satisfied that all items have been removed from the side in which the search started. Without losing positive control, the search man conducts the search on the other side of the subject.

Figure 3-11. Performing the Standing Search

NOTE: Changing sides is accomplished by moving the outside hand (the searching hand) to the subject's uncontrolled handcuffed hand located on the outside. The outside hand and forearm will cross over the top of the reverse-handshake grip of the inside hand (see *Figure 3-12,* page 16).

- Crosses over the reverse-handshake grip with the outside hand and grips the subject's uncontrolled handcuffed hand in a reverse-handshake grip.
- Releases (quickly) the first reverse-handshake grip moving that hand to the subject's elbow as he sidesteps to the side yet to be searched. The search man repeats the standing search technique used on the first side of the subject.
- Gives strong verbal commands and uses only the force necessary to control the subject if he resists or becomes aggressive at any time during the search.
- Removes all items found during the search to be evaluated as potential evidence or a weapon. Items not considered a weapon or of evidentiary value will be returned to the subject at a later time.

Prone Search

3-33. Safety is the primary concern when dealing with a person under apprehension. Therefore, the search man must protect himself by restraining subjects under apprehension before conducting this search. Remember that all subjects under apprehension have the propensity to turn violent or hostile. That is the reason a prone search would be necessary. After performing an arm bar takedown on a noncompliant subject, it is especially important for the search man to put him at a physical disadvantage to provide for the subject's safety and his own. By doing so, the subject's ability to hide, destroy, or remove any evidence is limited.

Figure 3-12. Performing the Standing Search (Changing Sides)

3-34. The safest way to handcuff or Flex-Cuf a noncompliant subject is in the prone position, when the search man has greater control of the subject. It is usually the safest position in which to conduct a search

3-35. Handcuff the subject quickly during the apprehension. It is impossible to predict what a subject will do when put under apprehension. Often a subject may initially be cooperative and compliant; however, as the he becomes aware of his situation, he may become overcome with emotion and begin to struggle or pull away.

3-36. There is no legal or regulatory prohibition against opposite-gender searches. Common sense prevails and whenever possible, searches should be done by the same gender to avoid obvious questions and complaints. Once a subject is at a physical disadvantage by cuffing and is more easily controlled, it affords the necessary time to get someone of the same gender there to conduct the search.

NOTE: The subject is put into a physical disadvantage when he is taken facedown to the ground using an arm bar takedown and cuffed.

- During the preparation phase, the search man—
  - Ensures that the Flex-Cufs are not too tight.
  - Rotates his entire body around the subject's shoulder and center at the outside of the subject's hip with his shin over the subject's shoulder and one knee on the ground. At least one knee should remain over the subject's back (near side), with the other on the ground.
  - Pins the subject down by pressing his knee on the side of the subject's back or down on his closest elbow if resistance or aggression is encountered. He simultaneously grips the subject's closest handcuffed hand, grabs it with a reverse handshake, and applies pain for compliance.

- Uses strong, forceful verbal commands, such as "Stop resisting, stop moving around."
- Ensures that the subject's feet and ankles are crossed.

CAUTION

Stay alert. If the subject uncrosses his feet and ankles, it is usually a sign of preattack.

- Removes all items found during the search so they can be evaluated as potential evidence. Items not considered a weapon or of evidentiary value will be returned to the subject at a later time.

• During the performance phase, the search man—

NOTE: The search man must be systematic from head to foot while conducting the prone search. If interrupted during the search, it is important to start from the beginning to ensure that nothing is missed. It is better to go over the same area two or more times than to miss something that can be used as a weapon.

NOTE: The search man searches only one side (the front, back, or side) of the subject at a time until that side is complete.

- Divides the subject (visually) straight down the middle, using the spine as the centerline.

NOTE: He does not cross this line while conducting the search.

- Searches with an open hand, using a rub down or squeezing method for best results. He does not just pat down the subject like in the frisk search.
- Works at an arm's reach, searching only the area of the subject's body that is not in contact with the ground. The sense of touch, not the sense of sight, is being used.
- Stops the search as the subject's knees are reached. He taps one of the subject's legs and firmly commands him to raise his foot up and back, allowing for an inspection of the bottom of his foot. He then searches from the knee to the ankle area. This is performed on both the subject's legs and feet.

NOTE: At this point, the entire back side of has been searched. What remains to be searched is the front side of the subject (the chest and front of the hips, waist, groin, and thighs).

- Pivots back and faces the subject. Using the hand closest to the subject's head, the search man reaches across the subject's back and firmly grips the shoulder, rolling the subject up on the his side toward him to expose the areas not yet searched.
- Rolls the subject back down on his chest once all areas have been searched.

3-37. The search man must be completely satisfied that he has found everything. He must be especially sensitive to areas where weapons are often concealed like the waistband, the small of the back, ankles, pockets, shoes, the tongue of the shoes, and the groin area.

NOTE: Before removing subjects from the area, a brief summary must accompany them. The summary must include the subject's name, what he is suspected of doing, the time and date of the incident, and who saw him do the action.

DETENTION

3-38. Civil authorities must provide adequate detention facilities for all subjects. Authorities must be prepared to detain large numbers of people. They may choose to expand existing detention facilities or set up temporary facilities to accommodate the extra load. If possible, large-scale arrests are delayed until sufficient detention facilities have been set up.

3-39. If US military forces are committed to support local authorities, commanders should coordinate with them to ensure that adequate detention facilities are available and to learn their locations and capacities. If there are more detainees than civil detention facilities can handle, civil authorities may ask the military to provide support by setting up and operating temporary facilities. Army correctional facilities cannot be used to detain civilians. A temporary military detention facility can be set up if—

- Federal troops have been employed according to provisions.
- The TF commander has verified that available civilian detention facilities can no longer accommodate the number of prisoners awaiting arraignment and trial by civil courts.
- The Army Chief of Staff has granted prior approval.

3-40. Use of the temporary facility ends as soon as civil authorities can take custody of the detainees. Military forces are responsible for the custody, health, comfort, and sustenance of all detainees in their facilities until custody is transferred to civil authorities. Temporary facilities cannot be used to confine people arraigned or convicted in civil courts.

3-41. Women are detained in their own facility. Temporary facilities are supervised and controlled by officers and NCOs trained and experienced in military correctional operations. Females guard females. If two female guards are unavailable, use two male guards or one male and one female guard (see *FM 3-19.40*).

3-42. The same operational procedures that apply to the management of an installation confinement facility (see *FM 3-19.40*) apply to the management of temporary detention facilities, except for training, employment, and administrative discipline. Guards and support personnel under direct supervision and control of military officers and NCOs need not be trained or experienced in military correction operations. They must be specifically instructed and closely supervised in the proper use of force, custodial procedures, and the completion of military and civilian forms and reports that may be used. Soldiers that may be tasked to operate a temporary detention facility should be familiar with the forms and reports used for civilian apprehensions too.

3-43. Temporary facilities are set up on the nearest military installation or on suitable property under federal control. Ideally, the facility should be close enough to the disturbance area to minimize transportation and escort needs. However, it should be far enough away not to endanger those being detained. Whenever possible, existing structures are adapted for this use, but construction may be needed to provide segregation for ensuring effective control and administration.

3-44. The basic structure must include the following:

- Search areas.
- Holding areas for incoming men and women.
- A processing area.
- Holding areas for men and women who have committed misdemeanors.
- Holding areas for men and women who have committed felonies or are violent.
- A holding area for detainees' property and evidence obtained during a search.
- A holding area for administrative support and records.
- A medical station.
- Latrines.

For more information on these structures see *FM 3-19.40.*

3-45. Facility personnel must ensure that proper sanitation is maintained. When large numbers of people are detained or processed through a facility, sanitation becomes a problem. Medical personnel must conduct regular health inspections to detect unsanitary practices and conditions.

3-46. Facilities must be organized for a smooth flow of traffic. Processing stations must be set up so there is a linear or circular sequence of movement. These stations are set up out of sight of the holding areas. If possible, they are separated by a door to reduce noise. To reduce distractions, each station may be partitioned. Detainees may be more cooperative if they are out of sight of each other.

3-47. Injured people are given prompt medical treatment and transportation to medical facilities when necessary. A medical aid station for screening detainees and treating minor injuries is set up inside or next to the detention facility. Treatment areas, however, must be out of sight of the processing and holding areas. If possible, access to the medical facility bypasses the holding and processing areas. Facility personnel may also consider setting up a separate holding area for injured detainees.

3-48. Detention facility operation plans must contain emergency procedures. As a minimum, the following areas must be addressed:

- Fire evacuation.
- Disturbance control.
- Facility security.

NOTE: Depending on the situation, other emergency procedures may be needed.

3-49. Upon arrival at the detention facility, the detainee is logged in and searched, regardless of whether a complete search has already been conducted by the apprehending team in the field. Separate search areas are set up for men and women. Weapons, contraband, flame-producing devices, suspected evidence, money, and high-value items are confiscated. Medications are also confiscated. Medical personnel screen detainees that have had medications confiscated. Receipts are provided for any property or evidence that is confiscated. Confiscated items are tagged, and the items are stored in a controlled-property area.

3-50. As a detainee is brought to the facility, a file is initiated. Assigned detainee case numbers are used on all paperwork, such as logs, evidence tags, reports, and visual documentation. All paperwork (including photographs) that began outside the detention facility is marked with the case number. Facility personnel may also use hospital identification tags. Using indelible ink, the case number is written and attached to the tag on the detainee's wrist. Different colors may be used to identify different subject classifications, such as misdemeanors, felonies, or violent offenders. If opposing factions are involved, procedures are established to ensure that members of opposing factions are not processed together or detained in the same holding area.

3-51. After a detainee has been searched and classified, he is taken to a processing station where his paperwork is processed. If subjects passively resist by going limp, they may be moved by wheelchair. This reduces the number of escorts needed and the fatigue to facility personnel. Paperwork is reviewed to ensure that all information is complete, including charges, witnesses, and reasons for the apprehension.

3-52. Equipment such as height charts, scales, fingerprinting equipment, and cameras must be available for completing the police report and identifying the subject. Detainees may use aliases and not carry identification, so pay special attention to obtaining as much information as possible about the detainee's physical characteristics.

3-53. After processing is complete, the detainee is placed in the appropriate holding area. Paperwork for the detainee is forwarded to the administrative section. Files are reviewed for completeness and to determine the disposition of the detainee. Information from the file may be placed in a computer to find repeat subjects. Computers may be used to file criminal information only.

3-54. Custody transfers and release procedures must be coordinated with civil authorities and the appropriate legal counsel. Every effort must be made to arraign subjects quickly. The purpose of a detention facility is not to keep people off the streets, but to aid in processing subjects through the legal system. To speed up the release process, planners should consider issuing citations or subpoenas for minor offenses.

3-55. Military force members may have to respond to writs of habeas corpus. Writs are court orders addressed to a prisoner's custodian. Such a writ directs the custodian to bring the prisoner to court to determine the legality of the prisoner's apprehension and detention. Custodians must ensure that all documentation concerning the case is brought to court with the detainee.

3-56. Military personnel must obey writs issued by federal courts. For writs issued by a state court, the custodian or his legal advisor should respectfully reply that US authority is holding the prisoner. The Office of the SJA can answer any questions and explain the correct procedures.

Chapter 4

Riot Shield and Riot Baton Techniques

A soldier may be required to be part of a riot control formation where his primary weapon is a riot baton and his only protection a riot shield. A riot baton in the hands of a well-trained soldier is an invaluable weapon in a crowd control situation and may also be the most appropriate weapon. A riot baton is a very versatile weapon that can be used as an offensive and/or defensive weapon. The riot shield offers protection from the top of the head to just below the knees.

OVERVIEW

4-1. Use of the riot baton and riot shield is based on METT-TC and the commander's choice-of-force option. Formations committed to a riot with riot batons must have lethal overwatch, lethal weapons, and NLW employed within the formation and reserve forces with additional NL and lethal capabilities. These forces are positioned to rapidly reinforce the control force formation.

4-2. During the termination phase of the disturbance (when the violence has subsided and is expected to remain that way), soldiers are placed in a defensive posture. They may be equipped with a riot shield and a riot baton to perform their tasks. The riot shield is primarily a defensive weapon, which may be used in an offensive mode. It is primarily used as an offensive weapon when the formation is in close contact with rioters. The riot baton can also be used when formations are on the offensive and in contact with the crowd. However, crowd control elements employed with riot shields and riot batons must have lethal weapons and NLW available.

WARNING

Riot control formations are never employed with riot shields and riot batons only. There must always be lethal overwatch elements in position and lethal and NL capabilities within the riot control formation with a ready reserve force armed with lethal and NL capabilities.

RIOT SHIELD

4-3. In the hands of a properly trained soldier, the riot shield has proven to be the first line of defense for a formation. The riot shield may be used as an offensive weapon when in contact with an aggressive crowd.

4-4. The riot shield may be held with one arm. It is secured to the weak side arm with a self-sticking strap system and a handle. This system is designed to make the riot shield an extension of the soldier's arm. The riot shield is held at a slightly inward angle. This allows for debris thrown at the soldier to fall harmlessly to the ground (see *Figure 4-1*).

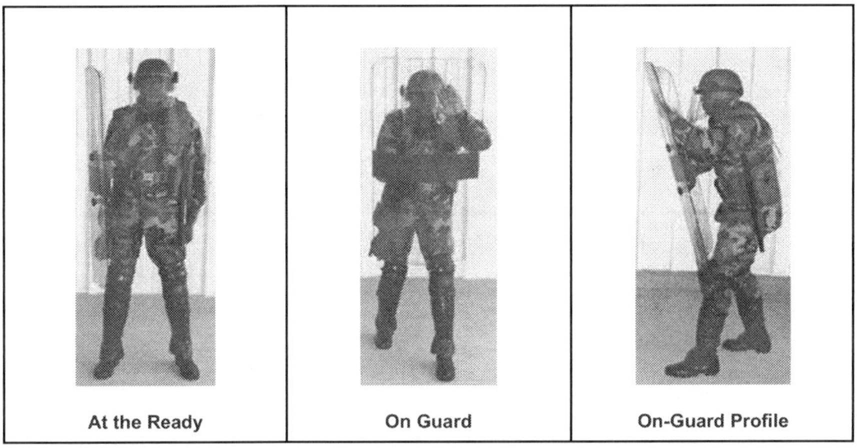

Figure 4-1. Riot Shield Positions

RIOT SHIELD CARRIES

4-5. The two basic riot shield carries are as follows:

- **At the ready.** The at-the-ready position is used while in a column formation or on the move toward hostilities. This position is designed for rapid movement. It should not be used when in contact with a crowd.
- **On guard.** The on-guard position is used when the riot control formation is in contact with the crowd or contact is eminent. It is used primarily in the following formations:
  - Line.
  - Wedge.
  - Echelon.
  - Diamond.
  - Circle.

DEFENSIVE AND RETENTION TECHNIQUES

4-6. The primary purpose of the riot shield is for defense. However, each riot shield holder must be proficient in riot shield retention techniques.

DEFENSE

4-7. The first line of defense for the formation is the riot shield. It is designed to withstand strikes from objects thrown by the crowd. In addition to protecting the riot shield bearer, it also protects other members of the formation from thrown objects.

RETENTION

4-8. When the control force is in contact with the crowd, the crowd will attempt to strip the riot shield from the bearer. When a subject grabs the top of the riot shield (see *Figure 4-2*), the riot shield holder slaps the riot shield with his strong hand near the subject's hands and gives the following commands (while continuing to slap the riot shield until the subject releases it):

- "Get back."
- "Get away."
- "Stop."

Figure 4-2. Riot Shield Retention Technique (Top Attempt)

4-9. If the riot shield is grabbed from the bottom (see *Figure 4-3,* page 4-4), then the riot shield holder forcefully drops to one knee, pinning the subject's fingers and/or hands to the ground. This must be done quickly and with force.

RIOT BATON

4-10. A riot baton in the hands of a properly trained soldier is a formidable weapon. However, it must be used in conjunction with other measures to be most effective. Soldiers must be trained with the riot baton to the point that its various techniques are automatic to them. This training must also include

Figure 4-3. Riot Shield Retention Technique (Bottom Attempt)

learning the vulnerable points of the human body so they can avoid areas that may cause permanent injury or death when struck.

WARNING

The riot baton is never raised above the head to strike a subject in a club fashion. Not only is it likely to cause permanent injury; but it also gives an unfavorable image of the control force. The soldier is also vulnerable to an attack on his rib cage when his arm is raised.

TARGET AREAS OF THE HUMAN BODY

4-11. The target areas of the human body are divided into three areas that represent the primary-, secondary-, and final-target areas. These areas are color coded as green (shown as gray in *Figure 4-4*), yellow (shown as white in *Figure 4-4*), and red (shown as patterned gray in *Figure 4-4*).

Primary-Target Areas

4-12. The primary-target areas are those areas which, when struck or restrained, cause the least amount of trauma to the body. These include the meaty or muscular areas of the body, such as the—

- Foot, shin, and instep.
- Inside and outside of the thigh.
- Lower abdominal region.
- Fore- and upper-arms.

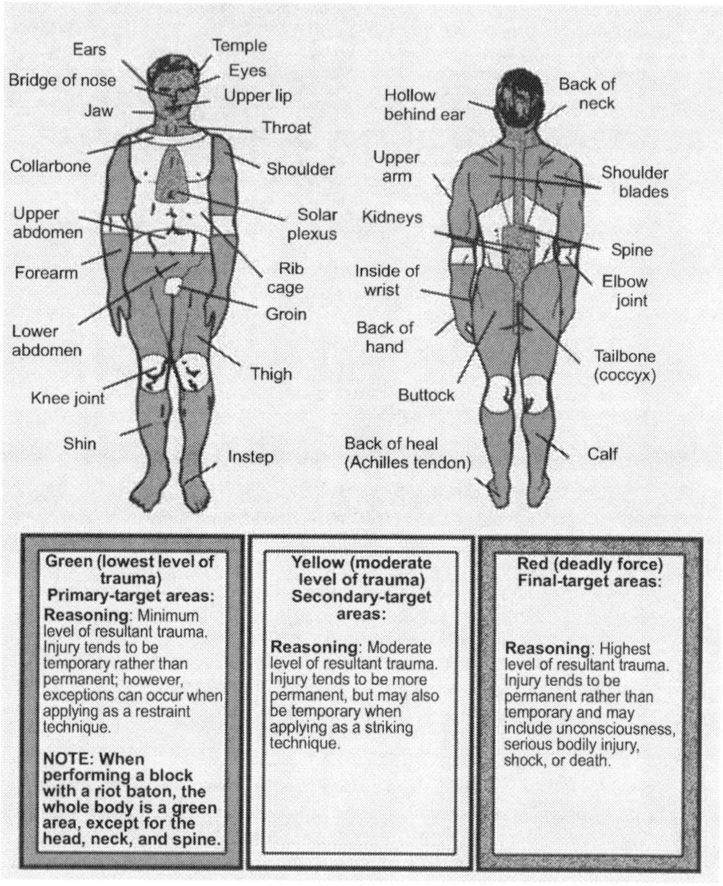

Figure 4-4. Escalation of Trauma Chart

Secondary-Target Areas

4-13. The secondary-target areas are those areas of the body which, when struck or restrained, cause a moderate level of trauma to the body. This type of trauma tends to be longer lasting because the time to heal is longer than with other parts of the body. If these areas are struck with a riot baton, serious injury may occur. However, some of these areas are used for control or restraining holds that tend to lessen the threat of injury. Some secondary targets are the—

- Collar bone and shoulder blades.
- Elbows and knees.
- Upper abdominal region.

Final-Target Areas

4-14. The final-target areas are those areas which, when struck with a riot baton, cause serious trauma that tends to be fatal or critical. However, there are some areas that (when used for restraining only), will not cause death or serious injury. Some final-target areas are the—

- Front and back of the head (eyes, ears, nose, upper lip, and hollow behind the ear).
- Neck and throat.
- Upper chest.
- Xiphoid process.
- Spinal column.
- Groin area.

JOINTS OF THE HUMAN BODY

4-15. The joints of the human body are often described as hinges and, when manipulated in a certain fashion, allow for better control of the subject being restrained, especially if he is aggressive. When applied correctly, minimal pressure is needed to control the subject. When a subject attempts to resist, he creates additional pain for himself. By anatomically locking out these hinges, the subject will be forced to comply. When restraining a subject, emphasis must be placed on locking out two of the three joints of the upper body or two of the three joints of the lower body. The joints are as follows:

- **Shoulder.** Apply pressure to the shoulder to prevent the subject from freely moving the upper portion of his body.
- **Elbow.** Apply pressure on the elbow or just above the elbow on the ulner nerve.
- **Wrist.** Apply pressure by pancaking the hand to twist or lock out the wrist. Apply pressure at the break of the wrist and push the subject's thumb toward a stationary object (the wall or the ground) to lock the hand out.
- **Hip.** Apply pressure to the hip to knock the subject off balance or alter his center of gravity.

- **Knee.** Apply pressure to the knee by pushing outward on it to cause the subject to be placed off balance, or use the knee as a hinge for pain compliance.
- **Ankle.** Apply pressure to the ankle by clasping and turning it in a counterclockwise manner to lock the hinge out.

TYPES OF RIOT BATONS

4-16. The Army is currently fielding two types of riot batons, wooden and expandable. The most common riot baton is the 36-inch hickory riot baton with thong. There is also the 24- to 36-inch expandable riot baton, which has been added to the nonlethal capabilities set (NLCS). Regardless of which riot baton the soldiers are equipped with, all techniques for blocking and striking are identical. Similar to the riot shield, the riot baton offers soldiers both offensive and defensive capabilities.

4-17. Soldiers must be properly trained in all blocking and striking techniques. Improper use of the riot baton by an untrained soldier has the potential for creating a greater problem than what already exists.

NOMENCLATURE OF THE RIOT BATON

4-18. The riot baton is divided into four basic parts: the long end, long portion, grip portion, and grip end (see *Figure 4-5*). Although the expandable and wooden riot batons differ, their nomenclature is quite similar.

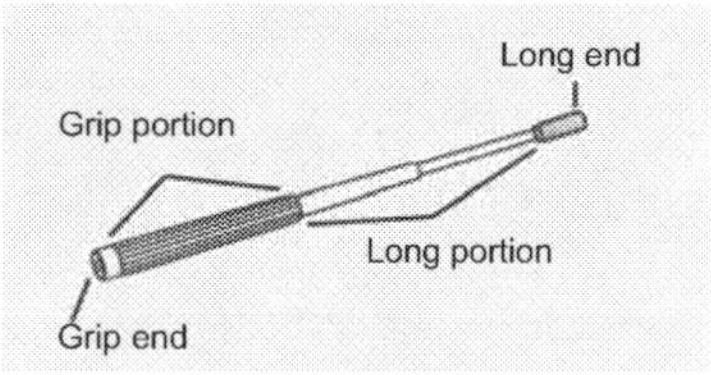

Figure 4-5. The Riot Baton

TYPES OF RIOT BATON CARRIES AND POSITIONS

4-19. As with any other weapon that the soldier will use, there are certain carries and positions from which the riot baton may be used offensively or defensively.

Two-Hand Carry

4-20. The two-hand carry is the most common carry for the riot baton. It is intended for executing all blocks and returning with strikes. When the expandable riot baton is in the collapsed position (at 24 inches), the long end of the riot baton is angled downward approximately 45°. When it is extended to 36 inches the long end is angled upward approximately 45°.

4-21. The wooden riot baton is secured with the soldier's hand by means of a leather thong. He hooks the thumb of his strong hand (palm down) in the thong, pulls the thong across the back of his hand, and grips the riot baton. This provides him with a very secure grip (see *Figure 4-6*).

4-22. The two-hand carry is intended for executing all blocks and returning with strikes. The soldier grips the riot baton with both hands. His strong hand grips the short end of the riot baton, and his weak hand grips the long portion of the riot baton, 2 to 4 inches from the end. His palm is down with his strong hand pulled in close against his strong side hip.

4-23. Wherever the long portion of the riot baton is pointed is where the soldier will strike or jab. The riot baton should be kept at a 45° angle to the ground. All blocks and two-hand jabs and strikes are executed from the two-hand carry.

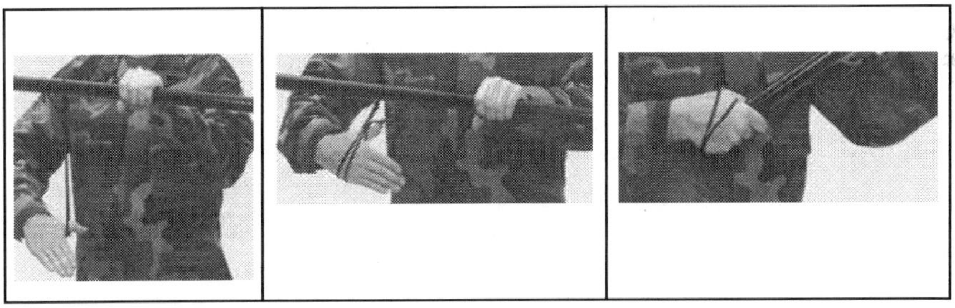

Figure 4-6. Two-Hand Carry

Outside-Arm Carry (Expandable Riot Baton Only)

4-24. The outside-arm carry (see *Figure 4-7*) is used with the expandable riot baton in the collapsed (24-inch) position only. Assuming a wide-based defensive stance (knees slightly bent and feet shoulder width apart), a soldier draws the grip portion of the riot baton with his strong hand. Without striking out or widely swinging the riot baton, he brings the long portion up against the outer part of his strong arm (the tricep area). The soldier's weak hand should be raised in a position to protect and block strikes from an aggressor. It is from this carrying position that a soldier can effectively use both the front and rear strikes. Front and rear strikes should never be performed with an expandable riot baton in the expanded position or with a 36-inch wooden riot baton.

CAUTION

The expandable and 36-inch wooden riot batons provide a greater risk of injury to the soldier's wrist.

Figure 4-7. Outside-Arm Carry

Vertical High-Profile Carry

4-25. Assuming a wide-based defensive stance, the soldier draws the riot baton with his strong hand by the grip portion. Once the riot baton has been drawn from the carrier, the soldier swings the riot baton in a downward motion, fully extending the arm and holding the riot baton down and parallel to the strong side leg. The soldier's weak hand is raised in a position to protect and block strikes from an aggressor (see *Figure 4-8*, page 4-10). Although this carry can be used with both riot baton systems, it is best suited for the expandable riot baton collapsed to 24 inches. The vertical high-profile carry clearly shows the soldier with a drawn riot baton in an effective defensive posture and ready for attack. A drawn riot baton in the hands of a trained soldier may offer just the kind of deterrence needed to stop the activity of an aggressor.

Vertical Low-Profile Carry

4-26. Assuming a wide-based defensive stance, the soldier draws the riot baton with his strong hand by the grip portion. Once the riot baton has been drawn from the carrier, the soldier swings the riot baton in a downward motion, fully extending his arm and holding the riot baton down and parallel to his strong side leg. His weak side hand is raised in a position to protect and block strikes from an aggressor (see *Figure 4-9*, page 4-10). Although this carry can be used with both types of riot batons, it is best suited for the expandable riot baton collapsed to 24 inches. This position portrays the soldier in an effective defensive position ready for an attack, but with a possible hidden weapon. The major difference between the vertical low-profile and high-profile carry is that in the low-profile carry the soldier hides the riot baton behind his strong side leg, allowing for some element of surprise to a potential aggressor. A drawn riot baton in the hands of a trained soldier may offer the deterrence needed to stop the activity of an aggressor.

Figure 4-8. Vertical High-Profile Carry

Figure 4-9. Vertical Low-Profile Carry

RIOT BATON BLOCKING TECHNIQUES

4-27. There are five basic types of riot baton blocking techniques that are executed from the two-hand carry position.

High Block

4-28. The high block is very effective in blocking a downward, vertical strike directed at the top of the head and shoulders. This type of downward, vertical strike may be from a club, pipe, or similar object. The high block is a two-step movement. To perform a high block the soldier performs the following steps:

Step 1. Raises the riot baton straight across his body until it is parallel to the ground (see *Figure 4-10*).

Step 2. Brings the riot baton straight-up in front of his body until the long portion of the riot baton is in front of his forehead (approximately 2 inches from his head). The riot baton should be in a horizontal position above his head and slightly angled down toward his weak hand, as it absorbs the shock from the strike (see *Figure 4-10*). The fingers of his weak hand should be open and behind the long portion of the riot baton when blocking the object to protect his fingers from being smashed by the strike.

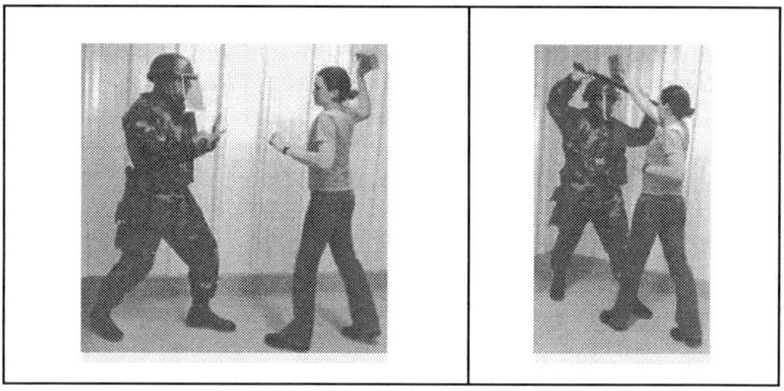

Figure 4-10. High Block (Steps 1 and 2)

Low Block

4-29. A low block is very effective in blocking an upward vertical strike directed at the groin, lower abdomen, chest, or chin. This upward vertical strike may be from an individual's foot, knee, or fist (for example, an upper cut to the chin). To perform a low block, the soldier performs the following steps:

Step 1. Raises the riot baton straight across his body until it is parallel to the ground (see *Figure 4-11*, page 4-12).

Step 2. Brings the riot baton straight down in front of his body, bending at his knees and keeping his body upright until the long portion of the riot baton is just below his knees (see *Figure 4-11*). The riot baton should be horizontal and parallel to the ground as it absorbs the shock from an upward strike. His weak hand fingers should be open and behind the long portion of the riot baton when blocking the object to protect his fingers from being smashed by the strike.

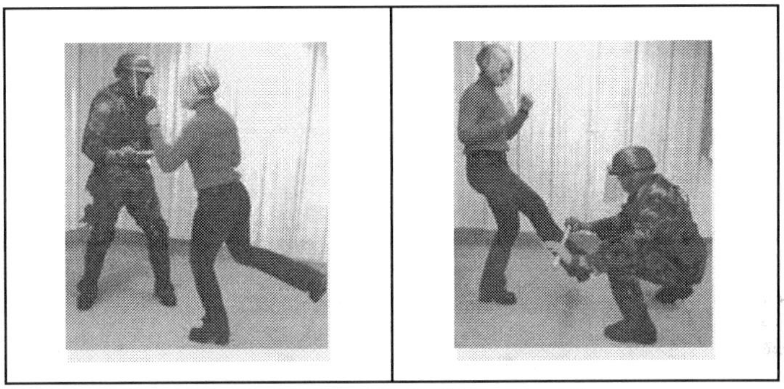

Figure 4-11. Low Block (Steps 1 and 2)

Strong Side Block

4-30. A strong side block is very effective in blocking a horizontal strike directed at the strong side of the head, neck, chest (or flank), or hip area. The horizontal strike may be from an individual's foot, fist, elbow, knee, or weapon. To perform a strong side block, the soldier performs the following steps:

Step 1. Raises the riot baton straight across his body in a vertical position where the grip portion is in his strong hand and the long portion is in his weak hand with the fingers straight up (see *Figure 4-12*).

Step 2. Turns his body laterally by rotating his hips toward his strong side and moving the riot baton across his body, keeping the riot baton vertical (see *Figure 4-12*).

4-31. The strong side block provides protection for a soldier from the front to the rear of his body. A riot baton held in the vertical position absorbs the shock of a strike coming from the strong side. The fingers of his weak hand should be open and behind the long portion of the riot baton, when blocking the object to protect his fingers from being smashed by the strike.

Weak Side Block

4-32. A weak side block is very effective in blocking a horizontal strike directed at the weak side of the head, neck, chest, or hip area. The horizontal strike may be from an individual's foot, fist, elbow, or knee. To perform a weak side block, the soldier performs the following steps:

Step 1. Moves the riot baton in a vertical position toward his weak side, ensuring that the grip portion is pointing straight down and the long portion is pointing straight up.

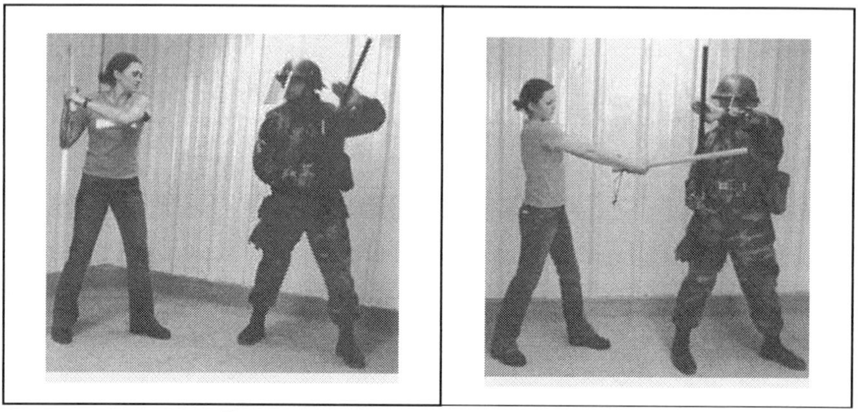

Figure 4-12. Strong Side Block (Steps 1 and 2)

Step 2. Keeps the fingers of his weak hand open and behind the long portion of the riot baton when blocking the object to protect his fingers from being smashed by the strike. At the moment of contact with the opposing strike, the blocking surface of the long portion should be at a right angle to the opposing object.

Step 3. Assess the situation following a weak side block, and take appropriate follow-up action, as needed.

Middle Block

4-33. A middle block is very effective in blocking a front horizontal strike directed at the face, throat, chest, abdomen, or hip. The horizontal strike may be from an individual trying to tackle near the waist or shove or punch at the face. To perform a middle block, the soldier performs the following steps:

Step 1. Moves the riot baton toward the middle of his body at a 45° angle, keeping the long portion pointed up and slightly forward of the grip portion (see *Figure 4-13*, page 4-14).

Step 2. Keeps the fingers of his weak hand open and behind the long portion of the riot baton when blocking the object to protect his fingers from being smashed by the strike. However, his weak hand may have to grip the long portion when pushing an individual away or forcing an individual down on the ground if the individual is trying to tackle.

RIOT BATON STRIKING TECHNIQUES

4-34. Seven strikes can be executed using the riot baton. Two of these strikes come from the outside-arm or vertical carry. The other five strikes come from the two-hand carry.

Figure 4-13. Middle Block

One-Hand Forward Strike

4-35. The one-hand forward strike is a very quick and effective offensive strike. It is usually employed as a countermeasure to an attack from the front that has been blocked effectively.

4-36. Target areas can be as high as the outside upper arm, down to the outer thigh region. An effective one-hand forward strike must be aggressively quick, with the purpose of distracting, disabling, or altering behavior. Therefore, those employing this technique must be sensitive to what part of the riot baton they use to strike the individual. Strike the target with the last 2 to 4 inches of the long end for optimal results.

4-37. A one-hand forward strike may be employed in one of the following three ways. When selecting available targets, use the principles in *Figure 4-4, page 4-5*.

- A horizontal manner.
- A downward diagonal manner to destabilize an individual.
- A vertical manner in front of the body to clear an aggressor's hands.

NOTE: When the one-hand forward strike is used in a vertical manner, keep the long portion straight up, not angled forward. By keeping the long portion up rather than angled forward, the chance of striking the individual in the face or head is significantly reduced. For the counterstrike to be effective, follow through with the technique when the baton makes contact with the aggressor and/or his object to destabilize the him. A pattern of movement such as a

forward shuffle, forward pivot, strong side step, or rear pivot may enhance this technique.

> **WARNING**
> **Do not execute the one-hand forward strike with the wooden or expandable riot baton in the extended position. It has been shown to cause injury to the wrist of the user.**

4-38. When using the vertical or outside-arm carry position (see *Figures 4-7,* page 4-9 and *4-8,* page 4-10), the soldier quickly moves the riot baton across his body using the strength and power of his hips, moving from his strong side to his weak side (see *Figure 4-14,* page 4-16). Ending this strike leaves the riot baton positioned under the weak side armpit in preparation for the one-hand reverse strike (see *Figure 4-14*). When delivering this strong side strike technique (in a horizontal or downward diagonal manner), ensure that the palm is facing up. In a vertical delivery, the palm will be toward the chest rather than up. Pause following the one-hand forward strike and assess the situation. Take appropriate follow-up action, as needed. If the situation does not call for additional strikes, avoid striking again.

One-Hand Reverse Strike

4-39. The starting position for the one-hand reverse strike is under the weak side armpit (see *Figure 4-14*). The one-hand reverse strike is used as a follow-up strike and is quick and effective.

4-40. Target areas can be as high as the outside upper arm, down to the outer thigh region. An effective one-hand reverse strike must be aggressively quick, with the purpose of distracting, disabling, or altering behavior. Therefore, those employing this technique must be sensitive to what part of the riot baton they use to strike the individual. Strike the target with the last 2 to 4 inches of the riot baton for optimal results.

4-41. A soldier may employee a reverse strike in one of the following three ways:

- A horizontal manner.
- A downward diagonal manner to destabilize an individual.
- A vertical manner.

4-42. In front of the body to clear an individual's hands. When the one-hand reverse strike is used in a vertical manner, keep the long portion straight up, not angled forward. By keeping the long portion up rather than angled forward, the chance of striking the individual in the face or head is significantly reduced. When selecting available targets, use the principles in *Figure 4-4,* page 4-5.

4-43. The one-hand reverse strike is considered a follow-up strike to the one-hand forward strike. As a follow-up strike, it is important that soldiers are taught the discipline of assessing the target. An aggressor may not require another strike, based on his condition and actions after enduring the one-hand forward strike.

Forward Strike	Target Assessment	Reverse Strike

Figure 4-14. One-Hand Forward and Reverse Strikes

4-44. When using the one-hand reverse strike, the solder quickly moves the straight baton across his body, using the strength and power of his hips (from weak side to strong side). When the strike is completed, the soldier's arm will be back in the outside-arm carry position (see *Figure 4-7,* page 4-9). The soldier ensures that his palm is facing down when delivering the counterstrike in a horizontal or diagonal manner. This will prevent injury to the user's wrist. In a vertical delivery, the palm will be toward the chest rather than up. Following the one-hand reverse strike, assess the situation and take appropriate follow-up action, as needed. If the situation does not call for additional strikes, the soldier should avoid striking again.

4-45. When the one-hand reverse strike is used in a vertical manner, keep the long portion straight up, not angled forward. By keeping the long portion up rather than angled forward, the chance of striking an individual in the face or head is significantly reduced. For the counterstrike to be effective, follow through with the technique when the baton makes contact with the aggressor and/or his object to destabilize him. A pattern of movement such as the forward shuffle, forward pivot, strong side step, or rear pivot may enhance this technique.

WARNING

This strike will not be executed with a 36-inch wooden or expandable riot baton in the extended position. It has been shown to cause injury to the wrist of the user.

Two-Hand Strong Side Horizontal Strike

4-46. Starting from the two-hand carry position, the two-hand strong side horizontal strike can be aggressively quick. The purpose of this strike is to create distance, distract, disable, or alter behavior. Strong side horizontal strikes are quick and effective offensive strikes, usually employed as a countermeasure to close frontal attacks that have been effectively blocked. When selecting available targets, use the principles in *Figure 4-4,* page 4-5.

4-47. The two-hand carry position enables soldiers to use riot batons at the 24- or 36-inch length. To effectively employ this strike, use the strength and power of the hips to thrust the riot baton in a horizontal manner. Simultaneously, pull back with the weak hand as the strong hand drives the grip end toward the target, striking a rib or the abdominal region of the aggressor (see *Figure 4-15*, page 4-18). Following the two-hand strong side horizontal strike, assess the target before following up with another strike. An aggressor may not require another strike, based on his condition and actions after enduring the one-hand forward strike.

NOTE: A pattern of movement such as a forward shuffle or forward pivot may enhance this technique.

Two-Hand Weak Side Horizontal Strike

4-48. An effective two-hand weak side horizontal strike must be aggressively quick to create distance, distract, disable, or alter behavior. The weak side horizontal strike is a quick and effective offensive strike usually employed as a follow-up strike to the strong side horizontal strike. It is a countermeasure designed to close frontal attacks that have been effectively blocked. When selecting available targets, use the principles in *Figure 4-4*. Maintaining a strong defensive stance after employing a strong side horizontal strike puts the soldier in position to execute a two-hand weak side horizontal strike.

4-49. Use of the two-hand carry position enables soldiers to use riot batons at 24 or 36 inches. To effectively employ this strike, use the strength and power of the hips to thrust the riot baton in a horizontal manner. Simultaneously, pull back with the strong hand as the weak hand drives the grip forward toward the aggressor, striking a rib or the abdominal region of the target (see *Figure 4-15*). Following the two-hand weak side horizontal strike, it is important to assess the aggressor. An aggressor may not require another strike, based on his condition and actions after enduring the one-hand forward strike.

NOTE: A pattern of movement such as a forward shuffle or forward pivot may enhance this technique.

Two-Hand Front Jab

4-50. The two-hand front jab starts from the two-hand carry position (see *Figure 4-16*, page 4-18). It is aggressively quick and creates distance, distracts, disables, and/or alters behavior. Front jabs are effective offensive strikes, usually employed as a countermeasure to charging and overpowering frontal attacks. When selecting available targets, use the principles in *Figure 4-4*.

Figure 4-15. Two-Hand, Strong Side, and Weak Side Horizontal Strikes

4-51. Use of the two-hand carry position enables soldiers to use riot batons at 24 or 36 inches. Use the strength and power of the hips to effectively employ this strike in a slightly downward or horizontal manner. Simultaneously, use both arms to shoot the long end of the riot baton straight out from the body to the target area. Immediately, pull back the riot baton to the two-hand carry position (see *Figure 4-16*).

Figure 4-16. Front Jab

4-52. Forward jabs drive the long end of the riot baton toward the aggressor, striking a rib, a hip flexor, or the abdominal region (see *Figure 4-4*, page 4-5). Following the two-hand front jab, it is important that soldiers assess the aggressor before following through with another strike, as the aggressor may not require another strike.

NOTE: A pattern of movement such as a forward shuffle will enhance the power of this technique.

Two-Hand Rear Jab

4-53. Starting from the two-hand carry position, the two-hand rear jab is aggressively quick and creates distance, distracts, disables, or alters behavior. Rear jabs are quick and effective offensive strikes usually employed as a countermeasure to charging and overpowering attacks from the rear. When selecting available targets, use the principles in *Figure 4-4,* page 4-5.

4-54. Use of the two-hand carry position enables soldiers to employ riot batons at 24 or 36 inches. To effectively employ this strike, use the strength and power of the hips to thrust the riot baton in a slightly downward or horizontal manner. Simultaneously, turn the head to the rear to see and assess the target. Use both arms to shoot the riot baton straight back from the body to the targeted area and immediately pull back the riot baton to the two-hand carry position (see *Figure 4-17*).

NOTE: A pattern of movement such as a rear shuffle will enhance the power of this technique.

Figure 4-17. Rear Jab

Two-Hand Middle Strike

4-55. The two-hand middle strike starts from the two-hand carry position. It is aggressively quick and creates distance, distracts, disables, or alters behavior. It is proven to be an effective follow-up technique to a two-hand weak or strong side block or to destabilize an individual by pushing him back and away. The two-hand middle strike is used as a two-count movement. When selecting available targets, use the principles in *Figure 4-4*.

4-56. To perform a two-hand middle strike (see *Figure 4-18,* page 4-20) the soldier does the following:

Step 1. Brings the riot baton up to chest level and slightly off parallel to the ground.

Step 2. Steps forward aggressively, simultaneously thrusting the long portion of the riot baton forward in a horizontal manner. Both arms are fully extended at the end of the movement with a quick snappy return to the step 1 position. Use the entire body to generate the power.

NOTE: It is important that soldiers assess the aggressor before applying a follow-up strike. If the soldier steps forward as he employs the middle strike, it will afford greater impact and power.

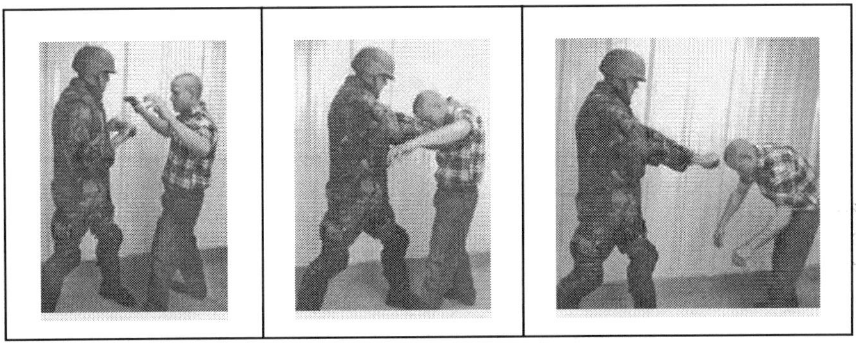

Figure 4-18. Middle Strike

Riot Baton Retention

4-57. A common technique an aggressor may use is grabbing for a soldier's riot baton or, in some cases, even taking the riot baton and using it against the soldier. This causes a disruption or distraction.

4-58. Maintaining physical control of the riot baton is essential for the safety of the soldier and the control force as a whole. Aggressors will generally attempt to grab the riot baton where they can get the strongest hold. Soldiers must resist the instinct of getting into a tug-of-war match over the riot baton, which often proves to be ineffective.

4-59. Using the momentum of the aggressor to pull the riot baton away, the soldier simply steps into or closer to the aggressor with the weak side foot. With the momentum created, the aggressor's hold on the riot baton is lessened. The soldier then rapidly uses the tracing-C technique to recover the riot baton.

4-60. A soldier using the tracing-C technique does the following:

Step 1. Pushes up on the riot baton.

Step 2. Pulls the riot baton around in an upward swing as if tracing a C in the air from bottom to top with the long end.

NOTE: This movement seems to wrap the aggressor's arms around themselves (see *Figure 4-19*).

Step 3. Drives (immediately with a quick and forceful movement) the long end of the riot baton straight down, as if slicing the C in half (see *Figure 4-19*).

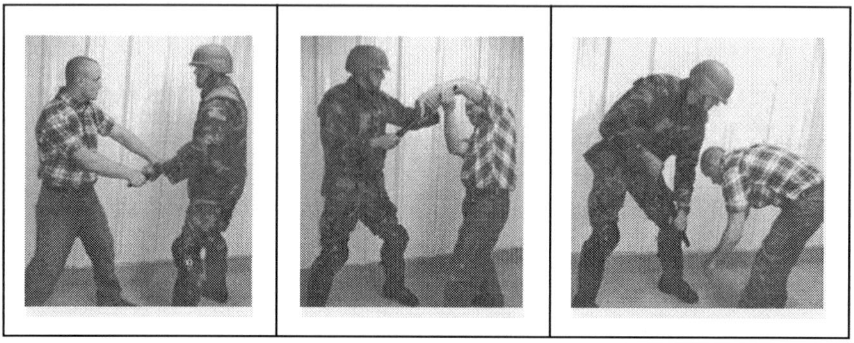

Figure 4-19. Riot Baton Retention

4-61. Once the soldier reaches the top of the C and the aggressor is tied up, it is impossible for the aggressor to maintain hold of the riot baton. Further actions such as a strike may or may not be necessary. Soldiers must be able to assess the situation and react accordingly.

WORKING AS A TEAM

4-62. Soldiers never have to rely solely on their own skills or abilities in riot control operations. Although individual skills were discussed in this chapter, it is important to remember that well-trained soldiers employing offensive, defensive, riot shield, and riot baton techniques will perform the collective task of crowd control well.

4-63. Riot control operations are dynamic and often chaotic. Soldiers armed with riot shields provide a first line of defense while those armed with riot batons and firearms provide overwatch as the formation moves toward engaging a hostile crowd. To work effectively and in concert, soldiers must be properly trained, drilled, and exercised. If an aggressor approaching a line of skirmishers violently grabs a riot shield, help is required to maintain control (see *Figure 4-20*, page 4-22). A soldier armed with a 36-inch wooden riot baton or an expandable riot baton (extended to 36 inches) can be of assistance. The soldier simply moves forward into a gap between the riot shields where he can quickly strike the top or side of the riot shield, slicing the long end across or straight down onto the hands and fingers of the aggressor (see *Figure 4-18*).

4-64. Riot batons are not the only way to effectively knock an aggressor's hands from a riot shield. Soldiers armed with 12-gauge shotguns or M16/M203s can just as effectively smash the hands or fingers of a persistent aggressor with the barrel of their weapons using the technique described above.

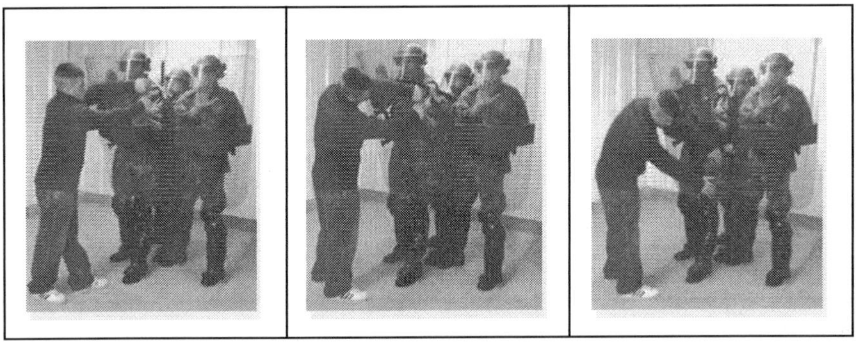

Figure 4-20. Clearing a Riot Shield

NONLETHAL MUNITIONS

4-65. With the inclusion of NL munitions, commanders now have additional tools that afford excellent standoff capabilities. This is critical because commanders now have intermediate options for dealing with a hostile crowd. No longer will a commander issue a proclamation, have it ignored by the crowd, and then be forced to use lethal munitions to backup the proclamation. Now he can use NL munitions to gain the same standoff distance but without the catastrophic results of lethal munitions.

4-66. NL munitions include (but are not limited to) a 12-gauge shotgun and 40-millimeter rounds. These blunt munitions have the capability of effectively engaging targets as close as 5 meters and as far as 50 meters. Range is dependent upon each round. These munitions are used from a distance to dissuade, discourage, and disperse the crowd. If the crowd disperses from 50 meters and soldiers are in close contact with the crowd, then the mission has been successful.

TAP-DOWN TECHNIQUE

4-67. Commanders must be familiar with the characteristics and capabilities of all rounds to use them effectively. When 12-gauge and M203 weapons are used within a line of skirmishers (usually located behind the riot shields), it is imperative that a method is developed, trained, and practiced to effectively operate as a team. Operators of these weapons may encounter problems in riot control situations when trying to identify and effectively engage legitimate targets. The tap-down technique is used to overcome target identification and engagement with NL munitions. If ordered to engage targets or the situation authorizes an engagement, soldiers armed with NL munitions may step forward, level their weapons, and fire upon legitimate targets (see *Figure 4-21*).

4-68. A soldier that is armed with a NLW and is using the tap-down technique must first tap on the shield holder's shoulder. Tapping on the shield holder's shoulder alerts him that the gunner is preparing to fire. In response to the tap, the shield holder drops to one knee while keeping the riot shield firmly affixed to their front for protection. The gunner takes a step forward and fills the gap. He then leans into the riot shield bearer's back with his knee and fires. This technique ensures that the long end of the weapon is extended beyond the riot shield.

> **WARNING**
>
> **The long end of the weapon must be projected beyond the riot shield before firing. This is a critical safety step, which will prevent NL projectiles from hitting the riot shield and injuring soldiers.**

4-69. Once soldiers firing munitions have fired, they raise their weapon back to a high port arms. The soldiers clear their weapons and tap the riot shield bearer a second time, signaling them to rise to a standing position.

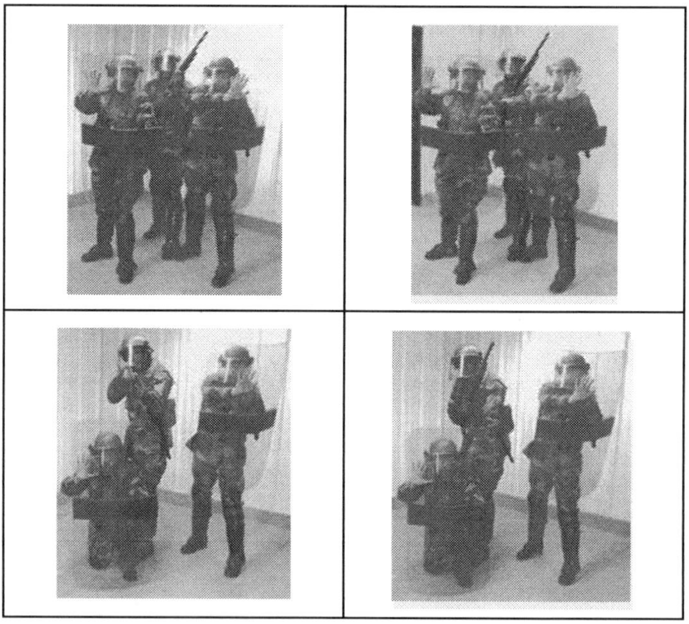

Figure 4-21. Tap-Down Technique

Chapter 5

Nonlethal Capabilities Set and Employment Considerations

During the past several decades, US forces have regularly performed peacekeeping and peace enforcement missions at a greater tempo than in the past. In the peacekeeping and peace enforcement environment, the use of conventional firearms or the threat of their use may not be the solution to a situation where US forces must separate two belligerent, hostile ethnic groups or prevent a similar group from entering an area that is off-limits to them. However, a show of force using NLW and NL munitions will cause crowds to disperse, separate, or leave the area with minimal casualities. The NLCS is a well-rounded, versatile package of both commercial off-the-shelf (COTS) and government off-the-shelf (GOTS) NL equipment and munitions. NLCS contents are divided into four distinct categories: personnel protectors, personnel effectors, mission enhancers, and training devices.

PERSONNEL PROTECTORS

5-1. Personnel protectors provide the individual soldier with added protection to the sensitive and vital areas of his body. They provide excellent protection for the individual soldier from trauma often inflicted by thrown objects (such as rocks, bricks, sticks, and bottles).

NONBALLISTIC RIOT FACE SHIELD

5-2. A nonballistic riot face shield (see *Figure 5-1,* page 5-2) has two adjustable positions and is constructed of hard durable plastic that securely attaches to the Kevlar® helmet. It may be worn with an M17 or M40 protective mask and is designed to protect the soldier's head, face, and neck from thrown objects. Nonballistic riot face shields will not protect from attacks with ballistic weapons, such as firearms. All soldiers operating in a civil disturbance environment should be outfitted with riot face shields.

NONBALLISTIC FULL-LENGTH SHIELD

5-3. A nonballistic full-length riot shield (see *Figure 5-2,* page 5-2) is a body shield constructed of hard durable plastic. Left- and right-handed personnel can easily use the nonballistic full-length riot shield. It is designed to protect the soldier's face, torso, and upper legs from thrown objects. Nonballistic full-length riot shields will not provide protection from ballistic weapons. Only the front line of skirmishers and those providing lateral protection to both flanks of the formation will be outfitted with nonballistic full-length riot shields.

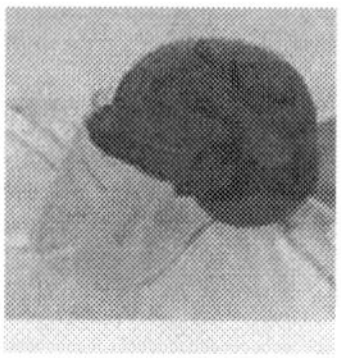

Figure 5-1. Nonballistic Riot Face Shield

Figure 5-2. Nonballistic Full-Length Riot Shield

NONBALLISTIC SHIN GUARD

5-4. The nonballistic shin guards (see *Figure 5-3*) are constructed of hard durable plastic and are easily fastened to a soldier's legs. They are designed to protect the soldiers, feet, shins, and knees from thrown objects. They will not provide protection from ballistic weapons, such as firearms. All soldiers operating in a civil disturbance environment should be outfitted with shin guards.

BALLISTIC RIOT FACE SHIELD

5-5. The ballistic riot face shield (see *Figure 5-4*) has two adjustable positions constructed of acrylic and bullet-resistant materials that provide the wearer with Level IIIA protection. The ballistic riot face shield easily attaches to the Kevlar helmet and is capable of being worn with either an M17 or M40 protective mask. It will protect the soldier from hand-thrown objects. Because of its excessive weight (3.4 pounds), the ballistic riot face shield should not be worn for standard riot control operations. Ballistic riot face shields are best suited for use within specialized teams associated with riot control operations; for example, teams responsible for the search and apprehension of identified criminals or building- and alley-clearing operations.

Figure 5-3. Nonballistic Shin Guards

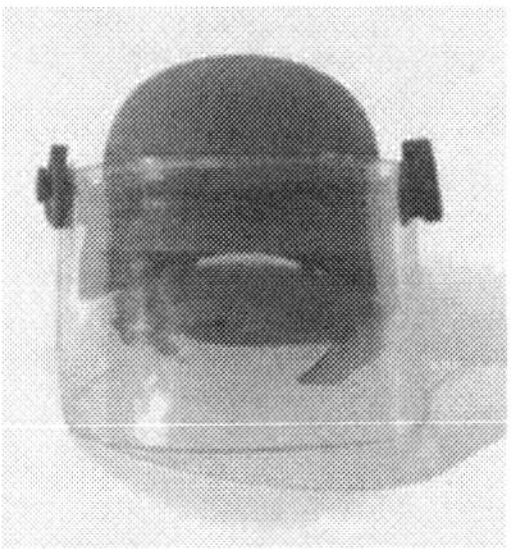

Figure 5-4. Ballistic Riot Face Shield

BALLISTIC FULL-LENGTH RIOT BODY SHIELD WITH LIGHT KIT

5-6. The ballistic full-length riot body shield (see *Figure 5-5,* page 5-4) is constructed from Spectra Shield® bullet-resistant material that provides a soldier with Level IIIA protection. It will protect the soldier from thrown objects. Because of its excessive size (18 pounds and 20 by 36 inches), the ballistic full-length riot body shield should not be used as part of a standard riot control operation. Ballistic full length riot body shields are best suited for use within specialized teams associated with riot control operations; for example, teams responsible for the search and apprehension of identified criminals or building- and alley-clearing operations.

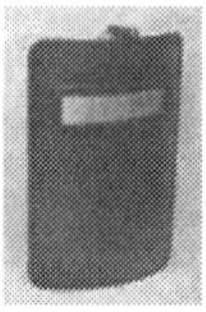

Figure 5-5. Ballistic Full-Length Riot Body Shield

BALLISTIC SHIN GUARDS

5-7. Ballistic shin guards (see *Figure 5-6*) are constructed of Kevlar KM2 materials and provide a soldier with Level IIIA protection to the feet, shins, and upper legs. They protect the soldier from hand-thrown objects. There are three different sizes, and because of their excessive weight (7 to 10 pounds), ballistic shin guards should not be deployed as part of a standard riot control operation. Ballistic shin guards are best suited for use within specialized teams associated with riot control operations; for example, teams responsible for the search and apprehension of identified criminals or building- and alley-clearing operations.

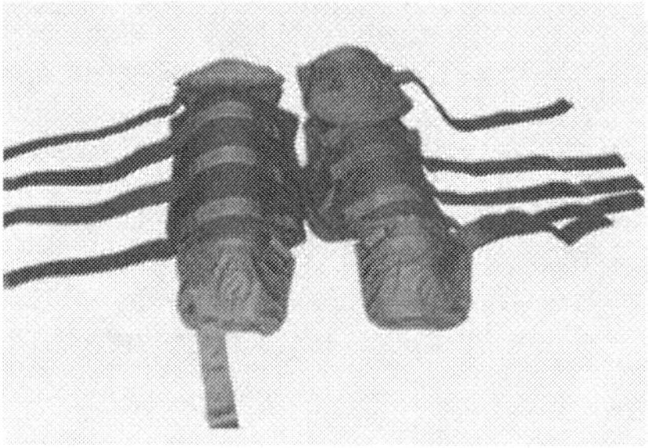

Figure 5-6. Ballistic Shin Guards

PERSONNEL EFFECTORS

5-8. Personnel effectors are those items that provide soldiers with an NL response without the necessity of always coming into direct physical contact with aggressors. Personnel effectors are 12-gauge and 40-millimeter ammunition that inflict blunt trauma. These munitions are compatible with the shotgun and the M203 grenade launcher already in the Army inventory. There are also diversionary and rubber ball grenades, which distract and inflict blunt trauma.

WOOD RIOT BATON WITH BELT RING (36 INCHES)

5-9. The 36-inch wood riot baton (see *Figure 5-7*) is made from solid oak. When held properly, the riot baton thong provides greater retention capabilities. Intended primarily for self-defense, the 36-inch wood riot baton can be used effectively as an offensive tool if it becomes necessary to keep rioters out of arm's reach of soldiers conducting crowd control operations.

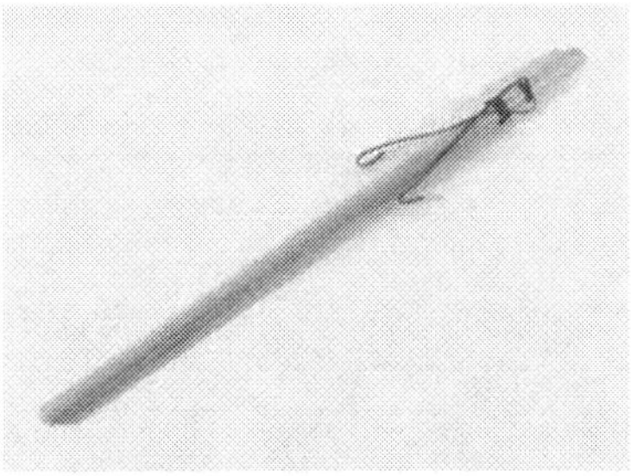

Figure 5-7. Wood Riot Baton

EXPANDABLE RIOT BATON WITH CARRIER (24 TO 36 INCHES)

5-10. The expandable riot baton (see *Figure 5-8*) provides the soldier with an NL means of crowd control and self-defense. The riot baton comes with a mounting device that attaches the riot baton to a belt. The riot baton is 24 to 36 inches in length and is primarily intended for self-defense. The expandable riot baton can be used effectively as an offensive tool if it becomes necessary to keep rioters out of arm's reach of soldiers conducting crowd control operations.

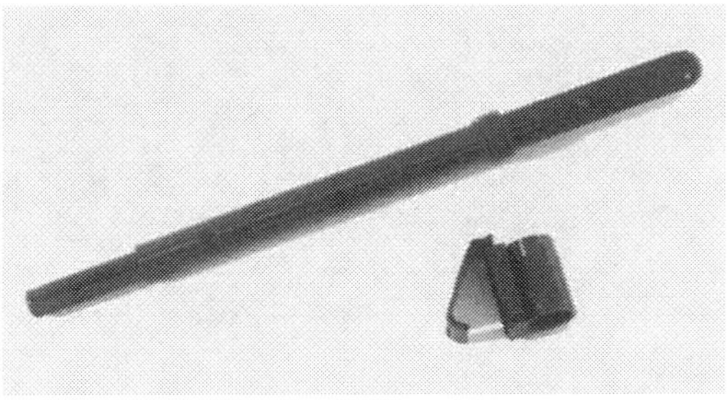

Figure 5-8. Expandable Riot Baton and Carrier

500 (3 INCH) 12-GAUGE SHOTGUN

5-11. The 12-gauge shotgun (see *Figure 5-9*) is a pump action shotgun currently in NLCS inventory. The pump action shotgun is chambered to take up to 3-inch shells. The 3-inch chamber allows for the use of M1012 and M1013 NL munitions. This shotgun also provides a visually distinct alternative to standard military weapons that may be desired based on mission considerations (force signature and acceptability).

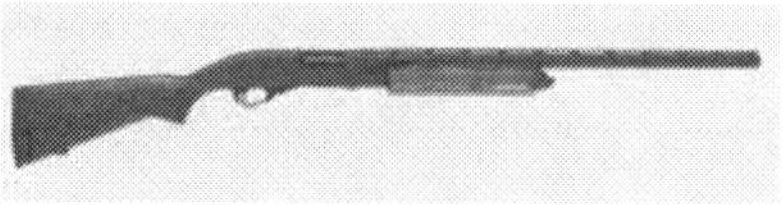

Figure 5-9. 12-Gauge Shotgun

M7 66-MILLIMETER LAUNCHER WITH M315 INSTALLATION KIT

5-12. The M7 (see *Figure 5-10*) is a 66-millimeter vehicle-mounted, NL, grenade-launching device that is mounted on a HMMWV. It is an indirect fire support system that can deliver the M98 distraction grenade that creates a flash-bang effect, L96A1 antiriot grenade, or M99 blunt trauma grenade that creates a sting-ball effect. The M315 installation kit is used to install an M7 discharger on the turret ring of appropriate HMMWV variants. An adjustable bracket allows the launch angle to be depressed for engaging targets at ranges of 50, 75, and 100 meters. The system enforces standoff distances and deters potential threats.

Figure 5-10. M7 Launcher

MODIFIED CROWD CONTROL MUNITION-GROUND EMPLACEMENT

5-13. Similar in nature to the claymore mine, the modified crowd control munition-ground emplacement (MCCM-GE) (see *Figure 5-11*) is an NL munition that can deliver six hundred .32-caliber rubber pellets and a flash-bang effect. The MCCM-GE will stop, confuse, disorient, and/or temporarily incapacitate area targets and/or personnel at ranges out to 30 meters.

> **WARNING**
> **Avoid employing the MCCM-GE within 10 meters of personnel as it may cause serious bodily harm or death.**

Figure 5-11. MCCM-GE

12-GAUGE, FIN-STABILIZED POINT, NONLETHAL CARTRIDGE (M1012)

5-14. The M1012 (see *Figure 5-12*) is a single projectile round made of hard rubber that is shaped like a bomblet and designed to be fired a at a single target. With a muzzle velocity of 500 feet per second, the M1012 has the effective range of no closer than 5 meters and no further than 30 meters. Engagement inside 5 meters could cause serious injury or death. Beyond 30 meters the kinetic energy dissipates to the point where the round becomes ineffective.

Figure 5-12. M1012

12-GAUGE, CROWD DISPERSAL, NONLETHAL CARTRIDGE (M1013)

5-15. The M1013 (see *Figure 5-13*) is a multiple projectile round with .23 caliber hard rubber pellets that is designed to be fired at and employed with the purpose of affecting multiple targets. With a muzzle velocity of 900 feet per second, the M1013 has an effective range of no closer than 5 meters and no further than 30 meters. Engagement inside 5 meters could cause serious injury or death. Beyond 30 meters, the kinetic energy dissipates to the point where the rubber pellets become ineffective.

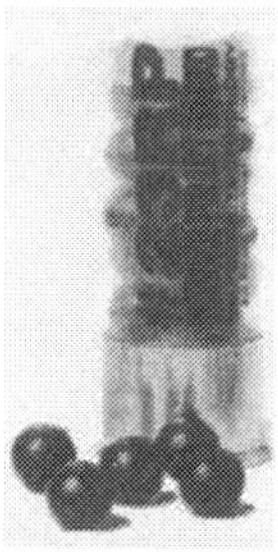

Figure 5-13. M1013

40-MILLIMETER SPONGE GRENADE (M1006)

5-16. The M1006 (see *Figure 5-14*) is a single projectile round made of pliable foam rubber with a hard plastic back. The M1006 is designed to be aimed and fired at a single target. With a muzzle velocity of 265 feet per second, the M1006 has an effective range of no closer than 10 meters and no further than 50 meters. Engagements inside 10 meters could cause serious injury or death. Beyond 50 meters, the kinetic energy dissipates to the point where the round becomes ineffective.

Figure 5-14. M1006

40-MILLIMETER CROWD DISPERSAL CARTRIDGE

5-17. The crowd dispersal cartridge (CDC) (see *Figure 5-15*) is a multiple projectile round with .48 caliber hard rubber balls that is designed to be fired and employed with the purpose of affecting multiple targets. With a muzzle velocity of 450 feet per second, the CDC has an effective range of no closer than 10 meters and no further than 30 meters. Engagement inside 10 meters could cause serious injury or death and beyond 30 meters the kinetic energy dissipates until the rubber balls become ineffective.

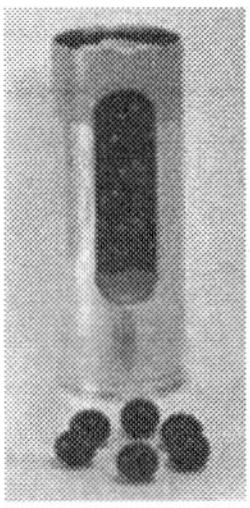

Figure 5-15. CDC

RUBBER BALL GRENADE (GG04)

5-18. The GG04 (see *Figure 5-16,* page 5-10) is a multiple projectile, flash-bang grenade with 100 .25-caliber hard rubber pellets. Each grenade has a fuse delay of 2.8 to 3 seconds with a flash measuring approximately 1 million CP and 180 decibels at 3.5 feet. At detonation, rubber pellets are dispatched at 360° with an effective range of 2 to 3 meters and a maximum engagement range of 15 to 20 meters. The rubber ball grenade is designed to be hand-thrown or muzzle-launched from a 12-gauge shotgun.

RUBBER BALL GRENADE LAUNCHING CUP AND AA30 12-GAUGE LAUNCH ROUND

5-19. The launch cup (see *Figure 5-17,* page 5-10) firmly attaches to the 12-gauge shotgun without any additional tools. It works in concert with the AA30 12-gauge launch round to propel the rubber ball grenade approximately 100 meters. The shooter simply inserts the full body and the safety lever of the grenade into the launching cup, pulls the pin, loads a 12-gauge round, and fires (approximately) at a 30° angle. The desired effect is to have the grenade explode approximately 7 to 10 feet above the target.

RELOADABLE FLASH-BANG GRENADE (M84)

5-20. The M84 (see *Figure 5-18,* page 5-10) is a reloadable flash-bang thrown grenade. Each grenade has a 1- to 2-second fuse delay, with a flash capability of 1.5 to 2.5 million CP and a bang of 168 to 175 decibels. Although it is part of the NLCS, use of the M84 is not recommended for use in crowd control situations. Specialty teams, such as security response teams, and search teams may use this device effectively.

Figure 5-16. GG04

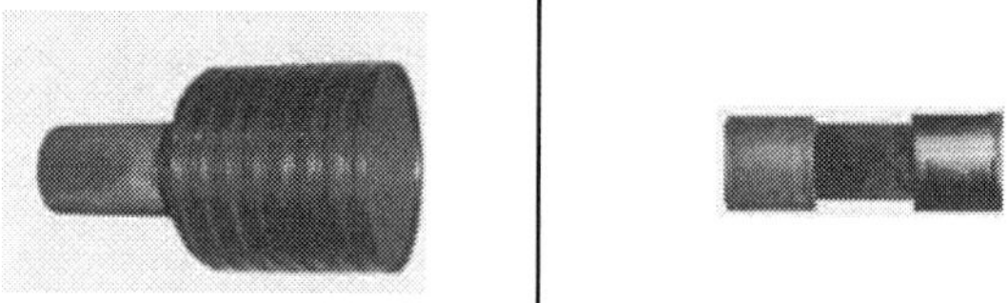

Figure 5-17. Launching Cup and AA30

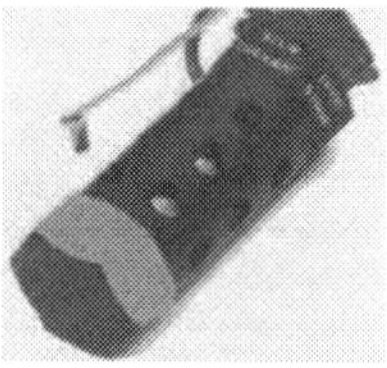

Figure 5-18. M-84

FLEX-CUFS

5-21. Flex-Cufs consist of a tough pliable plastic band with a self-locking mechanism in the center for each end. When the Flex-Cuf (see *Figure 5-19*) is threaded the restraint bands circle around the wrists or ankles, impeding movement and securing the individual. They are lightweight and disposable and require a cutting tool or other instrument to remove.

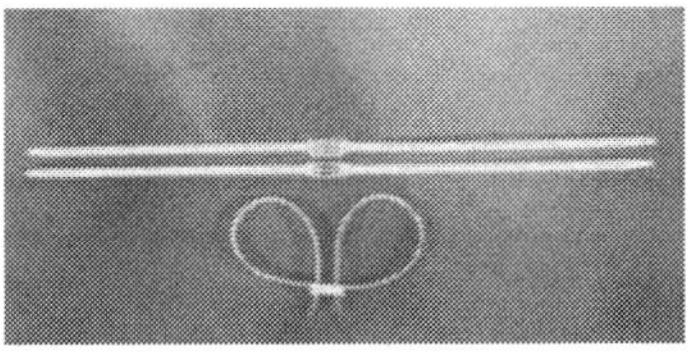

Figure 5-19. Flex-Cufs

INDIVIDUAL RIOT CONTROL AGENT DISPERSER (M36)

5-22. The M36 (see *Figure 5-20*) contains a Dibenz (F, -1 4-oxazepine [CR] solution. It can deliver 25 one-second bursts out to 12 feet. Individual disposable RCA dispersers are intended primarily for self-defense or to keep rioters out of arm's reach of soldiers conducting crowd control formations or soldiers engaged in missions where a noncombatant exists.

Figure 5-20. M36

MIDSIZE RIOT CONTROL AGENT DISPERSER (M37)

5-23. The M37 (see *Figure 5-21*, page 5-12) is the size of a standard fire extinguisher that uses compressed air to force the RCA out to a range of 30 feet. It has the capacity to employ 18 bursts for 3 seconds per charge. It is excellent for providing a wide coverage of RCAs onto a hostile crowd while maintaining excellent standoff capability. The M37 can be refilled and is rechargeable. It can be filled with CR solution (liquid agent) or CS1 (dry agent). For training purposes, CR can be substituted with water and CS1 can be substituted with talcum powder.

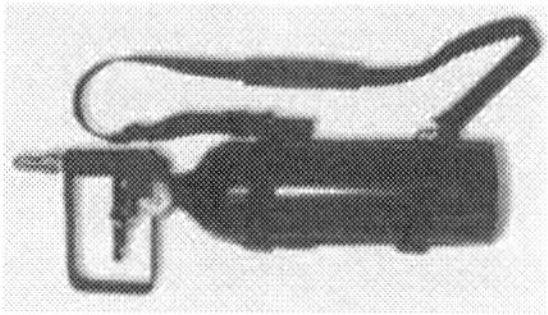

Figure 5-21. M37

SQUAD RIOT CONTROL AGENT DISPERSER (M33A1)

5-24. The M33A1 (see *Figure 5-22*) is designed to provide crowd control and protection at the squad level. It is capable of projecting a ballistic stream of RCAs beyond 25 feet in up to 25 half-second bursts. It consists of a frame and harness assembly, compressed-gas cylinder (agent container assembly), air pressure assembly, gun and hose assembly, multijet spray unit, and check valve assembly. The M33A1 can be refilled and is rechargeable. For training purposes, CR can be substituted with water and CS1 can be substituted with talcum powder.

Figure 5-22. M33A1

MISSION ENHANCERS

5-25. Mission enhancers aid the commander in various tactical situations. Mission enhancers include various types of devices used for illumination, vehicle denial, traffic control missions, and force protection enhancement.

PORTABLE BULLHORN

5-26. The portable bullhorn (see *Figure 5-23*) provides squad leaders with the capability to project their voices above the noise and commotion created by crowds and mobs. The megaphone has a built-in siren for alarm purposes that allows the operator to use the device as an ordinary megaphone and siren alarm during emergencies. The bullhorn is portable and easy to carry. The acoustic range of the megaphone for a clear voice output is ¾ to 1 mile. The power output of the megaphone is rated at 15 watts with a maximum rating of 20 watts. The dimensions of the megaphone are 14.5 inches with a 9-inch diameter bell. It weighs 3.5 pounds without batteries. Battery requirements for the megaphone are eight dry-cell size C batteries.

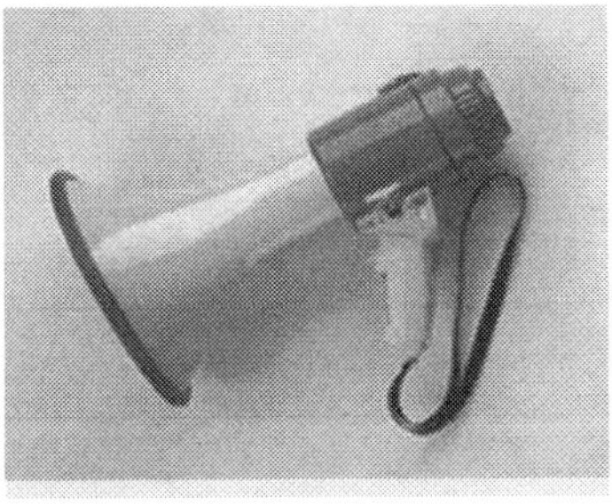

Figure 5-23. Portable Bullhorn

GROUND-MOUNTED BULLHORNS

5-27. Ground-mounted bullhorns (see *Figure 5-24,* page 5-14) are a critical communication enhancement device for conducting crowd control tactics. The bullhorn can facilitate communication with the crowd in conjunction with PSYOP support and assistance with the communication of commands to troops engaged in the crowd control process. The power output of the ground-mounted bullhorn is rated at 22 watts. The dimensions of the bullhorn are 13.75 by 9 by 14.75 inches, and it weighs 15 pounds.

INDIVIDUAL VOICE AMPLIFICATION SYSTEM (M7)

5-28. The M7 (see *Figure 5-25,* page 5-14) is a critical communication enhancement device for conducting crowd control tactics using RCAs while wearing the M40 protective mask. The M7 is fitted to the M40 protective mask to facilitate oral communication and increase the user's ability to communicate using radios and other devices.

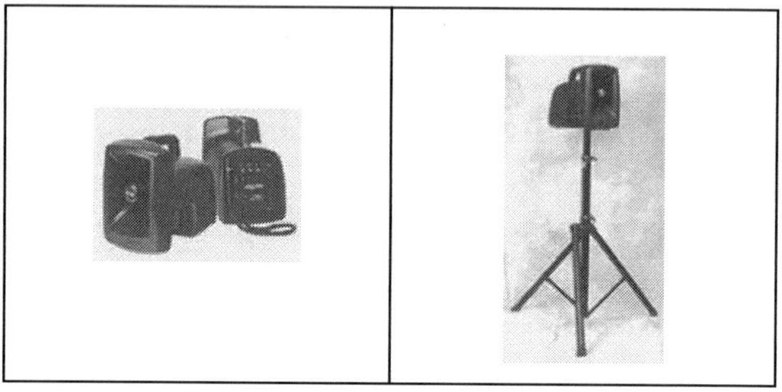

Figure 5-24. Ground-Mounted Bullhorns

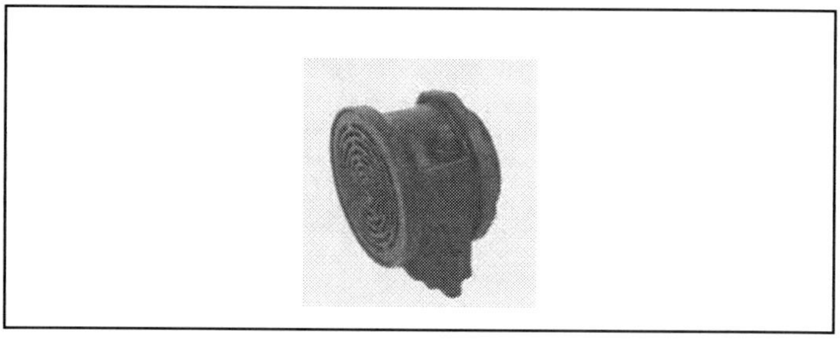

Figure 5-25. Individual Voice Amplification System

INDIVIDUAL HIGH-INTENSITY SEARCHLIGHT WITH HOLSTER

5-29. The individual high-intensity searchlight (see *Figure 5-26*) is used primarily for illumination in crowd control operations at night. An individual can use it for general illumination of the operational area to pinpoint agitators and threats, reduce the ability of the rioters to see troop formations and actions, enhance tactical deception and actions, and enhance tactical deception techniques for units conducting crowd control operations.

HIGH-INTENSITY LIGHT

5-30. A high-intensity light (see *Figure 5-27*) is intended for use in low light or night conditions. A high-intensity light can project a beam that will enable the user to identify an individual person up to 1,900 yards away. It has an adjustable beam spread of 1° to 15° and will run continuously at maximum power for up to 45 minutes. The internal power supply requires 4 hours for a full charge.

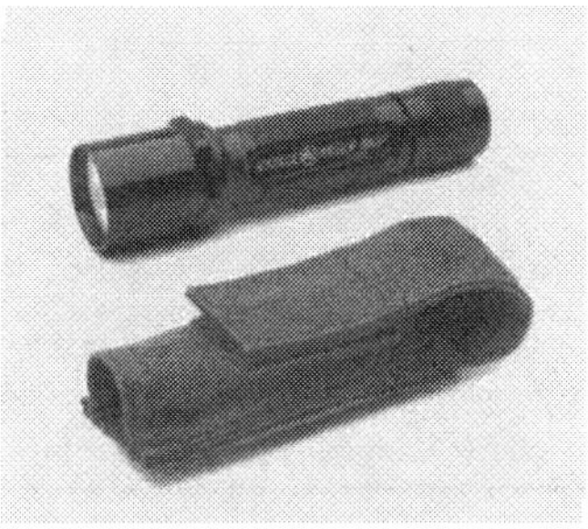

Figure 5-26. Individual High-Intensity Searchlight with Holster

Figure 5-27. High-Intensity Light

HIGH-INTENSITY LIGHT SYSTEM WITH 12-GAUGE SHOTGUN ATTACHMENT KIT

5-31. This 12-guage high-intensity light system (see *Figure 5-28,* page 5-16) is used to illuminate possible targets in crowd control situations where there is little or no light. It serves to blur or blind the aggressor's ability to see. It mounts easily to the end of the 12-gauge shotgun barrel.

40-MILLIMETER CARRYING POUCH

5-32. The 40-millimeter carrying pouch is made from durable canvas material, (see *Figure 5-29,* page 5-16) and can be easily affixed to load-bearing equipment (LBE) and/or a load-bearing vest (LBV) using Velcro straps. It is capable of carrying up to six 40-millimeter rounds.

Figure 5-28. 12-Gauge High-Intensity Light System With 12-Gauge Shotgun Attachment Kit

Figure 5-29. 40-Millimeter Carrying Pouch

12-GAUGE UTILITY POUCH

5-33. The 12-gauge utility pouch is made from durable canvas material (see *Figure 5-30*). It can be easily affixed to an LBE and/or LBV using metal clasps. The 12-gauge utility pouch is capable of carrying up to twenty-five 12-gauge shotgun rounds.

Figure 5-30. 12-Gauge Utility Pouch

12-GAUGE BUTTSTOCK CUFF

5-34. The 12-gauge buttstock cuff (see *Figure 5-31)* firmly attaches to the shotgun by sliding over the stock. Elastic straps aid in preventing the buttstock cuff from sliding or rolling. It holds up to seven 12-gauge cartridges, which give the soldier quick access to additional rounds.

Figure 5-31. 12-Gauge Buttstock Cuff

DIVERSIONARY AND/OR RUBBER BALL GRENADE POUCHES

5-35. The diversionary and/or rubber ball grenade pouch (see *Figure 5-32*) is an ammunition-carrying pouch made of durable canvas material. It is partitioned in a manner to provide the soldier with a means of holding up to six rubber ball grenades or six flash-bang grenades in separate compartments.

Figure 5-32. Diversionary and/or Rubber Ball Grenade Pouch

CALTROPS

5-36. The caltrop (see *Figure 5-33,* page 5-18) is a field-expedient area denial system. Securing approximately 50 caltrops to 550-type military cord allows a soldier to quickly toss and recover the system. Caltrops are used to deny vehicle and pedestrian movement in designated areas. The four prongs are approximately 2 inches tall and will puncture tires or boots. A soldier can easily employ the caltrop by scattering it in a designated area hidden by camouflage.

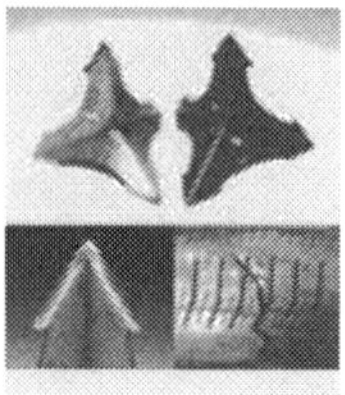

Figure 5-33. Caltrop

PORTABLE VEHICLE ARREST BARRIER

5-37. A portable vehicle arrest barrier (PVAB) (see *Figure 5-34*) is a lightweight, portable, manually-emplaced and recoverable-barrier entrapment device that is used (on command) to create a barrier that will stop vehicles being driven by known or suspected hostile forces with minimal damage to the vehicle and its occupants. Two individuals can easily emplace it in approximately 4 to 8 hours. It has the capability of stopping a light truck (up to 7,500 pounds) at speeds of up to 45 miles per hour with a stopping distance of less than 200 feet. It is not to be used as a stand-alone system. The PVAB may be used at dismount points, checkpoints, and roadblocks to prevent unauthorized access by wheeled vehicles into or out of areas.

Figure 5-34. PVAB

TRAINING DEVICES

5-38. Training devices that commanders can use to effectively prepare their soldiers are included in the NLCS. Training devices allow soldiers to learn and build confidence in the development of personal defensive skills. These are critical pieces of equipment because they allow soldiers to train at almost full speed, which provides reinforcement of realistic motor skills. Training in this manner has proven to be more effective than just striking out at the air.

IMPACT TRAINING SUIT

5-39. Impact training suits (see *Figure 5-35*) are made of closed-cell shock-absorbent foam that allows soldiers to hone their riot baton skills through realistic hand-to-hand training. Impact training suits absorb strikes inflicted by the current expandable and wooden riot batons. They provide protection for the head, face, hands, and legs without significant degradation of the wearer's mobility.

Figure 5-35. Impact Training Suits

TRAINING STRIKE BAGS

5-40. Training strike bags (see *Figure 5-36*) give the soldier a durable, yet absorbing surface to strike or kick for training. It also helps develops proficiency in open-hand control and riot baton techniques.

Figure 5-36. Training Strike Bag

PRACTICE GRENADE BODY (GG05)

5-41. The GG05 is blue in color and is easily identified as a practice grenade (see *Figure 5-37*). It is designed to replicate the live GG04 rubber ball grenade. It is made of durable rubber and can be hand-thrown or muzzle-launched from the 12-gauge shotgun. It can be reloaded.

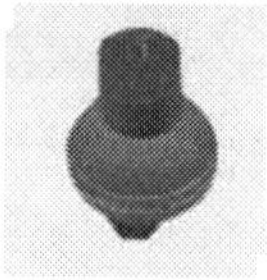

Figure 5-37. GG05

PRACTICE GRENADE FUSE

5-42. The practice grenade fuse (see *Figure 5-38*) is identified by its blue spoon and brown stripe. It easily screws into the body of the GG05. Once used, unscrew the practice grenade fuse from the body and reload it with a new fuse.

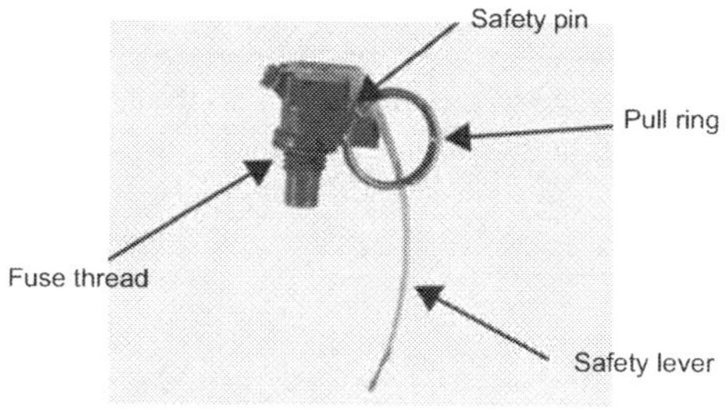

Figure 5-38. Practice Grenade Fuse

CORE CAPABILITIES

5-43. Core capabilities are those fundamental competencies that enable the commander to achieve the desired operational outcome. In the case of NLW, this includes providing a flexible means of response in order to protect friendly forces and/or influence the actions of potential adversaries and noncombatants. With the availability of the NLCS (see *Figure 5-39*) these goals can be achieved without resorting to lethal force and in a manner that will minimize collateral damage. The core capabilities associated with NL effects fall into the following two major categories:

NOTE: There are nine platoon equivalents within a light infantry battalion.

PLT Organization	Personnel and Weapons	Nonlethal Weapons & Equipment
PLT	7 personnel (5 M16/M4 and 2 M60/M240)	3 riot face shields 3 sets of shin guards 3 sets of full-body restraints 3 individual voice amplification systems 3 individual RCA dispensers with carriers 1 portable bullhorn 1 ground-mounted bullhorn 1 high-intensity searchlight 1 set of caltrops 3 ballistic riot face shields 3 sets of ballistic shin guards 3 ballistic riot body shields 1 riot baton training suit 1 riot baton strike pad
SQD SQD SQD	3 personnel (3 M16/M4)	3 riot face shields 3 sets of shin guards 3 riot body shields 3 expandable riot batons with carriers 9 sets of cinch-strap restraints 3 individual voice amplification systems 3 individual RCA dispensers with carriers 3 individual high-intensity search lights with carriers (see *Figure 5-26*)
TM TM TM TM TM TM	24 personnel (12 M16/M4 and 6 M16/M4 with M203 6 SAW)	24 riot face shields 24 sets of shin guards 100 sets of disposable flex-cuffs 24 individual voice amplification systems 24 individual RCA dispensers with carriers 6 expandable riot batons with carriers 11 grenade pouches 12 forty-millimeter grenade pouches 6 twelve-gauge ammunition pouches 6 twelve-gauge bitstock ammunition carriers

Figure 5-39. NLCS Distribution

- **Counterpersonnel.** NL counterpersonnel capabilities enable the application of military force with a reduced risk of fatalities or serious casualties among noncombatants or, in some instances, among enemy forces. There are several specific NL counterpersonnel capabilities to be explored. These include the means to influence the behavior and activities of a potentially hostile crowd, and the capability to bring a mob engaged in a riot under control. While there are many similarities in these two scenarios, each involves unique challenges, which may require radically different solutions. For more information, see *FM 3-22.40*.
- **Countermaterial.** NL countermateriel capabilities enhance operations by rendering equipment and facilities unusable without complete destruction. An NL countermateriel capabilities enable the

application of military force to defuse potentially volatile situations under circumstances in which more destructive conventional military means might prove counterproductive. For example, pre-emptive strikes against troublesome and aggressive nations may be politically unacceptable when only conventional weapons with the high risk of casualties are involved. With NL countermateriel capabilities, the enemy's ability to threaten its neighbors could be curtailed with less political risk by attacking only their weapons of war and their supporting infrastructure. For more information, see *FM 3-22.40.*

ADDITIONAL CAPABILITIES

5-44. NLW are only NL when fired within their parameters.

DANGER

If NLW are fired outside their parameters, they may cause serious injury or death.

5-45. If a situation turns from NL to lethal, an NL gunner only needs to change his point of aim and distance to inflict serious bodily harm and/or death to the threat. The closer the soldier is to the target, the greater the kinetic energy and the more severe the effects are to the threat.

EMPLOYMENT CONSIDERATIONS

5-46. NL munitions and equipment are not completely NL; they are NL by intent. The term NL does not guarantee zero mortality or nonpermanent damage. NLW can add flexibility to combat operations and enhance force protection by providing an environment in which friendly troops can engage threatening targets with a reduced risk of noncombatant casualties and collateral damage.

5-47. If the tactical situation dictates a NL response, a soldier who levels a weapon loaded with NL munitions must be trained in distance to the target and appropriate points of aim. Based on the soldier's understanding and training on the characteristics and capabilities of the round, they can easily discern and engage targets to the desired effect. If the tactical situation dictates a lethal response, soldiers can easily adjust the point of aim to a more vulnerable target location on the body and move in closer to the target. NL munitions can be used as a lethal response if the situation dictates and the intent of the soldier is to inflict death.

NONLETHAL WEAPONS CAPABILITIES IN FORMATIONS

5-48. The mix of NLW within the formation is based on METT-TC. For example, the first rank of the formation will have riot shields, riot batons, and their personal weapon (slung across their back [left to right] with the butt up and muzzle down or holstered). The second row will have a mix of shotguns and M203s. Shotguns and M203s may be used as area denial or point target weapons depending on the specific munitions used. The shotgun provides cover for the slower loading M203. The commander may move these weapons within the formation to meet the needs of the mission and to create distance between the formation and the crowd (see *Appendix D*).

Chapter 6

Civil Disturbance Formations

Control force formations, when properly employed and executed against a crowd, are some of the most practical methods of crowd control. The riot baton has been the weapon of choice for the control force. With the development of NLW, the commander now has more options at his disposal. Before NLW, the only other choices were chemical irritants, MWDs, and lethal force. Now commanders can create a standoff distance with NLW that reduces the risk of close fighting with the mob. During any crowd control operation, the commander must protect the control force with lethal overwatch. This overwatch may be in the form of marksmen and observer teams (sniper with observer) and a response force with lethal capabilities.

OVERVIEW

6-1. With advances in NLW, it is possible to create an effective standoff distance of 15 to 100 meters between the crowd and the control force formation. It is now possible to advance and disperse the crowd without coming into direct contact with it. Some NLW may also be used to assist in the identification and apprehension of instigators.

6-2. Control force formations are used to disperse, contain, and/or block a crowd. These formations are more effective in urban areas than in open areas, such as parks or fields. When control force operations are employed in urban areas, it is easier to split a crowd into smaller segments, isolate instigators, or funnel the crowd into the desired location by using buildings and other man-made structures. Commanders must be mindful of the fact that well-organized crowds may attempt to move to the flanks of a formation and/or get behind the formation to gain a tactical advantage.

6-3. Commanders must be aware of the limitations of control force formations. They are not the answer to all civil disturbance situations. Do not expose the formation to sniper fire and/or unnecessary violence simply for a show of force. Control force formations should be used when it is decided that a crowd poses a threat and must be dispersed or moved to a specified area. Before dividing a large crowd, the commander must consider that this might not solve the problem. It may worsen the problem by creating smaller elements that may engage the control force in small-mob tactics, such as sniping, looting, burning, and attempting to envelop the control force. Commanders must anticipate that breakaway crowds may engage in small-mob tactics. These tactics can be defeated by area control measures, such as building searches, saturation patrolling, and other tactics. After dispersing the crowd, the control force must not assume that there will be an immediate return to peaceful activities. The use of formations is only part of the total dispersal effort.

6-4. Commanders must assess and secure the areas of operations (AOs). This may be accomplished with the use of helicopter overflights and by visually securing rooftops, high elevations, and advantage points. Designated marksmen teams may be used to secure the rooftops and provide lethal overwatch for the control formation. The control force commander must be aware that designated marksmen teams are positioned on rooftops.

6-5. Crowd control formations and their support teams have more capabilities than just crowd dispersal. They also have the capability to apprehend and detain certain members of the crowd that the commander feels may instigate the crowd to further violence. Commanders must analyze the threat, determine the mission of the control force, and decide which control force formations will accomplish the mission according to METT-TC.

6-6. During CONUS operations, there are usually no enemy forces. However, commanders must analyze and prepare for any potential opposition. The formations discussed in this chapter are guidelines and may be adapted to fit any mission or situation. Whatever the modification, soldiers must stay in position and on line. Only through training and rehearsals will soldiers become proficient in basic formation movements and be able to adjust to changing situations. Soldiers must also be prepared to employ NLW. *Figure 6-1* is used as a key to identify soldiers within crowd control formations.

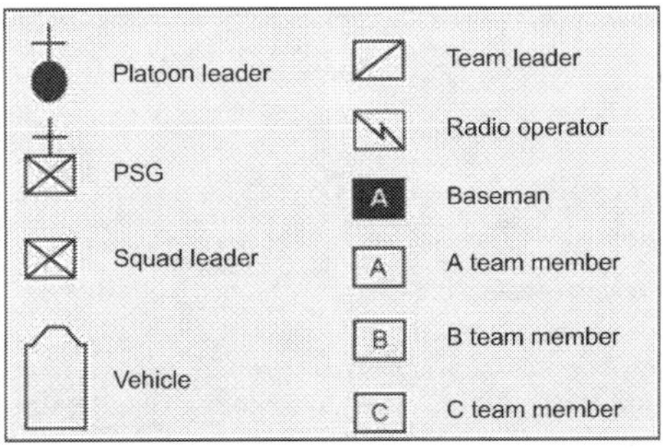

Figure 6-1. Symbols for a Formation Element

ACTIONS BEFORE MOVEMENT

6-7. Before moving the control force to the civil disturbance area and into a potentially hostile urban environment, the commander and his staff must review the preincident plans and expand them (as needed) based on current information. Because of the employment of forces against ambiguous threats,

IPB becomes critical. Information is the key to developing an appropriate response to civil disturbances.

6-8. Civil disturbance training must be part of the annual unit training plan. If a soldier is deployed to an area where civil disturbances are a recurring event, sustainment training must occur more frequently. Before moving the control force to the rally point, an intense rehearsal must be conducted by the leadership and the control force. A top-down review of the rules of engagement (ROE/rules for the use of force (RUF) must be accomplished. The ROE/RUF must be part of the unit training plan and each soldier must be trained to standard (see *Chapter 8*). The use of NLW (type specific) must be granted to the on-site commander from the approval authority.

6-9. Information is the key to developing plans for appropriate responses to a civil disturbance, and IPB is crucial to the success of the mission. During the planning phase (see *Chapter 2*), the information provided by the IPB will aid the staff in selecting various routes and rally points. The selection of ingress and egress routes must include a variety of ways in and out of the area. Direct and indirect routes and those with cover and concealment are necessary. The rally point must be cleared and secured by an advanced party capable of controlling the area.

RALLY POINT ACTIONS

6-10. Once at the rally point, the commander makes contact with the local police, civil authorities, or military officer. He then assesses the situation, decides the next COA, and issues orders, as appropriate. The platoon sergeant (PSG) forms the platoon into a column formation and places team leaders directly in front of their teams (see *Figure 6-2,* page 6-4). Team and squad leaders make last minute checks of their soldiers and await orders.

6-11. The commander places his lethal overwatch (designated marksmen [DM] teams) in areas that overlook the control force and the mob. They may be to the flanks and slightly to the front of the formation. It is their responsibility to protect the formation from lethal fire by constantly scanning the crowd.

6-12. When the commander receives word that DM teams are in place, he forms the control force into a crowd control formation and moves them quickly from the rally point to where the crowd is assembled. Depending on the situation, movement should be in a column formation or one of the three basic control force formations. Reserve forces are left at the rally point until ordered forward as additional lethal overwatch or to reinforce the formation.

CONTROL FORCE FORMATIONS

6-13. Control force formations, when properly employed and executed against a crowd of limited size, are one of the most practical methods of crowd control. Experience has indicated that the most frequently used control force

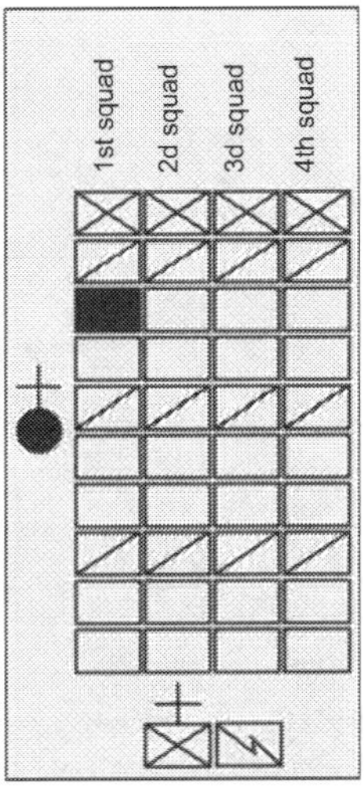

Figure 6-2. Column Formation

formations are the line, echelon, and wedge. The two lesser-used formations are the diamond and circle. Descriptions of all the formations are as follows:

- **Line formation.** The line is the basic formation and it is used more often because of its offensive and defensive applications. As an offensive formation, the line is used to push or drive crowds straight back, across an open area, or up a city street. As a defensive formation, the line is used to hold the crowd or to deny access to areas.
- **Echelon formation.** The echelon is an offensive formation used to turn or divert groups in open or built-up areas and to move crowds away from buildings, fences, and walls.
- **Wedge formation.** The wedge is an offensive formation that is used to penetrate and split crowds into smaller groups.
- **Diamond formation.** The diamond, when used as an offensive formation, is used to enter a crowd and is the formation of choice for extraction teams. As a defensive formation, the diamond is used when all-around security is required in open areas. The decision to use this formation is based on the conformation of the crowd.
- **Circular formation.** The circular formation is used for the same purpose as the diamond formation. The decision to use this formation is based on the conformation of the crowd.

6-14. There are many suitable variations of control force formations that may be employed, but appropriate commands and signals must be created to execute the formations. Because of their somewhat complicated nature and the coordination required for these formations, new variations must be trained, practiced, and rehearsed before they are used in a civil disturbance.

VEHICLES AND FORMATIONS

6-15. Vehicles may be employed with troops in control force formations, especially when a large riotous crowd is on hand. Although vehicles add strength to formations, certain precautions must be taken. Covering the windshield with sturdy, close mesh fencing and the standard safety glass will offer some level of protection to the occupants. Shields and/or mobile barriers may be built by mounting a wooden or metal frame strung with barbed wire across the front of a vehicle. Members of the formation should walk as near to the front corners of each vehicle as possible to keep rioters from attacking the sides and rear of the vehicles.

6-16. Up-armored HMMWVs, ASVs, or armored vehicles may be employed with the control force formation when determined by METT-TC. They present a strong psychological effect and offer protection for the occupants. Wheeled armored vehicles are best because they do not deface the pavement. Whenever vehicles are used in a crowd control formation, the commander must ensure that he is still able to see and control the formation. This may require him to occupy a position in a similar vehicle behind the formation or command subordinate leaders to move away from the front line and into line-of-sight (LOS) control. This will require additional hand-and-arm signals to allow the commander and subordinate leaders to communicate. Commanders can effectively use vehicles as primary communication tools by using horns and lights as signals to organize movement or actions. However vehicles are used, it takes a good measure of training and practice to be effective.

ELEMENTS OF A CONTROL FORCE FORMATION

6-17. Four elements make up the basic crowd control formation. They are as follows:

- **Base element.** The base element is made up of two ranks. The first rank is shield holders, while the second rank contains the NLW. This is the front line of the formation.
- **Support element.** The support element forms up in a column formation behind the base element and has a variety of uses. It may be used to replace base element members as needed, provide lateral or direct support, or perform extraction team operations. The support element helps the base element by performing the following three essential tasks:
  - **General support.** The general-support element is formed from an uncommitted squad in the platoon (generally the 2d squad). When a company is tasked as the control force element, one of the platoons becomes the general-support element. The element is in a column formation centered on and behind the main formation. From this formation, the general-support element can move to lateral or direct support, as needed.
  - **Lateral support.** The lateral-support element is used to protect the flanks of the formation. This is done by moving a set number of teams forward from the general-support element or by using the

end teams in the formation; however, this will make your formation smaller. Once in position, these teams become part of the formation, with their riot shields facing the flanks of the formation.

  - **Direct support.** When direct support is ordered, the general-support element moves forward as they form their own formation. While moving forward, the squad and team leaders from the control force formation step back and allow the riot shield holders from the support element to step behind riot shield holders of the original formation. This allows the support riot shield holders to form up between and behind the riot shield holders of the original formation. Now there are two lines of riot shield holders between the crowd and NLW firers. This formation is the strongest of the formations and requires more planning and practice to master.

- **Command element.** The command element contains several different members. A general configuration for the command element is the platoon leader, the PSG, a radio operator, a video recorder operator (if required), and an interpreter (if required). This element does not have a fixed location within the formation and moves about as needed.
- **Lethal overwatch element.** The lethal overwatch element is a team formed from reserve security forces.

6-18. The reserve support element is not part of the control force formation until it is brought forward from the rally point to join the formation. It remains until released by the formation commander. Lethal overwatch teams are formed from the reserve force. Once lethal overwatch teams are deployed and in position, they are under the control of the formation commander. Communication between the lethal overwatch team and the formation commander is a priority because he approves target selection and engagement.

WEAPON POSITIONS

6-19. At the commander's discretion, members of the control force formation may carry their weapons in any of the positions prescribed in *FM 3-21.5*. However, the primary method of carrying weapons in the control force formation is at the safe-port position (see *Figure 6-3*). It allows the soldier to control both ends of the weapon while moving in and out of the formation and advancing on the crowd.

6-20. High-port position (see *Figure 6-3*) is a position of complete readiness. It is used whenever troops are in contact with a crowd that is showing resistance or not withdrawing. This is a tiring position and is hard to maintain for extended periods. Commanders must rest their troops at every opportunity by using less tiring positions.

Figure 6-3. Weapons Positions

COMMANDS

6-21. Commands are given to the control force formation orally or with hand-and-arm signals. Oral commands for the on-guard position are given in one count. All other commands are given in two counts, a preparatory command followed by a command of execution. However, oral commands cannot be relied on completely, so commanders must plan to use hand-and-arm signals too. When vehicles are part of the formation, radios become an alternate method for relaying commands. To improve communication, commanders may use hand-and-arm signals (see *Figure 6-4,* page 6-8) with oral commands. The basic commands used in control force formations are outlined in *Table 6-1,* page 6-9. These commands can be added to or combined to form more complex commands, therefore forming more complex formations.

MOVEMENTS

6-22. Control force formations move in the same manner as regular formations. This allows the commander to more precisely maneuver the formation by commanding the number of steps it should move and in what direction, such as "Five Steps, Forward March." The normal rate of march for entering and leaving a control force formation is double-timing. The half-step march is used when the formation is in direct contact with the crowd. This slows the formation down and allows for better command and control. The squad and team leaders echo preparatory commands and provide cadence while in control force formations.

INTERVAL AND DISTANCE

6-23. Interval is the lateral space between elements. Distance is the space between elements in a column. The usual interval and distance between soldiers in control force formations is 30 inches. The interval and distance can be adjusted based on METT-TC. If the crowd has to be physically pushed back, a close interval is preferred. This allows the riot shields to be overlapped, creating a stronger wall. During peaceful demonstrations, the interval could increase to double arm width, allowing the formation to cover more area and still control the crowd. The interval and distance could change several times during the course of operations before the crowd is dispersed.

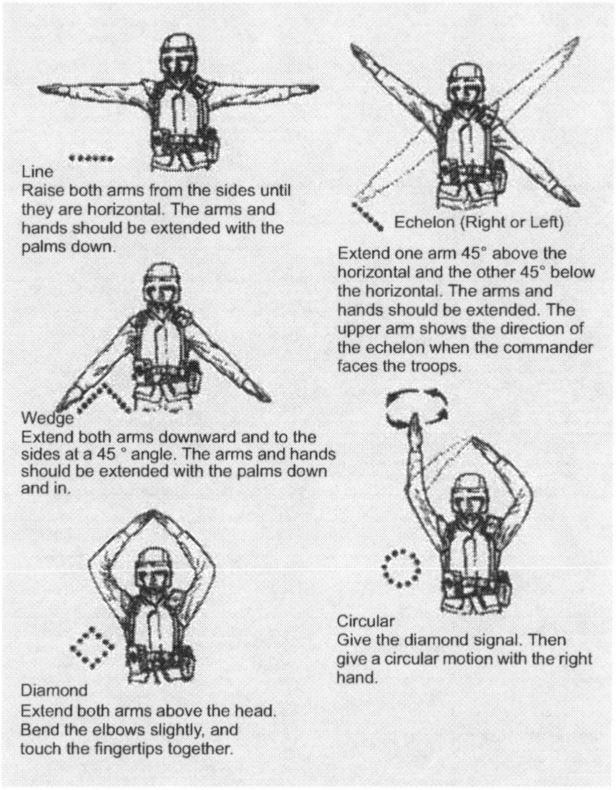

Figure 6-4. Hand-and-Arm Signals

Table 6-1. Basic Commands

Commands for	Command	Given by	Purpose
Formations	Platoon On Line	Platoon leader and/or commander	To form a line formation.
	Platoon Wedge	Platoon leader and/or commander	To form a wedge formation.
	Platoon Echelon	Platoon leader and/or commander	To form an echelon formation.
	Platoon Diamond	Platoon leader and/or commander	To form a diamond formation.
	Platoon Circle	Platoon leader and/or commander	To form a circular formation.
	Move	Platoon leader and/or commander	As a command of execution. **NOTE: The platoon leader and/ or commander also identifies the location for the formation by pointing his arm to the desired location for the formation.**
Support	General Support	Platoon leader and/or commander	To place a specified unit in the rear of the base element.
	Lateral Support	Platoon leader and/or commander	To place support elements on the left or right flank of the base element.
	Direct Support	Platoon leader and/or commander	To move support elements forward to strengthen the base element.
	Open	Extraction team leader	To open a space in the formation.
	Up	Extraction team leader	To inform the extraction team to start moving back to the main formation.
Fire	Weapon System Gunners, Number of Rounds, Type of Round, and Prepare to Fire	Platoon leader and/or commander	To ready NL gunners to fire weapons; for example, M203 gunners, three rounds, area target, prepare to fire.
	Shield Down	NL gunners	To tell riot shield holders to get down on one knee and lower the riot shield to the ground.
	Shield Up	NL gunners	To tell riot shield holders to stand and raise the riot shield.
Weapons	Port Arms	Platoon leader and/or commander	To use when not in contact with the crowd.
	High Port	Platoon leader and/or commander	To tell NL gunners what position to hold their weapons in while the formation is in physical contact with the crowd. This allows for rapid targeting and firing of NL munitions.

EXTRACTION TEAM OPERATIONS

6-24. The extraction team is a squad (see *Figure 6-5*) that is generally formed from the support element, but could also be formed from reserve security forces. Once the team is formed, it is OPCON to the formation commander until the team's mission is complete and it returns to the rear of the formation. This team provides the riot control formation with the means to employ NL and lethal cover forward of the formation. It also provides the formation with the ability to remove and detain rioters downed by munitions.

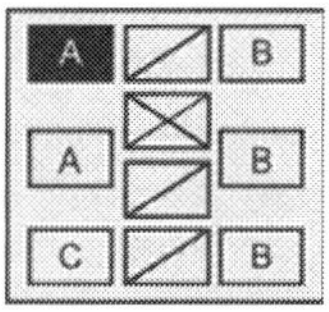

Figure 6-5. Extraction Team Formation

The extraction team—

- Extracts vehicles or personnel that are in immediate danger from the crowd.
- Detains and escorts downed rioters to the rear of the formation.
- Restrains and searches subjects.
- Covers confined or congested areas where a full riot control formation cannot be inserted.

6-25. The extraction team is usually a squad. While the recommended distance for deploying the extraction team from the control force formation is no farther than 10 meters, this may change based on METT-TC. The control formation protects the extraction team and provides immediate support, if necessary.

6-26. The extraction team may deploy from anywhere within the formation. When given orders to deploy, the squad leader of the extraction team gives the formation commands, identifies at least two personnel in the squad to conduct the search and apprehension of the subject, and leads the squad to the base line of the formation. He then extends his arm between the two riot shield holders and commands, "Open," while tapping the riot shield holders on the side of the shoulder. The two riot shield holders take one step backward and one step to the right or left. This clears a path for the extraction team. The extraction team leader then states the number of personnel leaving the formation, such as "Ten security personnel leaving." As the last extraction team member exits the formation he states, "Last Man." The two riot shield holders then return to their original position in the formation. The adjacent base element squad and/or team leaders pass the number of personnel in the team through the formation.

6-27. As the extraction team approaches an instigator or incapacitated demonstrator, the riot shield holders envelop the target and face in the direction of the nearest threat. NL gunners on the left and right cover their respective areas. The search and apprehension team immediately controls and restrains the target. To avoid confusion, one team member concentrates on controlling the subject while another team member applies a restraining system. When the search and apprehension team has the subject under control, the extraction squad leader sounds off with, "Up." Upon hearing the "Up" command, it is repeated by all extraction team members. Team members then grab a shoulder of the team member in front of them to backtrack into the formation. As the extraction team approaches the formation, the nearest base element leader will extend his arm and command, "Open," while tapping the riot shield holders on the side of the shoulder. The riot shield holders move one step backward and to the right or left to open the formation. As he enters the formation, the extraction team leader sounds off with the number of personnel entering the formation, such as "Ten security personnel and one demonstrator." The adjacent base element leader counts the number of personnel as they enter the formation. When the last man enters the formation, he states, "Last Man." The riot shield holders then return to their original places in the formation. The adjacent squad and team leaders in the base element pass information through formation that the extraction team has returned.

LETHAL OVERWATCH TEAMS

6-28. During a NL engagement, the use of a DM team provides confidence and safety to those facing the mob. If a lethal threat is presented, the DM team in an overwatch position (armed with a standard infantry rifle that is mounted with a high-powered scope) can scan the crowd, identify agitators and riot leaders for apprehension, or fire lethal fire if so ordered or warranted. It is also ideally suited for flank and countersniper operations.

6-29. The DM team sets the security overwatch and provides real-time intelligence and reporting that is vital to mission success. It provides coverage during the entire approach to the crowd, and its ability to select positions and provide cover fire (if and when warranted) is critical to the safety of the control force. Successful execution requires training, practice, and rehearsal with the control force formation.

6-30. The DM team is organized from a standard three- or four-man team. Each individual on the team is equipped with a standard-issue weapon. One individual has an infantry rifle with a scope, another has a set of binoculars, and two other individuals provide security for the team. Each team is equipped with a radio for communication with the HQ element of the control force formation.

6-31. The teams must have a visual advantage over the crowd to provide lethal protection to the formation. This is best accomplished by placing them on nearby rooftops, in the upper floors of buildings, and on hilltops. They must have the optical equipment to identify mob leaders, instigators, and/or individuals with weapons. Team integrity must be maintained. In organizations with three-man teams, one individual becomes the marksman, another the observer, and another the security. When teams have a 4th member, that individual becomes the video camera operator. This will help identify individuals for later apprehension and provides a record of events.

6-32. Two teams should be deployed to cover the flanks of the formation. Each team should be deployed to a position that is in front of the formation and in an overwatch position. As the formation moves forward, the teams leapfrog forward from the formation.

6-33. The reserve forces, located at the rally point, can provide another lethal force of some size that can be brought forward, as needed. Each soldier in the control force formation should have a sidearm (9-millimeter pistol) or an M16 or M4 rifle. The rifle is slung diagonally across his back with the rifle butt over his left shoulder and the muzzle below his right hip. Lethal munitions are in a separate ammunition pouch that is isolated from NL munitions. This is to prevent the mixing of NL and lethal ammunition.

6-34. Once the lethal overwatch teams are deployed, they are under the OPCON of the formation commander. The lethal overwatch element must have direct and secure communications with the formation commander. The number of teams deployed is based on METT-TC. Teams will leapfrog to keep ahead of the advancing control force formation with one team in the overwatch position at all times.

RESERVE FORCES

6-35. A large reserve of soldiers should be maintained during civil disturbance operations. Knowing that a large reserve force is available provides confidence and safety within the control force and helps prevent them from overreacting to provocative acts with disorderly and criminal elements in the crowd. When determining the number of reserve forces required, it should be resolved in favor of a large number. Lethal overwatch teams, forces to augment the control force, and apprehension teams come from the reserve forces.

6-36. Apprehension teams are a must when it is likely that a large number of people will be apprehended. These teams provide an organized response to what could be a chaotic situation. The teams can be organized at the team or squad level, depending on the number of apprehensions expected. These teams usually operate behind the base element. During peaceful demonstrations, these teams can be sent forward to remove demonstrators who refuse to disperse.

6-37. Each team consists of an apprehension, movement, and security element (added if the control force cannot provide security); a recorder; and a video or still photographer. The apprehension element is an officer or NCO and a civilian police officer (CONUS or OCONUS) or US marshal. It decides who will be apprehended, informs the subject of the reasons, supervises the handling of the subject, and ensures that the apprehension is properly conducted and documented. The apprehension team is responsible for the overall apprehension process, limiting the number of individuals appearing in court to document the apprehension. The movement element helps the apprehension element handle the subject. They move, restrain, and search the individuals being apprehended under the supervision of the apprehension element. The security element prevents the crowd from interfering with the mission. The recorder documents the apprehension on *DD Form 2708*. When possible, the video and/or still photographer recorder photographs each subject with the apprehension element. This provides the apprehension element with the necessary documentation for use in court. One individual may need to conduct many of the duties of an apprehension team based on METT-TC.

6-38. The video and/or still photographer record the situation before, during, and after the apprehensions. These images may also be used as pictorial documentation for use in court. It is possible that the presence of video and still cameras actively recording the situation can cause crowd members to be less prone to unruly or criminal acts.

6-39. The behavior of the selected individual to be apprehended is observed to decide if the subject is violent, cooperative, or passive. Unless there is good reason to believe otherwise, the subject must be considered dangerous. The mood of the crowd is observed to estimate what it might do. Is it only curious onlookers or does it support the individual targeted for apprehension? Will it interfere?

6-40. When apprehending an individual, the actions must be kept as simple as the situation allows. The element of surprise can give the apprehension team an advantage over the apprehended individual. It is best not to use surprise if the individual might panic and injure someone. Select an area for the apprehension that is to the advantage of the team and not the individual. Avoid apprehensions in crowded areas as these areas provide many avenues of escape and contain supporters of the apprehended individual.

6-41. All verbal commands given to the apprehended individual must convincing authority with short, clear, and distinct commands. Voice, attitude, and deportment must convey an expectation that the apprehended individual will comply.

6-42. The individual should be searched immediately; however, circumstances (violence or sympathizers for the individual being apprehended) may dictate temporarily postponing the search until the individual can be moved to a safer area. Flex-Cufs or hand irons may be applied.

SQUAD FORMATIONS

6-43. Civil disturbance training is a collective task. Using team and squad concepts, team leaders instruct their teams and squad leaders instruct their squads. Safety is key when using force on force training for the control force formation. Soldiers must understand that serious injuries may occur if they do not follow guidelines set by the commander.

6-44. The smallest formation is a squad formation, which is used in back of the main formation to cover side streets. Squad members must know their positions regardless of which control force formation they are in, squad to company.

6-45. Training must enforce soldiers' understanding of the need to stay informed for individual protection. Individual soldiers are much more vulnerable to attack when they break ranks and chase after crowd members. When an individual soldier breaks ranks he not only puts himself in danger, but the entire formation.

6-46. When moving a squad into a crowd control formation from a column formation, the squad leader takes one or more steps to the right of the squad and faces it. He gives the command and hand-and-arm signal for the formation he wants. Then he indicates where he wants the formation located by pointing in that general direction. The A team leader commands, "Follow Me." He places the baseman at the position indicated by the squad leader. The squad forms on the baseman position.

LINE FORMATION

6-47. The command for forming a squad in a line formation is "Squad Line Formation, Move." On the command of execution, the A team leader leads the baseman to the place indicated by the squad leader. The squad members align themselves in sequence with the baseman at normal intervals (see *Figure 6-6*). Based on the situation, the squad leader may designate a specific interval. He does this when he gives the preparatory command. If no interval is specified, the squad automatically forms using the normal interval.

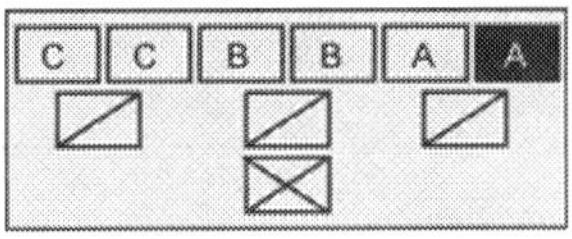

Figure 6-6. Squad Line Formation

ECHELON FORMATION

6-48. The command for having a squad form in an echelon formation is "Squad Echelon Right (or Left), Move" (see *Figure 6-7*). On the command of execution, the A team leader places the baseman at the location indicated by the squad leader. The squad members align themselves in sequence with the baseman, one pace to the side and one pace to the rear.

WEDGE FORMATION

6-49. The command for moving a squad formation into a wedge formation is "Squad Wedge, Move." On the command of execution, the A team leader moves the baseman to the place indicated by the squad leader. The A team lines up to the left of the baseman, one pace to the left and one pace to the rear of each preceding man. B team members align themselves with the baseman, one pace to the right and one pace to the rear of each preceding man. C team members align themselves in the same way to the left of the baseman (see *Figure 6-8*).

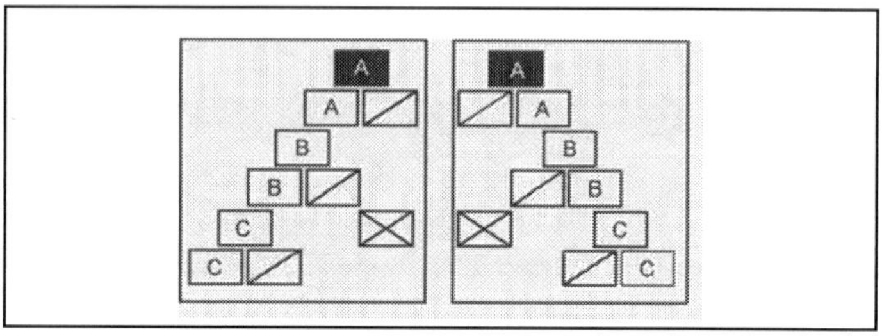

Figure 6-7. Squad Echelons Left and Right Formations

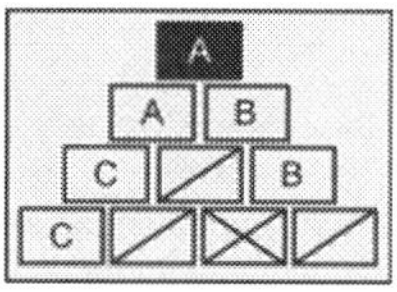

Figure 6-8. Squad Wedge Formation

ASSEMBLY

6-50. When assembling a squad from the line formation, the squad leader takes a position a sufficient distance to the rear of the squad and commands, "Squad Assemble." At the same time, he raises his right hand in the air and makes a circular motion. Squad members, who hear the command, automatically come to the port arms position. The A team leader and the baseman do an about-face movement.

6-51. The other squad members face toward the baseman. On the command of execution "Move," the squad leader points to the place where he wants the squad to assemble. The A team leader double-times to the designated spot, and the other members of the squad follow him. Team leaders fall into the file in their designated positions as they move to the AA. To assemble from a squad echelon formation, the steps are the same as a line formation.

6-52. To assemble from a squad wedge formation, the squad leader takes the same steps as for the line and echelon formations. B team members do a half left face, and the A team and C team members do a half right face. The C team pauses at the baseman position and allows the B team to clear the formation before moving to the AA.

DIAMOND AND CIRCLE FORMATIONS

6-53. The diamond and circle formations are used during extraction team operations. They are small formations used to penetrate the crowd or cover small areas. The decision on which formation to use is based on METT-TC and the ability of the squad to perform.

6-54. Commands for forming a squad in a diamond or circle formation is "Squad Diamond (or Circle), Move." On the command of execution, the A team leader moves the baseman to the place indicated by the squad leader. The A team lines up to the left of the baseman, covering the 8 to 12 o'clock positions. B team members align themselves to the right of the baseman, covering the 12 to 4 o'clock positions. C team members align themselves covering the 4 to 8 o'clock positions (see *Figure 6-9*).

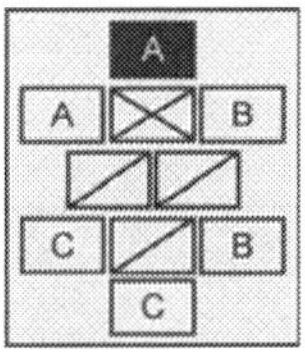

Figure 6-9. Squad Diamond or Circle Formation

6-55. To assemble from a squad diamond or circle formation, the squad leader takes the same steps as he would for a line formation. The A team leads the squad to the AA followed by the B team and then the C team. Teams fall into a column formation as they file back to the AA.

THREE-SQUAD PLATOON FORMATIONS

6-56. The platoon HQ consists of the following individuals: the platoon leader, the PSG, and a messenger or radio operator. When directed by the platoon leader, other individuals such as the selected marksman, interpreter, RCA disperser operators, and firefighters may augment the platoon HQ. In forming all control force formations from the column, the platoon leader moves to the right of the platoon and faces them when giving his commands. He gives the hand-and-arm signal and the preparatory command. On the command of execution "Move," he points to the approximate location where he wants the platoon to form. The platoon leader pauses between the preparatory command and the command of execution to allow each squad leader to issue a preparatory command to his squad.

Platoon Line Formation

6-57. The command for forming a platoon line is "Platoon on Line, Move." Immediately following the platoon leader's preparatory command, the squad leaders of the 1st and 3d squads give the command, "Follow Me." The leader of the 2d squad gives the command, "Stand Fast." On the platoon leader's command of execution, the 1st and 3d squads move forward to the point designated by the platoon leader. The 1st team member of the 1st squad is the baseman for the platoon formation. The 1st squad forms a squad line to the left of the baseman while the 3d squad forms a squad line to the right of the baseman and dresses with the 1st squad. After the 1st and 3d squads have cleared the platoon, the 2d squad leader gives the command, "Follow Me." The 2d squad splits and the A and B teams form a line to the right of the 3d squad.

The C team (with the squad leader) forms a line to the left of the 1st squad (see *Figure 6-10*).

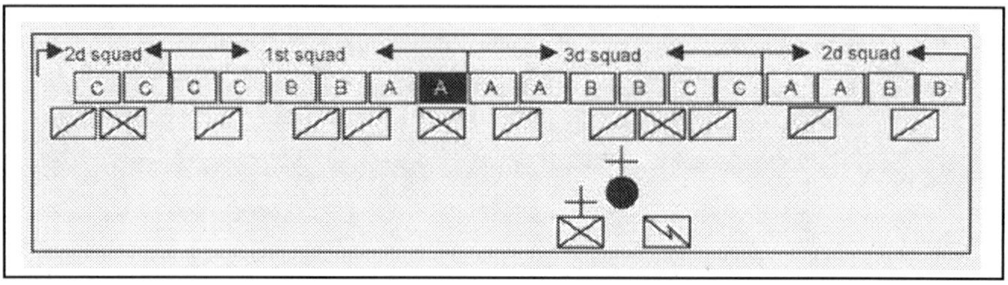

Figure 6-10. Platoon Line Formation

Assembling From a Platoon Line Formation

6-58. When assembling the platoon from the platoon line formation, the platoon leader and his HQ personnel take a position to the rear of the platoon. He then gives the command, "Platoon Assemble," while raising his right arm and making a circular motion above his head. Platoon members automatically assume the port arms position. The 1st squad leader and the baseman do an about face. The 3d squad leader moves to the head of his squad. All other members do an about-face movement toward the baseman. On the platoon leader's command of execution, the platoon leader points to where he wants the platoon to assemble. The 1st squad leader gives the command, "Follow Me," and moves his squad toward the AA. The squad moves to make a single file line to the baseman position followed by the C team of the 2d squad. As the 1st squad clears the baseman position, the 2d squad leader comes to a halt at the baseman position. The 3d squad leader gives the command, "Follow Me," and leads his squad to the AA followed by the A and B teams of the 2d squad. The A team leader of the 2d squad comes to a halt at the baseman position. Once the 3d squad has cleared the baseman position, the 2d squad leader gives the command, "Follow Me," and leads his squad to the AA. The 2d squad reassembles into their original column formation as they move to the AA. As the 1st squad comes to a halt (facing the crowd) in the AA, the 3d squad dresses with the 1st squad and leaves space for the 2d squad to return to its place in the formation. Once the platoon is formed into a column formation, HQ personnel take their position in the formation. The PSG then gives the command, "Countercolumn March," and moves the platoon to the rally point.

Platoon Line Formation With General Support

6-59. Whenever the command for establishing a crowd control formation contains the phrase "In Support" (without modification), it means that the support element is to remain in general support.

6-60. The command for forming a platoon line with the 2d squad in general support is "Platoon Line, 2d Squads in Support, Move." The 1st and 3d squads of the platoon line with general support execute the line, while the 2d squad remains in the column (see *Figure 6-11*). To assemble a platoon with a squad in general support, the procedure is the same as the platoon line, except that the 1st and 3d squads dress with the 2d squad.

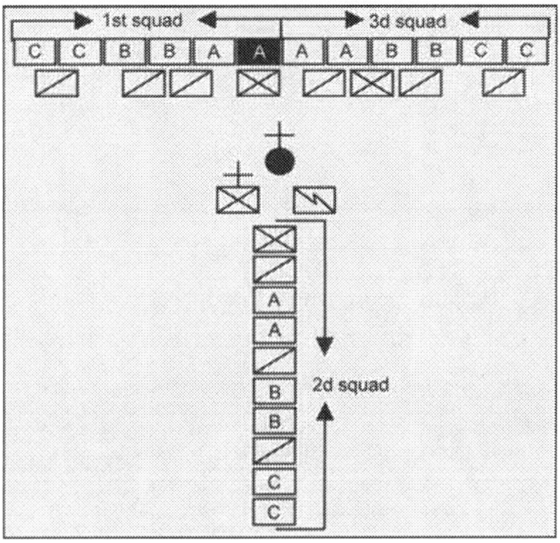

Figure 6-11. Platoon Line Formation With General Support

Platoon Line Formation With Lateral Support

6-61. The command for forming a platoon line with one squad in lateral support is "Platoon Line, 2d Squad in Lateral Support, Move." The 1st and 3d squads execute the line as before, and the 2d squad stands fast. After the line has been formed, the 2d squad leader gives the command, "Follow Me." The A and B teams move to the right flank while the squad leader and the C team move to the left flank of the formation. The 2d squad forms columns behind the last men on the line formation. If no direction is given, the 2d squad splits to cover both flanks. A squad may be used to cover one flank by giving the command, "2d Squad Right (or Left) Lateral Support" (see *Figure 6-12*).

6-62. The 2d squad may be moved from general to lateral support at any time by the platoon leader. The platoon leader commands, "2d Squad, Lateral Support, Move." To have the 2d squad join the line from either support position, the platoon leader commands, "2d Squad, Extend the Line, Move." The 2d squad leader commands, "Follow Me." On the command of execution, the squads set up individual lines, closing and dressing on the existing line. To assemble the platoon from a platoon line with lateral support, follow the same steps as for the platoon line.

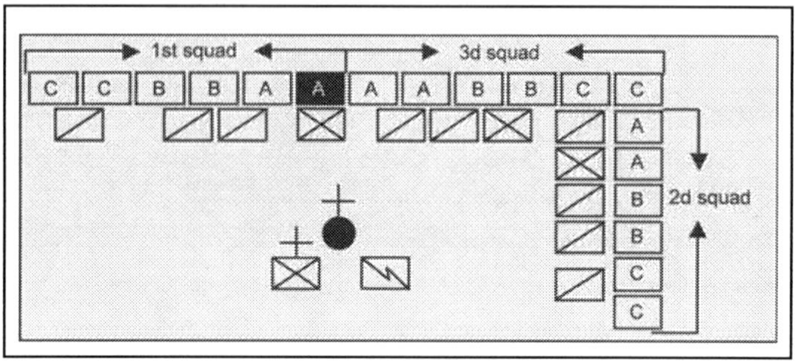

Figure 6-12. Platoon Line Formation With Lateral Support (Right)

Platoon Line Formation With Direct Support

6-63. The command for forming a platoon line with one squad in direct support is "Platoon Line, 2d Squad in Direct Support, Move." The 1st and 3d squads execute a line as before. The 2d squad executes a squad line directly to the rear and centered on the 1st line. The men in the supporting line cover the intervals between the men in the 1st line. To do this, the squad and team leaders of the 1st and 3d squads must take two steps to the rear to allow space for the 2d squad to move into place. After seeing the squad and team leaders move back from the line, the 2d squad leader commands, "Follow Me," and leads his squad into position (see *Figure 6-13*).

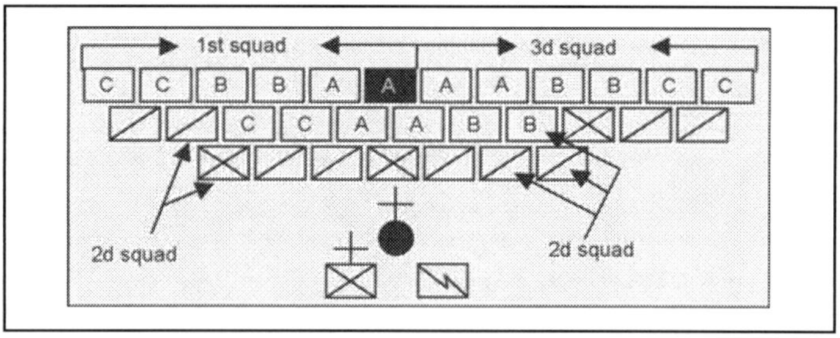

Figure 6-13. Platoon Line Formation With Direct Support

6-64. To assemble a platoon with direct support, the procedure is the same as the platoon line, except that the 2d squad moves to the AA first. To assemble the support squads from any position to general support, the platoon leader commands, "2d Squad in Support, Move." The 2d squad then returns to a column behind the line formed by the other two squads.

Platoon Echelon Right and Left Formations

6-65. The command for forming a platoon echelon right is "Platoon Echelon Right, Move." After the platoon leader's preparatory command, the squad leader of the 1st squad commands, "Follow Me." The squad leaders of the 2d and 3d squads command, "Stand Fast." On the command of execution, the 1st squad moves out and executes an echelon right at the location designated by the platoon leader. As the 1st squad clears the column, the 3d and 2d squads extend the echelon. The 2d squad performs support to this formation in the same ways it supports a line formation.

6-66. The command for forming a platoon echelon left is "Platoon Echelon Left, Move." The platoon echelon right is formed in the same manner as the echelon left (see *Figure 6-14*). *Figures 6-15; 6-16,* page 6-22; and *6-17,* page 6-23 show support formations in a left echelon.

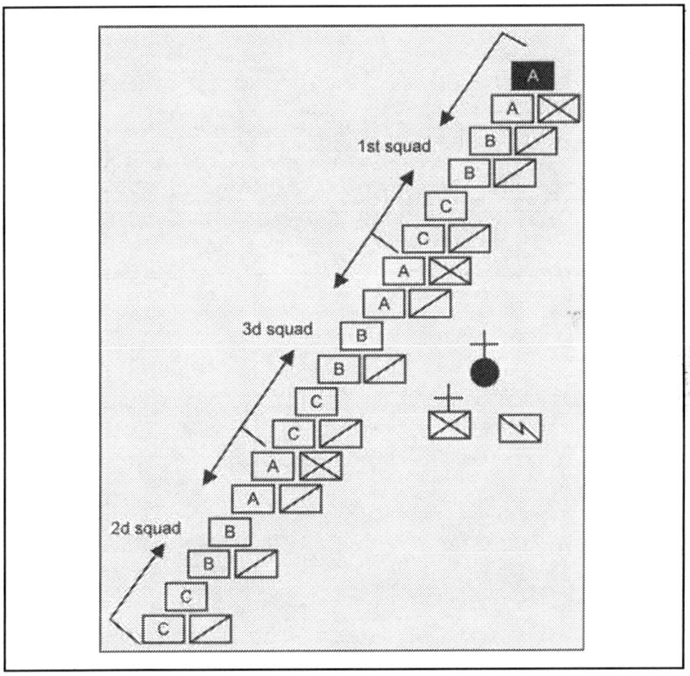

Figure 6-14. Platoon Echelon Left Formation

6-67. To assemble the platoon, the commander takes his position to the rear of the formation with the platoon HQ. On the preparatory command, "Platoon Assemble," the 1st squad leader does an about-face movement while all other

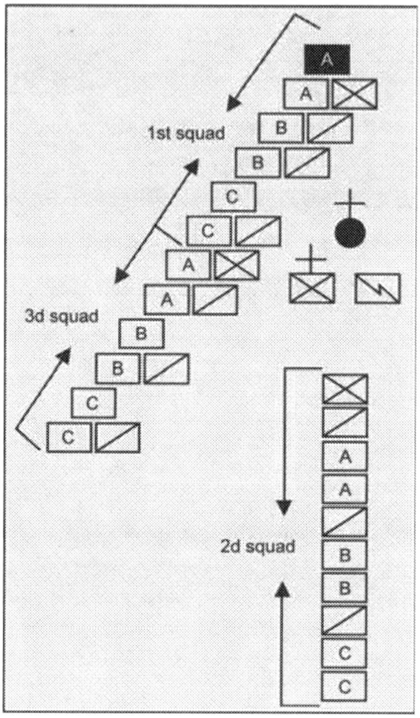

Figure 6-15. Platoon Echelon Left Formation With 2d Squad in General Support

members assume the port arms position and face the baseman. On the command, "Move," the 1st squad moves first, followed by the 3d and 2d squads. Each squad dresses with the 1st squad as it reaches the AA. The assembly of the echelon right formation is the same as the echelon left.

Platoon Wedge Formation

6-68. The command for forming the platoon wedge formation is "Platoon Wedge, Move." After the platoon leader's preparatory command, the squad leaders of the 1st and 3d squads command, "Follow Me." At the same time, the 2d squad leader commands, "Stand Fast." On the platoon leader's command of execution, the 1st and 4th squads move directly to the front. The 1st squad executes an echelon left on the baseman while the 3d squad executes an echelon right. When the last man of the 1st and 4th squads clear the 2d squad, the 2d squad leader commands, "Follow Me," and moves out to the left and right. The 2d squad splits and the A and B teams go to the right and the C

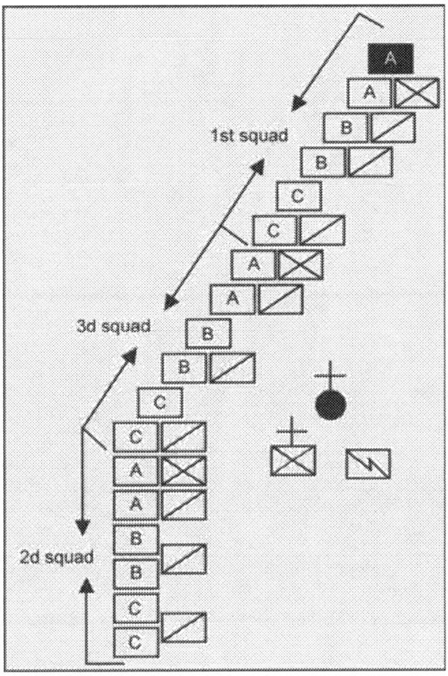

Figure 6-16. Platoon Echelon Left Formation With 2d Squad in Lateral Support

team goes to the left to extend the formation (see *Figure 6-18*). Assembly of the wedge formation is the same as the platoon line.

Platoon Wedge Formation With General Support

6-69. The command for forming a platoon wedge formation with support is "Platoon Wedge, 2d Squad in Support, Move." The 1st and 3d squads execute the wedge while the 2d squad remains in a column formation (see *Figure 6-19*, page 6-24).

Platoon Wedge Formation With Lateral Support

6-70. The command for forming a platoon wedge formation with lateral support is "Platoon Wedge, 2d Squad in Lateral Support, Move." The 1st and 3d squads execute the wedge while the 2d squad stands fast. After the 1st and 3d squads have cleared the column, the 2d squad leader commands, "Follow Me." If no direction is given, the 2d squad splits to cover both flanks. A squad

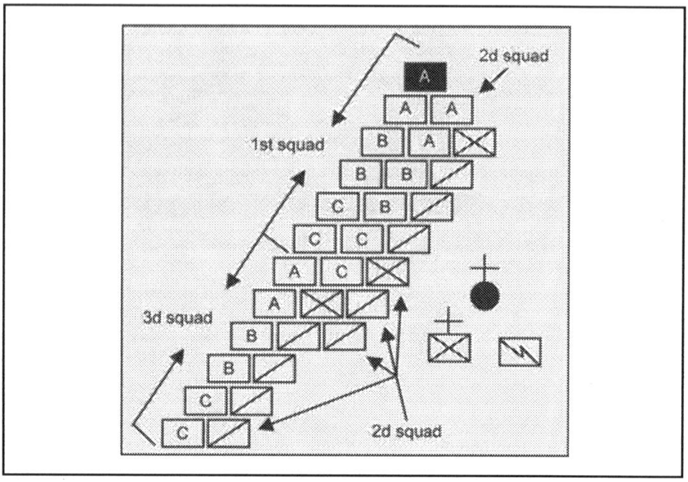

Figure 6-17. Platoon Echelon Left With 2d Squad in Direct Support

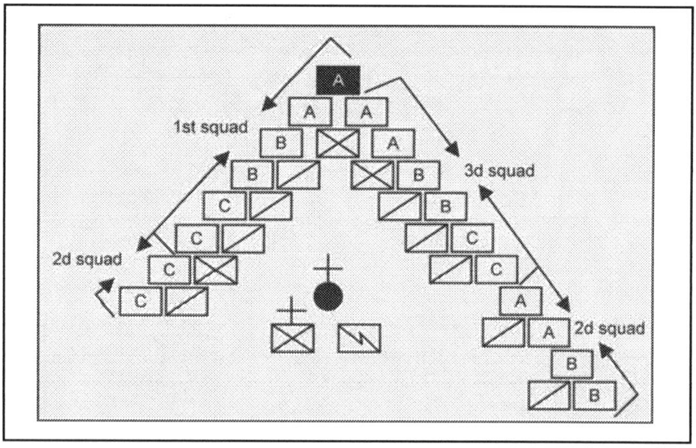

Figure 6-18. Platoon Wedge Formation

may be used to cover one flank by commanding, "2d Squad Right (or Left) Lateral Support" (see *Figures 6-20* and *6-21,* page 6-25).

6-71. To move the 2d squad from general support to lateral support, the platoon leader commands, "2d Squad, Lateral Support, Move." To have the 2d squad join the wedge from general or lateral support, the platoon leader

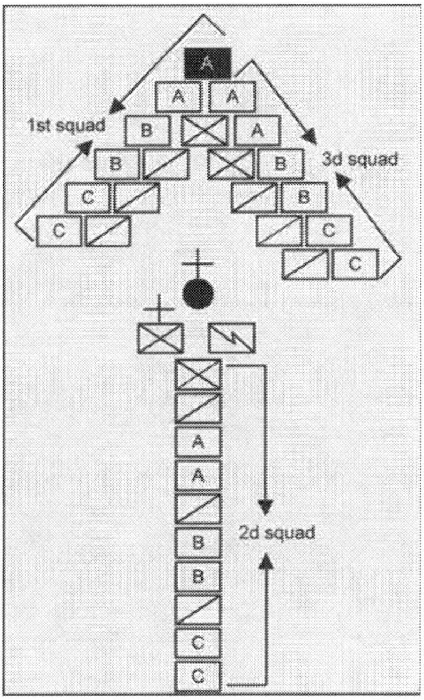

Figure 6-19. Platoon Wedge Formation With 2d Squad in General Support

commands, "2d Squad, Extend the Wedge, Move." To extend the wedge, the 2d squad splits and the teams move to the left and right, respectively.

Platoon Wedge Formation With Direct Support

6-72. The command for forming a platoon wedge formation with the 2d squad in direct support is "Platoon Wedge, 2d Squad in Direct Support, Move." The 1st and 3d squads execute a wedge. The 2d squad waits for the squad and/or team leaders to move to the rear, as in the line formation, before executing a wedge directly behind and centered on the main formation. The individuals in the supporting wedge cover the intervals between individuals in the leading wedge (see *Figure 6-22,* page 6-26).

Platoon Diamond and Circle Formations

6-73. These formations are used in open areas or where the crowd may be able to envelop the formation. The diamond and circle formations are primarily defensive in nature and difficult to maneuver. The decision about which formation to use is based on the ability of the platoon to perform them.

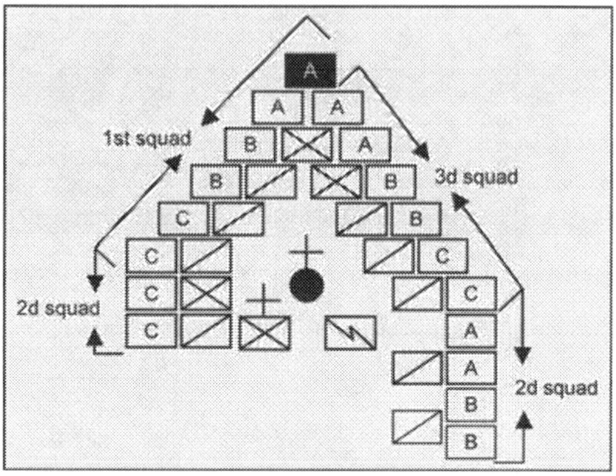

Figure 6-20. Platoon Wedge Formation With 2d Squad in Lateral Support

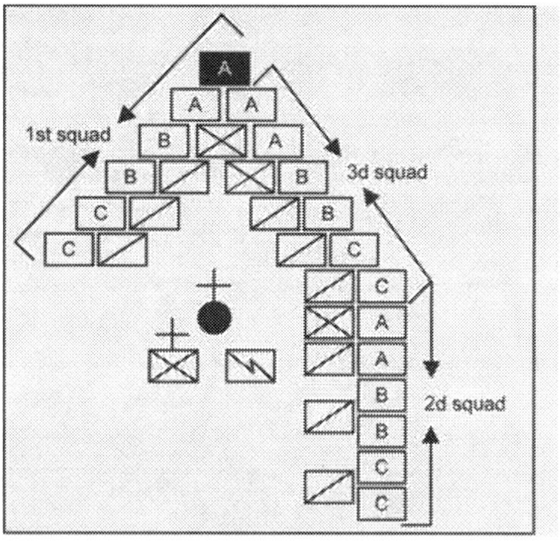

Figure 6-21. Platoon Wedge Formation With 2d Squad in Right Lateral Support

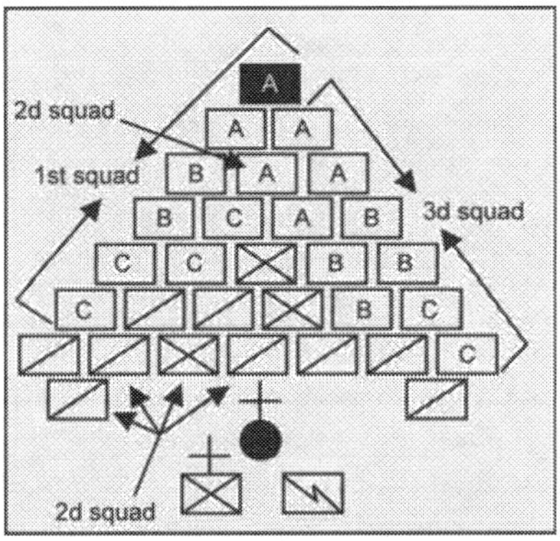

Figure 6-22. Platoon Wedge Formation With 2d Squad in Direct Support

The command for having a platoon form in a diamond or circle formation is "Platoon Diamond (or Circle), Move." On the command of execution, the 1st squad leader moves the baseman to the place indicated by the squad leader. The baseman becomes the 12 o'clock position while the 1st squad lines up to the left of the baseman covering the 8 to 12 o'clock positions. The 3d squad aligns itself to the right of the baseman, covering the 12 to 4 o'clock positions and the 2d squad aligns itself, covering the 4 to 8 o'clock positions (see *Figures 6-23* and *6-24*).

6-74. To assemble from a platoon diamond or circle formation, the HQ element takes the same steps as that of a line formation. The 1st squad leads the platoon back to the AA followed by the 2d and 3d squads. Squads fall into a column formation as they file back to the AA.

RELIEF ELEMENTS IN PLACE

6-75. During crowd control operations, the platoon leader may rotate his squads to give them a rest. He is not limited to using only those squads specified in the preceding examples as the lead elements of the formations. He can replace the base elements by commanding direct support to the 1st squad and then directing the 1st squad into general support. The support element (the 2d squad) then passes through the 1st squad. Once the 2d squad is formed directly behind the 1st squad, the 1st squad leader commands, "Right Face." The members of the 1st squad do a right-face movement and the 2d squad steps quickly though the 1st squad and forms the base formation. The 1st squad leader then commands, "Follow Me," and leads his squad to the general support position. This procedure is often necessary when protective masks are used and chemical irritants employed. The 1st squad then replaces the 3d squad in the same manner. Platoons participating in company operations are relieved in the same manner.

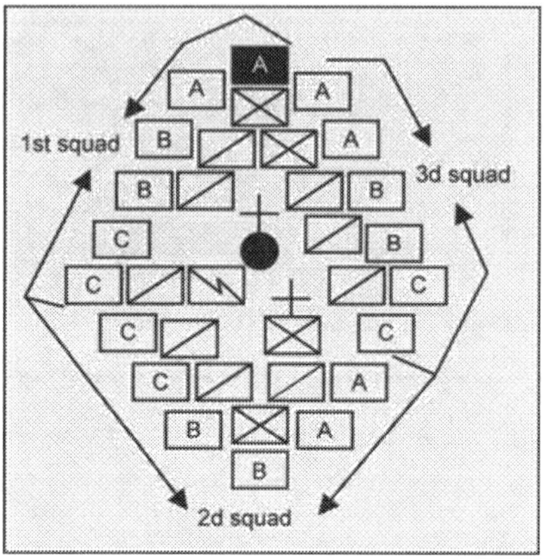

Figure 6-23. Platoon Diamond Formation

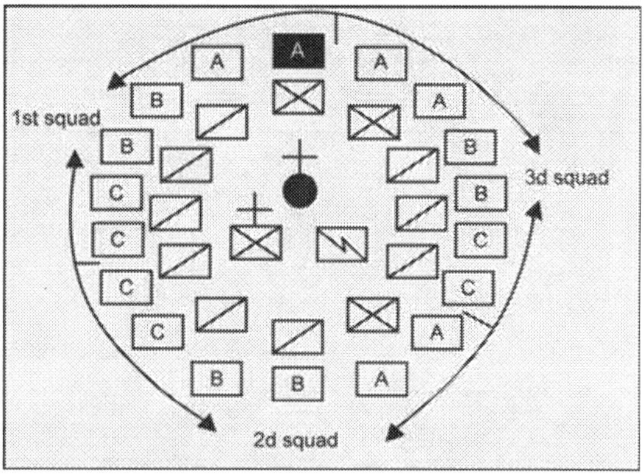

Figure 6-24. Platoon Circle Formation

FOUR-SQUAD PLATOON FORMATIONS

6-76. When forming control force formations from the column, the platoon leader moves to the right of the platoon and faces it when giving his commands. He gives the hand-and-arm signal and the preparatory command. On the command of execution, "Move," he points to the approximate location where he wants the platoon to form. The platoon leader pauses between the preparatory command and the command of execution to allow each squad leader to issue a preparatory command to his squad.

Platoon Line Formation

6-77. The command for forming a platoon line formation is "Platoon Line, Move." Immediately following the platoon leader's preparatory command, the squad leaders of the 1st and 4th squads command, "Follow Me." The squad leaders of the 2d and 3d squads command, "Stand Fast." On the platoon leader's command of execution, the 1st and 4th squads move forward to the point designated by the platoon leader. The 1st A team member of the 1st squad is the baseman for the platoon formation. The 1st squad forms a squad line to the left of the baseman and the 4th squad forms a squad line to the right of the baseman with all lines dressed to the right. After the 1st and 4th squads have cleared the platoon, the 2d and 3d squad leaders command, "Follow Me." The 2d squad forms a squad line to the left of the 1st squad. The 3d squad forms a squad line to the right of the 4th squad. All squads dress with the 4th squad (see *Figure 6-25*).

Platoon Line Formation Assembly

6-78. When assembling from the platoon line formation, the platoon leader and his HQ personnel take a position to the rear of the platoon. He then gives the command, "Platoon Assemble," while raising his right arm and making a circular motion above his head. Platoon members automatically assume the port arms position. The 1st squad leader and the baseman do an about-face movement. All other platoon members do an about-face movement toward the baseman. On the platoon leader's command of execution, he points to where he wants the platoon to assemble. He commands, "Follow Me," and moves his squad toward the AA. The 1st squad files to the baseman position, followed by the 2d squad. As the 2d squad clears the baseman position, the 4th squad leader then commands, "Follow Me," and leads his squad to the AA, followed by the 3d squad. As the 1st squad comes to a halt facing the crowd in the AA, the 2d squad assumes its position to the right of the 1st squad. As the 4th squad assembles, it dresses on the 1st squad, leaving space for the 3d squad to return to its place in the formation. Once the platoon is formed into a column formation, HQ personnel take their position in the formation. The PSG then commands, "Countercolumn, March," and moves the platoon to the rally point.

Platoon Line Formation With General Support

6-79. Whenever the command for establishing a crowd control formation contains the phrase "In Support Without Modification," it means that the support element is to remain in general support. The command for forming a platoon line with two squads in general support is "Platoon Line, 2d and 3d Squads in Support, Move." The 1st and 4th squads execute the line, while the 2d and 3d squads remain in the column (see *Figure 6-26,* page 6-30). To assemble a platoon with two squads in general support, the procedure is the same as that of the platoon line, except that the 1st and 4th squads dress on with the 2d and 3d squads.

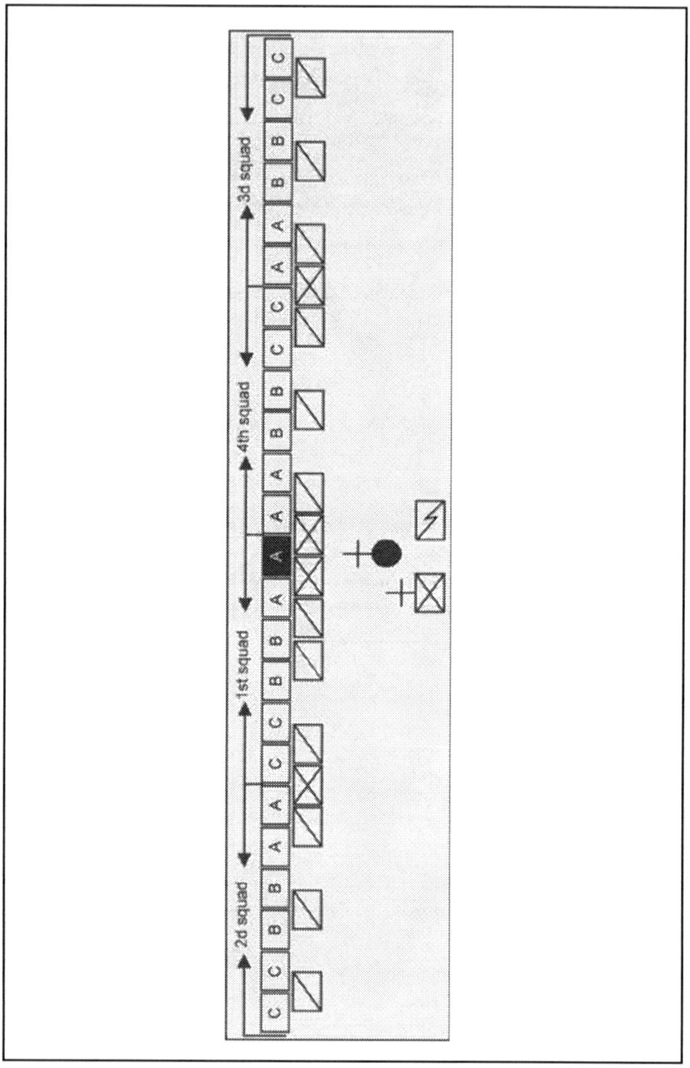

Figure 6-25. Platoon Line Formation With Four Squads

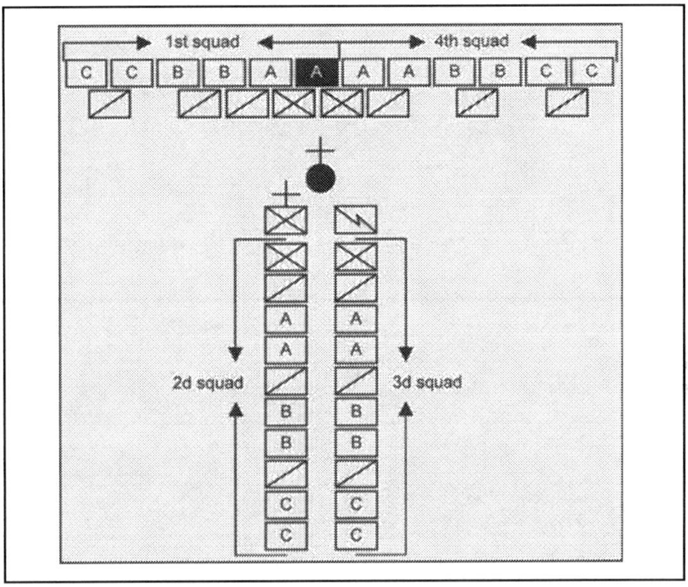

Figure 6-26. Platoon Line Formation With d and 3d Squads in General Support

Platoon Line Formation With Lateral Support

6-80. The command for forming a platoon line formation with two squads in lateral support is "Platoon Line, 2d and 3d Squads in Lateral Support, Move." The 4th and 1st squads execute the line as before, and the 2d and 3d squads stand fast. After the line has been formed, the squad leaders of the 2d and 3d squads command, "Follow Me." The 2d squad forms a column behind the last man of the 1st squad. The 3d squad forms a column behind the last man of the 4th squad (see *Figure 6-27*).

6-81. Squads may be moved from general support to lateral support at any time by the platoon leader. He commands, "2d and 3d Squads, Lateral Support, Move." To have the 2d and 3d squads join the line from either support position, the platoon leader commands, "2d and 3d Squads, Extend the Line, Move." The 2d and 3d squad leaders command, "Follow Me." On the command of execution, the squads set up individual lines, closing and dressing with the existing line. To assemble the platoon from a platoon line with lateral support, the procedure is the same as that of the platoon line, except lateral support squads face forward until they move to the AA.

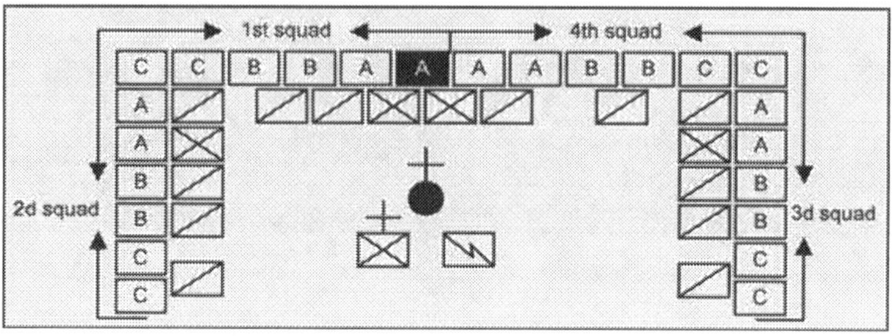

Figure 6-27. Platoon Line Formation With 2d and 3d Squads in Lateral Support

Platoon Line Formation With Direct Support

6-82. The command for forming a platoon line with two squads in direct support is "Platoon Line, 2d and 3d Squads in Direct Support, Move." The 1st and 4th squads execute a line as before. The 2d and 3d squads execute a 2d line directly to the rear and centered on the 1st line. The members in the direct support line cover the intervals between members in the baseline. To do this, the squad and team leaders of the 1st and 4th squads must take two steps to the rear to allow space for the 2d and 3d squads to move into place. After seeing the squad and team leaders move back from the line, 2d and 3d squad leaders command, "Follow Me," and lead their squads into position (see *Figure 6-28*).

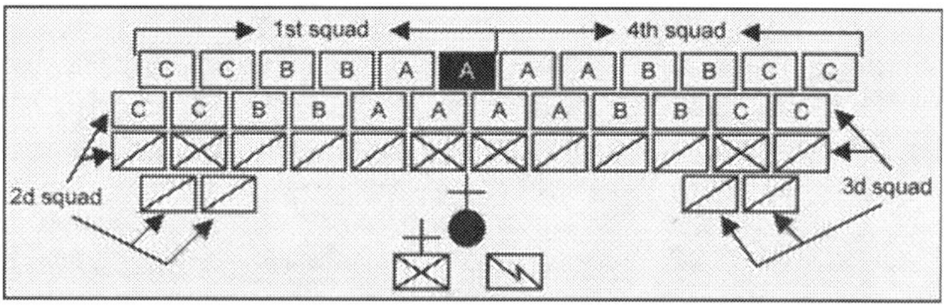

Figure 6-28. Platoon Line Formation With 2d and 3d Squads in Direct Support

6-83. To assemble the support squads from any position to general support, the platoon leader commands, "2d and 3d Squads in Support, Move." The 2d and 3d squads then return to a column behind the line formed by the 1st and 4th squads. To assemble the entire platoon, the procedure is the same as that of a platoon line, except that the 2d squad moves to the AA first, followed by the 3d, 1st, and 4th squads.

Platoon Line Formation With One Support Squad

6-84. The command for forming a platoon line formation with one squad in support is "Platoon Line, 3d Squad in Support, Move." The 1st squad moves out and executes a squad line. The 2d and 4th squads form a line on the side of the left and rights sides of the 1st squad. The 3d squad remains in column formation. The support squad may be used in lateral support on one or both sides of the formation or in direct support of any segment of the formation. To assemble the platoon, the procedure is the same as that of a platoon line, except that all squads dress with the 3d squad.

Platoon Echelon Right and Left Formations

6-85. The command for forming a platoon echelon right formation is "Platoon Echelon Right, Move." After the platoon leader's preparatory command, the squad leader of the 1st squad commands, "Follow Me." The squad leaders of the 2d, 3d, and 4th squads command, "Stand Fast." On the command of execution, the 1st squad moves out and executes an echelon right at the location designated by the platoon leader. As each squad clears the column, the next successive squad moves out and extends the echelon that has been formed. To assemble the platoon, the procedure is the same as that of a platoon line, except that all squads file back in order.

6-86. The command for forming a platoon echelon left formation is "Platoon Echelon Left, Move." The platoon echelon left is formed in the same manner as the echelon right, but in reverse order (see *Figure 6-29*). To assemble the platoon, the procedure is the same as that of an echelon right. See *Figures 6-30,* page 6-34; *6-31,* page 6-35; and *6-32,* page 6-36, for examples of echelon formations with support.

Platoon Wedge Formation

6-87. The command for forming the platoon wedge formation is "Platoon Wedge, Move." After the platoon leader's preparatory command, the 1st and 4th squad leaders command, "Follow Me." At the same time, the 2d and 3d squad leaders command, "Stand Fast." On the platoon leader's command of execution, the 1st and 4th squads move directly to the front. When the last members of the 1st and 4th squads clear the 2d and 3d squads, the 2d and 3d squad leaders command, "Follow Me," and move out to the left and right. The 1st squad executes an echelon left on the baseman and the 4th squad executes an echelon right. The 2d squad forms an echelon left on the last man of the 1st squad. The 3d squad forms an echelon right on the last man of the 4th squad. This completes the wedge formation (see *Figure 6-33,* page 6-36). To assemble the platoon, the procedure is the same as that of a platoon line.

Platoon Wedge Formation With General Support

6-88. The command for forming a platoon wedge formation with two squads in support is "Platoon Wedge, 2d and 3d Squads in Support, Move." The 1st and 4th squads execute the wedge while the 2d and 3d squads remain in column formation (see *Figure 6-34,* page 6-37).

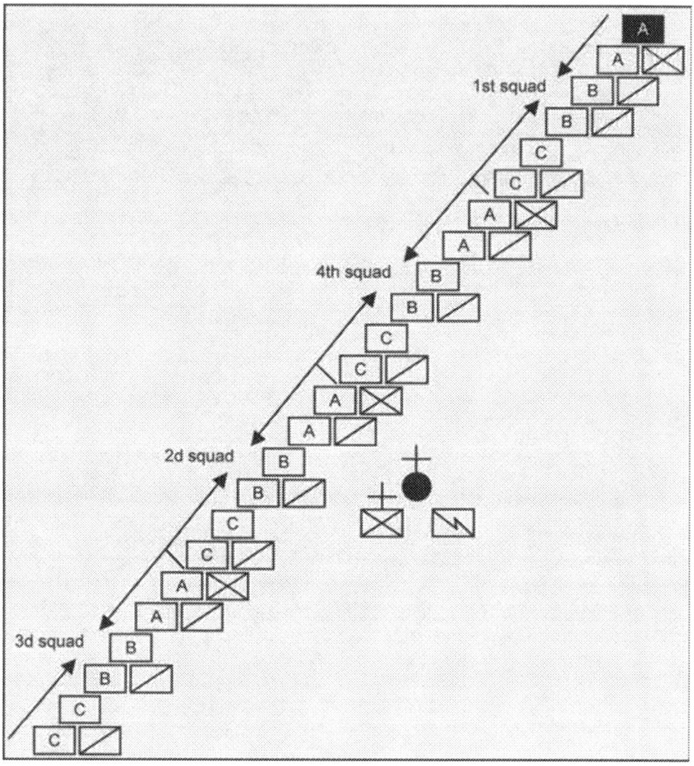

Figure 6-29. Platoon Echelon Left Formation With Four Squads

Platoon Wedge Formation With Lateral Support

6-89. The command for forming a platoon wedge formation with lateral support is "Platoon Wedge, 2d and 3d Squads in Lateral Support, Move." The 1st and 4th squads execute the wedge while the 2d and 3d squads stand fast. After the 1st and 4th squads have cleared the column, the 2d and 3d squad leaders command, "Follow Me," and move out to the left and right. The 2d squad forms a column formation behind the last man of the 1st squad. The 3d squad forms a column formation behind the last man of the 4th squad (see *Figure 6-35*, page 6-38).

6-90. To move the 2d and 3d squads from general to lateral support, the platoon leader commands, "2d and 3d Squads, Lateral Support, Move." To have the 2d and 3d squads join the wedge from general or lateral support, the

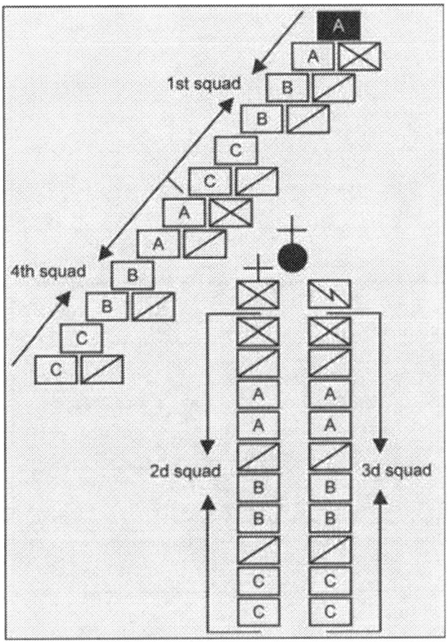

Figure 6-30. Platoon Echelon Left Formation With 2d and 3d Squads in General Support

platoon leader commands, "2d and 3d Squads, Extend the Wedge, Move." The 2d and 3d squad leaders command, "Follow Me," and move out to the left and right to extend the wedge on the 1st and 4th squads.

Platoon Wedge Formation With Direct Support

6-91. The command for forming a platoon wedge with two squads in direct support is "Platoon Wedge, 2d and 3d Squads in Direct Support, Move." The 1st and 4th squads execute a wedge. The 2d and 3d squads execute a wedge and close in on the leading wedge. The members in the supporting wedge cover the intervals between members in the leading wedge. To do this, the squad and team leaders of the 1st and 4th squads must take two steps to the rear to allow space for the 2d and 3d squads to move into place. After seeing the squad and team leaders move back from the line, the 2d and 3d squad leaders command, "Follow Me," and lead their squads into position (See *Figure 6-36,* page 6-38). To move the 2d and 3d squads back to general support, the platoon leader commands, "2d and 3d Squads Support, Move." Then the 2d and 3d squads move to their general support positions. To

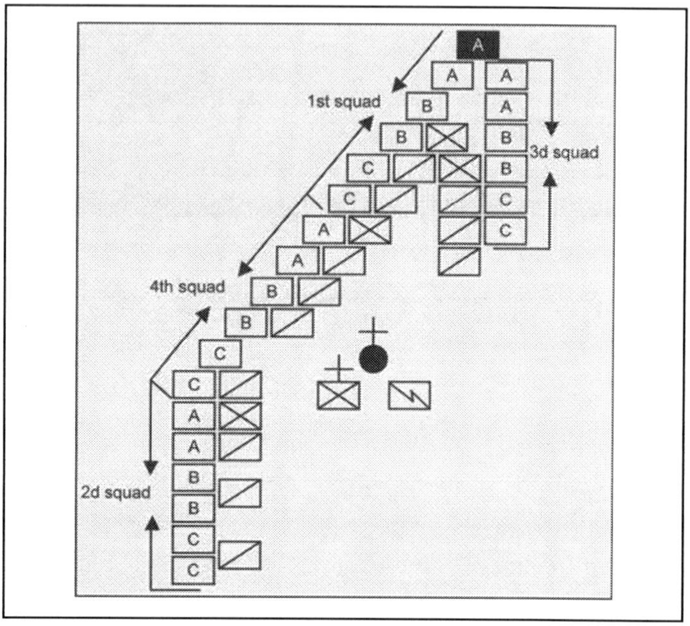

Figure 6-31. Platoon Echelon Left Formation With 2d Squad in Lateral Support

assemble a platoon from a platoon wedge with direct support, the procedure is the same as that of a platoon wedge, except that the 2d squad assembles first, followed by the 3d, 1st, and 4th squads.

Platoon Wedge Formation With One Support Squad

6-92. The command for forming a platoon wedge formation with one squad in support is "Platoon Wedge, 3d Squad in Support, Move." The 1st squad moves out and executes a squad wedge. The 2d and 4th squads form echelons left and right on the 1st squad. The 3d squad remains in the column formation.

Platoon Diamond and Circle Formations

6-93. The platoon diamond and circle formations are used in open areas or where the crowd may be able to envelop the formation. They are primarily defensive in nature and hard to maneuver. The decision concerning which formation to use is based on the ability of the platoon to perform them. The command for having a platoon form in a diamond or circle formation is "Platoon Diamond (or Circle), Move." On the command of execution, the 1st squad leader moves the baseman to the location indicated by the squad leader. The baseman becomes the 12 o'clock position. The 1st squad lines up to the

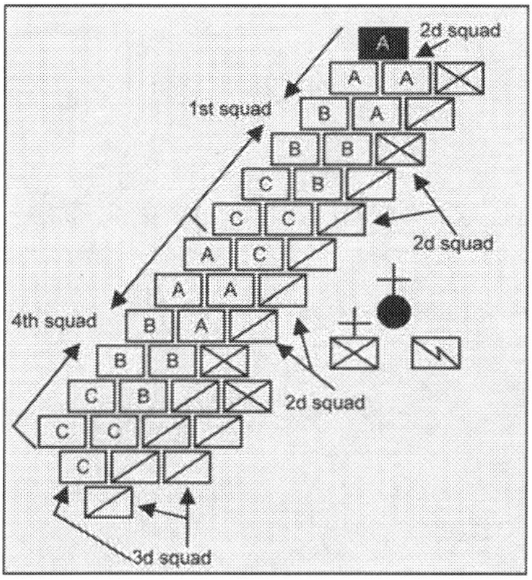

Figure 6-32. Platoon Echelon Left Formation With 2d and 3d Squads in Direct Support

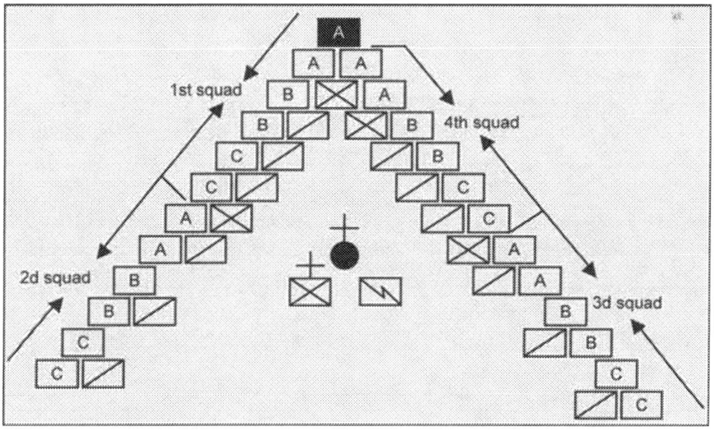

Figure 6-33. Platoon Wedge Formation With Four Squads

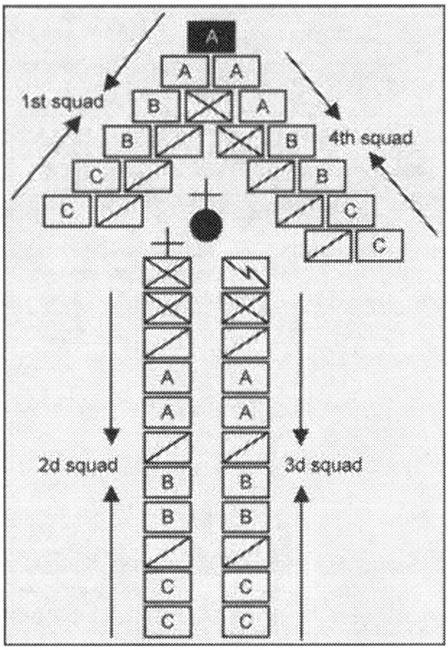

Figure 6-34. Platoon Wedge Formation With 2d and 3d Squads in General Support

left of the baseman covering the 9 to 12 o'clock positions. The 4th squad aligns themselves to the right of the baseman covering the 12 to 3 o'clock positions. The 3d squad aligns itself covering the 3 to 6 o'clock positions. The 2d squad aligns itself covering the 6 to 9 o'clock positions.

6-94. To assemble a platoon from a platoon diamond or circle formation, the HQ element uses the same procedures as that of a line formation. The 1st squad leads the platoon back to the AA followed by the 2d, 3d, and 4th squads. The squads fall into a column formation as they file back to the AA.

COMPANY FORMATIONS

6-95. When a company commander orders his company into control force formations from a column, he moves to the left near the head of the column so he can be seen by the platoon leaders. He then faces the company. As he gives his preparatory command, he points to where he wants the formation. If he does not indicate a location, the company forms immediately in front of the lead platoon.

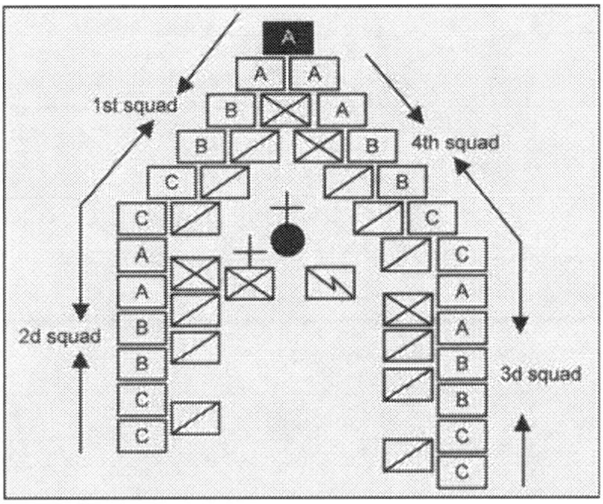

Figure 6-35. Platoon Wedge Formation With 2d and 3d Squads in Lateral Support

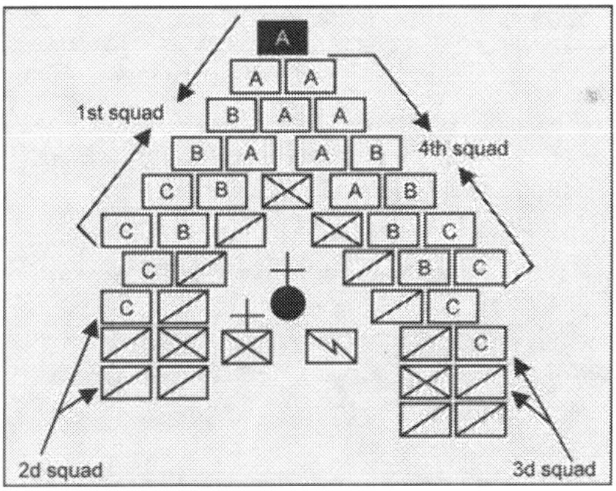

Figure 6-36. Platoon Wedge Formation With 2d and 3d Squads in Direct Support

Relieving Elements in Place

6-96. During crowd control operations, the platoon leader may rotate his squads to give them a rest. See *paragraph 6-76,* page 6-27.

Company Line in Depth Formation

6-97. The command for forming a company line in depth formation (see *Figure 6-37*) is "Company Line in Depth, Move." After the company commander's preparatory command, each platoon leader gives the command, "Platoon Line." The squad leaders follow with their commands to their squads to form the platoon line. On the command of execution, each platoon establishes a platoon line immediately to its front.

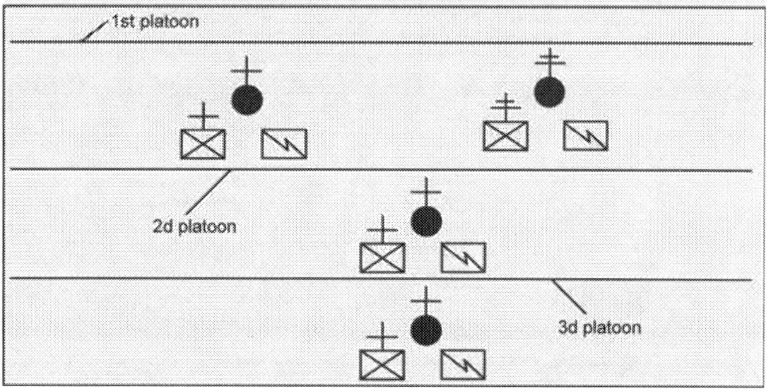

Figure 6-37. Company Line in Depth Formation

6-98. If the commander wants a more formidable formation, he commands, "2d Platoon, Direct Support, Move." The squad and team leaders from the 1st platoon take two steps backward. The 2d platoon moves forward and covers the intervals between the members of the lead platoon. The platoon leader of the 3d platoon then moves his platoon forward to occupy the position formerly held by the 2d platoon.

6-99. If the 3d platoon is called on to support the company line, it moves forward and takes up a position directly behind the 2d platoon in the same manner. Platoon leaders and PSGs help each other in controlling the company.

Company Line in Mass Formation

6-100. The command for forming a company line in mass formation is "Company Line in Mass, Move" (see *Figure 6-38,* page 6-40). The company line in mass is the same as a company line with both platoons in direct support. Instead of forming it in stages, the company commander indicates his desire for this formation in his initial command. The platoons each form a line, and the 2d and 3d platoons close on the 1st platoon without further command.

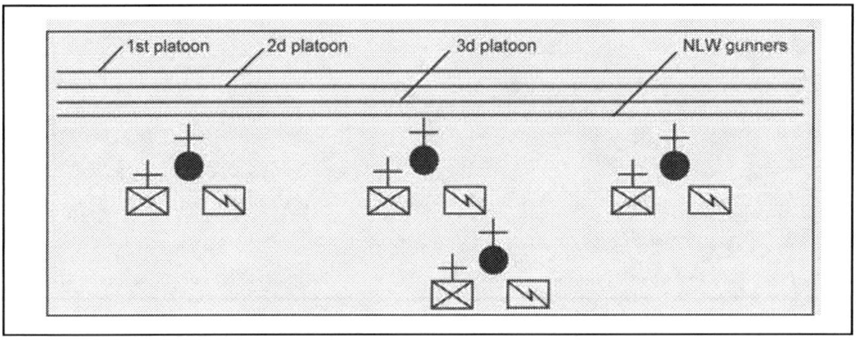

Figure 6-38. Company Line in Mass Formation

Company Line Formation With Support

6-101. In company formations, the 1st platoon usually forms the lead element and the 2d and 3d platoons provide support (see *Figures 6-39; 6-40;* and *6-41*). The support platoons can be employed in the same manner as the support squads in platoon formations. One platoon can be relieved from the lead element by another platoon while in formation by having a support platoon pass through the lead platoon.

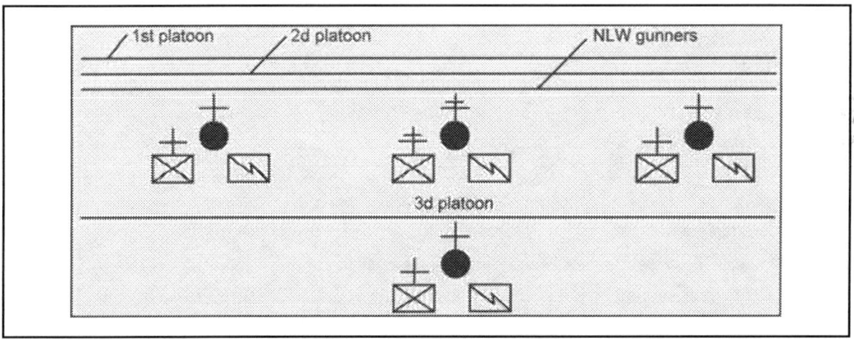

Figure 6-39. Company Line in Mass Formation With One Platoon in Depth

Company Echelons and Wedge Formations

6-102. Company echelons (see *Figures 6-42,* page 6-42*; 6-43,* page 6-43*; 6-44,* page 6-43*;* and *6-45,* page 6-44) and company wedge formations (see *Figures 6-46,* page 6-45; *6-47,* page 6-46; *6-48,* page 6-46; and *6-49,* page 6-47) are formed in the same manner and with the same variations as the company line formation. A company assembles from a crowd control formation in the same manner as a platoon or a squad does. However, due to the size of company

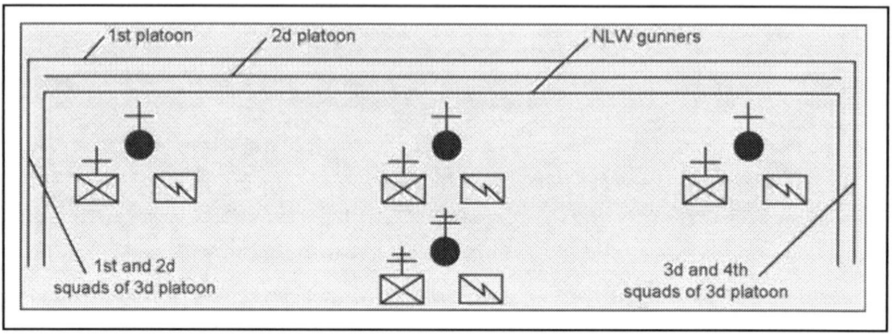

Figure 6-40. Company Line in Mass Formation With One Platoon in Lateral Support

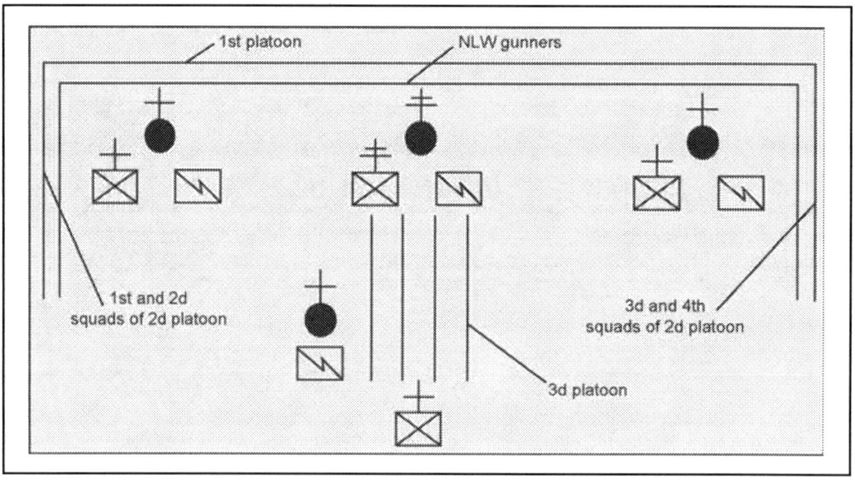

Figure 6-41. Company Line, One Platoon in Lateral Support and One Platoon in General Support

formations, the commander must consider the area that the company is operating in when he assembles the company. If he follows the guidelines of platoon assembly, he will have no difficulty. When assembling the company, the commander takes a position to the rear of the formation and commands, "Company Assemble." The platoon leaders take charge of their respective platoons. Usually, the 3d platoon assembles first, just behind the formation. After the 3d platoon is assembled, the 2d platoon assembles in front of the 3d platoon and the 1st platoon assembles in front of the 2d. All platoons assemble facing the commander. The 2d and 3d platoons then dress with the 1st platoon. As each platoon is assembled, they are ordered to port arms by the platoon leader to await further orders from the commander. If four platoons are used, the 4th platoon assembles first.

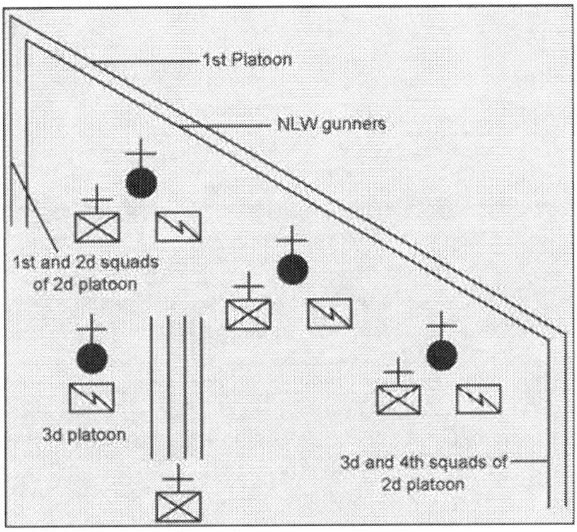

Figure 6-42. Company Echelon Right Formation With One Platoon in Lateral Support and One Platoon in General Support

Company and Platoon Formations With Vehicles

6-103. The commands for vehicles and foot troops are the same as those for foot troops alone. Because of the additional noise and distances involved, the company commander gives the hand-and-arm signal for the line. The command and the signal apply to the lead platoon only. The other platoons remain in general support until further directed (see *Figures 6-50,* page 6-48*; 6-51,* page 6-49*; 6-52,* page 50*; 6-53,* page 6-50*; 6-54,* page 6-51*; 6-55,* page 6-52*;* and *6-56,* page 6-53).

6-104. The motor section moves out with vehicle number two moving to the right. Vehicle number two passes the lead vehicle and establishes the position for the center of the line. At the same time, the 3d and 4th vehicles swing out to the left and right. They form to the left and right and slightly to the rear of vehicle number two. The 1st and 4th squads of the lead platoon then move forward and form lines to the left and right, on the front end of vehicle number two. As soon as the troops are in position, the 3d and 4th vehicles close in on the squads and tighten the formation. The 2d and 3d squads of the lead platoon then form a line to the left and right of the 3d and 4th vehicles to complete the company line. The command vehicle (vehicle number one) takes

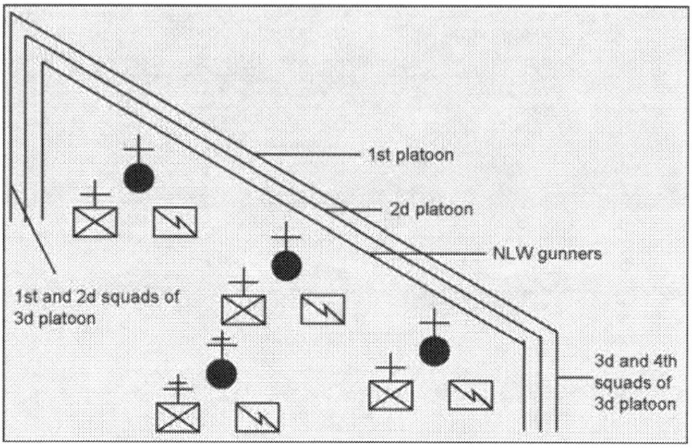

Figure 6-43. Company Echelon Right in Mass Formation With One Platoon in Lateral Support and One Platoon in General Support

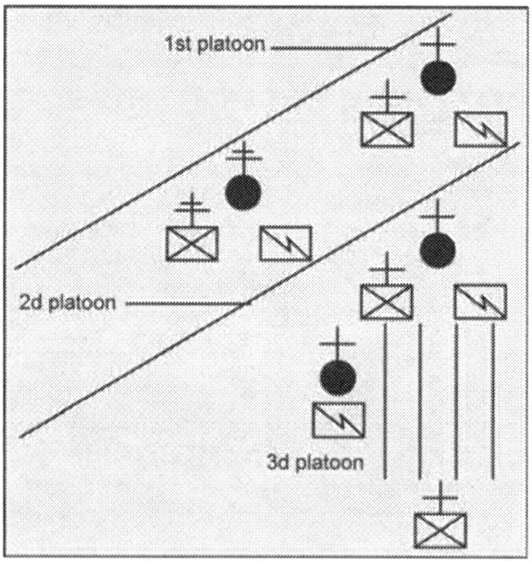

Figure 6-44. Company Echelon Right in Mass Formation With One Platoon in Lateral Support

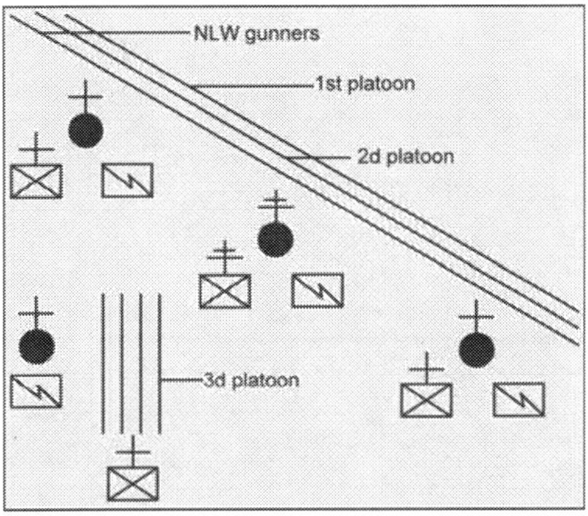

Figure 6-45. Company Echelon Right in Mass Formation With One Platoon in General Support

up a position behind the line where the commander can direct and control the unit. The 2d and 3d platoons move forward and are used in general, lateral, or direct support. In some instances, foot troops move into position first and vehicles join the formation, as needed.

6-105. To form a company echelon or a company wedge with vehicles, the company follows the same procedures used for company formations without vehicles. Vehicles are moved into the appropriate positions using procedures similar to those used for a company line.

6-106. To assemble troops and vehicles, the command vehicle moves to a position behind the formation. The commander faces the formation and gives the vehicle section the signal for assembly. Immediately, the vehicles return to their proper positions in column formation behind the command vehicle while the troops stand fast. The second in command then assembles the troops in the usual manner.

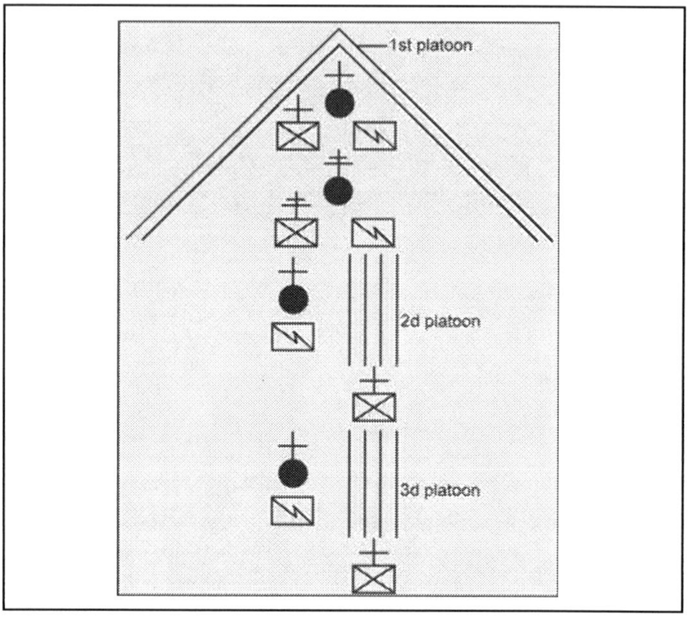

Figure 6-46. Company Wedge Formation With Two Platoons in General Support

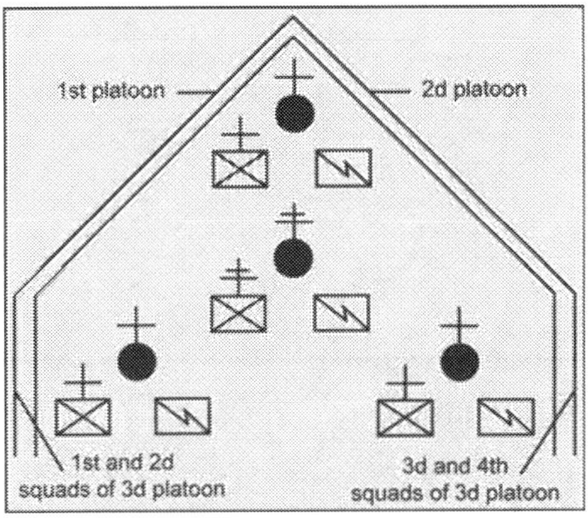

Figure 6-47. Company Wedge Formation With One Platoon in Lateral Support

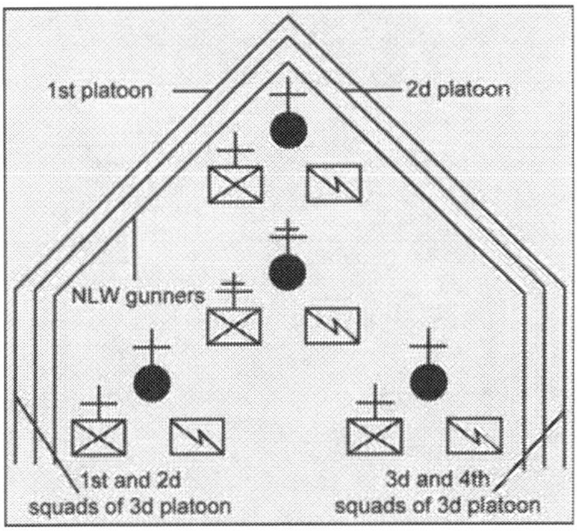

Figure 6-48. Company Wedge Formation With One Platoon in Lateral Support and One Platoon in General Support

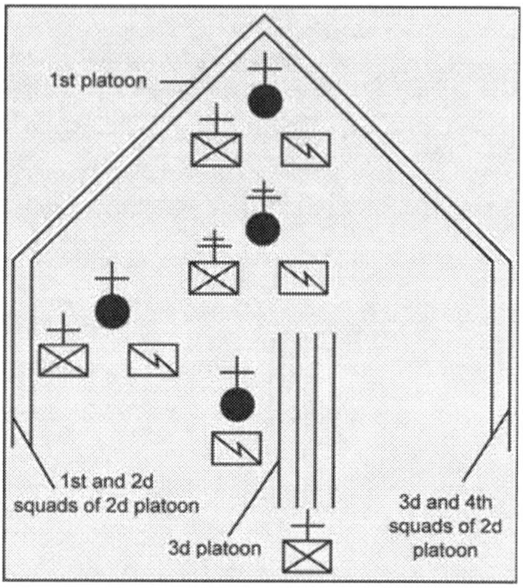

Figure 6-49. Company Wedge Formation With One Platoon in Lateral Support and One Platoon in General Support

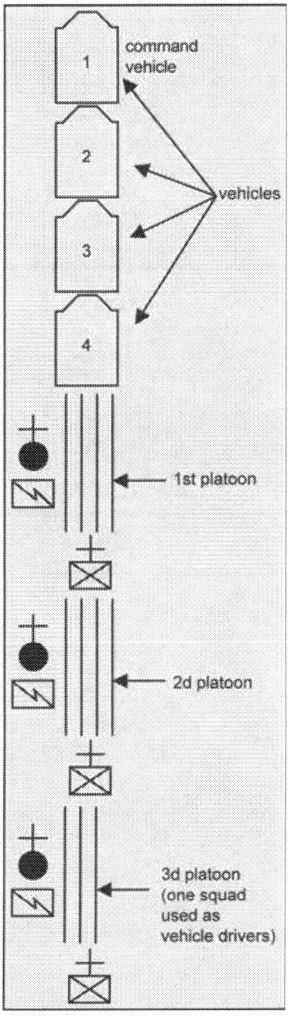

Figure 6-50. Company Column Formation With Vehicles

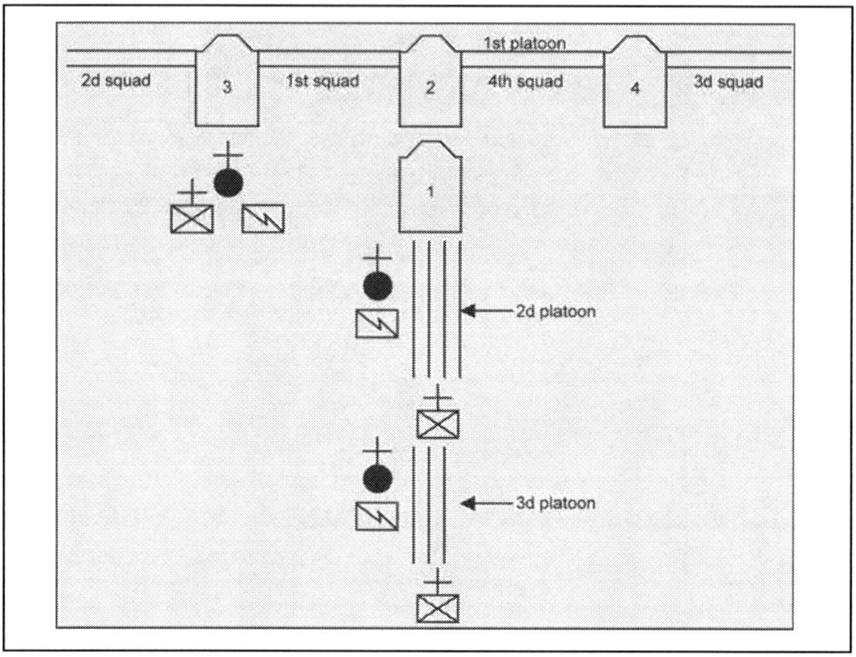

Figure 6-51. Company Line Formation With Vehicles and Two Platoons in General Support

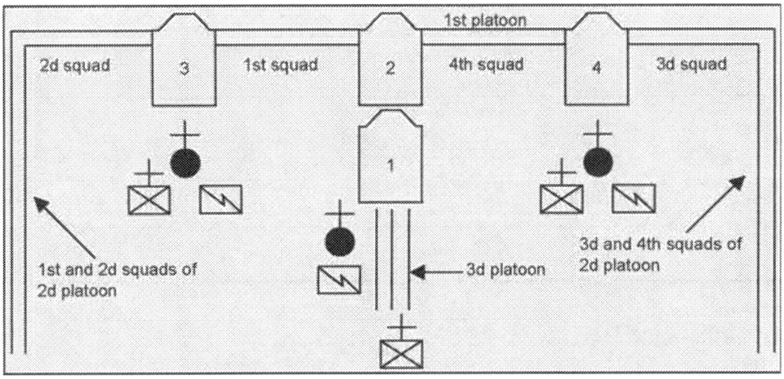

Figure 6-52. Company Line Formation With Vehicles, One Platoon in Lateral Support, and One Platoon in General Support

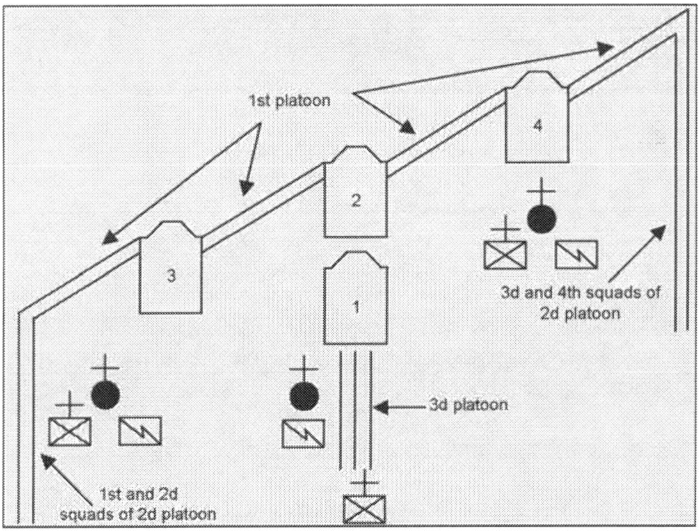

Figure 6-53. Company Echelon Right Formation With Vehicles, One Platoon in Lateral Support, and One Platoon in General Support

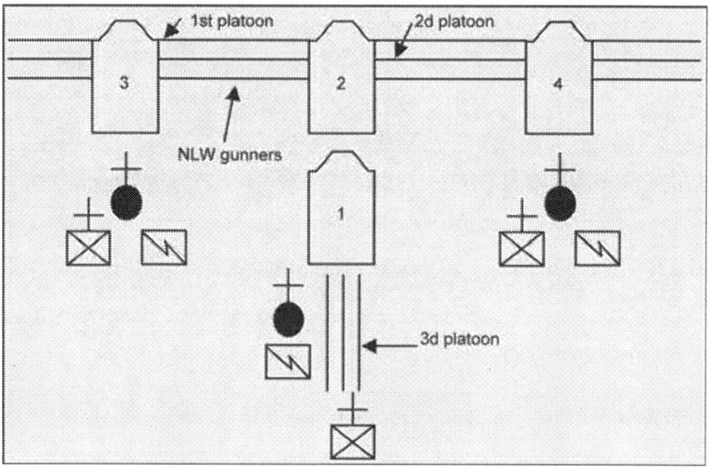

Figure 6-54. Company Line in Mass Formation With Vehicles and One Platoon in General Support

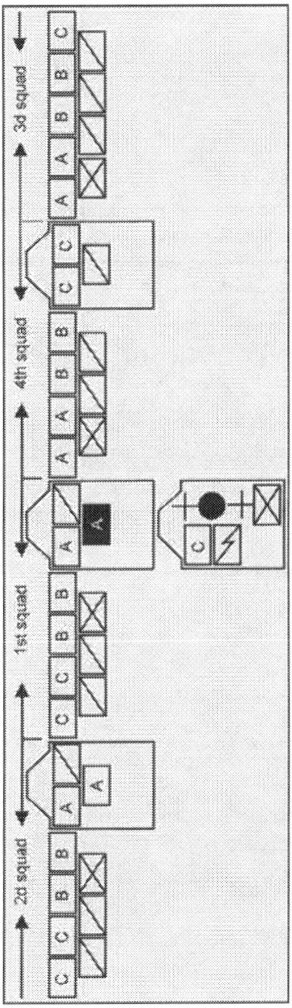

Figure 6-55. Platoon Line Formation With Vehicles

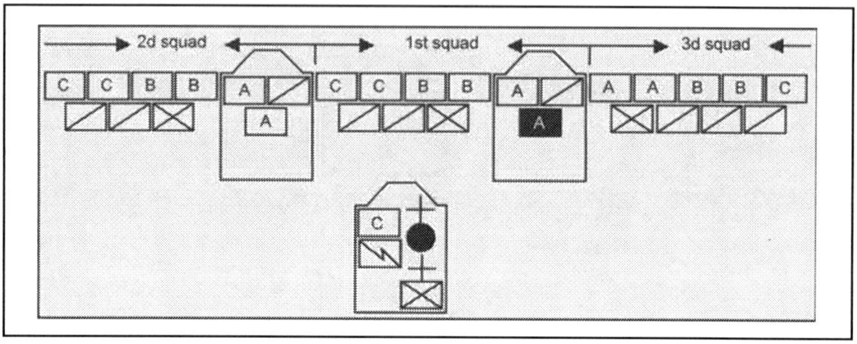

Figure 6-56. Platoon Line Formation With Direct Support and Vehicles

Chapter 7

Civil Disturbance Operations in Confinement Facilities

Civil disturbance doctrine is usually written for quelling disturbances that occur in large open areas where the size of the responding force or the type of formation employed is not hampered by a lack of space. However, in the military confinement environment, inmate disturbances often occur in small, closed areas, such as an inmate's housing unit, a gymnasium, a dining facility, or a chapel where maneuver space and equipment limit the size of the element that may be employed. The layout of each facility is different and presents its own unique problems when employing forces. This chapter covers the use of small military police teams, forced cell move teams (FCMTs), and the formations used to move an unruly inmate from one cell to another. Small generic riot control formations are discussed to provide the commander with a basic employment option that he can tailor to his facility to enable him to regain control of the affected area. The use of NL munitions and RCAs are incorporated into this discussion.

CROWD DYNAMICS WITHIN THE CONFINEMENT FACILITY

7-1. In confinement facility disturbances, inmates employ a number of tactics to resist control or to achieve their goals. Tactics may be unplanned or planned and nonviolent or violent. When a disturbance carries many purposes, it is likely that their tactics are well planned.

NONVIOLENT TACTICS

7-2. Nonviolent tactics range from name-calling to building barricades. Inmates may attempt to distract control force members by shouting at or ridiculing them or by using abusive language, obscene remarks, taunts, and jeers. The inmates' goals are to anger and demoralize the control force. They also want authorities to take actions that may later be exploited as acts of brutality.

VIOLENT TACTICS

7-3. Violent crowd tactics used by inmates are often extremely destructive and can include physical attacks on cadre, other inmates, and property. Use of violent tactics is limited only by the attitudes and ingenuity of the inmates, the training of their leaders, and the materials available to them. Inmates often commit violence with crude homemade weapons. If unplanned violence occurs, inmates will use mops, brooms, chairs, beds, or whatever else is on hand as weapons of violence. During planned violent disturbances, inmates can easily conceal makeshift weapons or tools to use against the control force. Inmates often erect barricades to impede troop movement and/or to prevent the control force from entering certain areas or buildings. Inmates can be expected to vent their emotions on individuals (cadre or inmates), the control force, their equipment, and/or the institution's property. Inmates direct dangerous objects like carts, barrels, liquids, and burning shoe polish at the control force.

GRIEVANCE PROTEST

7-4. A grievance protest may be organized as a riot. However, under normal circumstances, a riot of this type is not of an extremely violent nature. It may turn violent when leaders attempt to exploit any successes or weaknesses found in the control force.

UNORGANIZED RIOTS

7-5. Unorganized riots are characterized as being spontaneous in nature. However, they could be exploited and diverted by leaders at any subsequent stage of the riot, turning it into a different type. They may begin as an isolated assault against authority figures, or acts of civil disobedience. Under determined leadership, an unorganized riot could change to an organized riot.

MULTIPLE RIOTS

7-6. Multiple riots may happen when a disturbance occurs in one area of the confinement facility, causing others to riot in their areas. To control this situation with only one riot control force available, the force commander should first subdue the most violent riot. At the same time, he should employ some of his control forces to contain the other riots until the main control force is able to move to those locations to subdue them.

CADRE BEHAVIOR

7-7. The cadre is susceptible to crowd behavior. They become emotionally stimulated during tense confrontation with unruly and violent inmates. Commanders must be aware of the mood and attitude of the crowd and its effect on the control force cadre. To counteract the effects of crowd behavior on the cadre, the commanders must institute rigorous training and firm and effective leadership. This training must include a complete awareness and understanding of the use of force and the commander's intent. These are all necessary to offset the effect of crowd contagion upon the control force. Even with the best training and leadership, control force members must exercise individual and collective restraint.

PLANNING

7-8. Preplanning is the preparation and training conducted before a crisis occurs. The purpose of preplanning is to have plans and standing operating procedures (SOPs) in place so the cadre may react to an emergency and successfully contain and neutralize the situation. The preplanning process includes plan development, intelligence gathering, and training.

NOTE: Intelligence gathering is essential for the successful containment, neutralization, and/or prevention of a riot.

7-9. Once the control force team has been alerted to a disturbance within the confinement facility, leaders further develop preplans to fit the situation much the same way that DOD develops war plans.

7-10. During the planning process, leaders must be aware of the potential risks involved in quelling disturbances within a confinement facility. Careful planning minimizes collateral damage and risks to control force members and hostages.

7-11. The use of NLW and NL munitions must also be considered when developing plans. There must be strict accountability and control of RCAs, their employment, and other appropriate NL means (such as high-pressure water hoses). When using RCAs, plans must be flexible enough to accommodate changes in the situation and weather because it impacts the effectiveness of some agents. Planners must also consider the—

- Location of the disturbance.
- Estimated number of rioters.
- Accessibility of weapons, tools, or cleaning supplies.
- Inmate's military training.
- Hostage situation.

7-12. Based on an analysis of these factors, the commander makes an estimate of the situation. The estimate must be as thorough as time permits. Using the estimate, the commander considers COAs, selects RCAs, and determines the need for engineer support. In choosing a COA, consider the—

- Number of hostages.
- Number of inmates holding the hostages.
- Leader.
- Attitude and demeanor of the rioters.
- Location of the riot.
  - Inside areas of the confinement facility, such as the dining facility, work site, or housing wing.
  - Outside areas of the confinement facility, such as in exercise or work areas.
- Accessibility of weapons or material to make weapons.
- Accessibility of chemicals, such as cleaning supplies, cooking oils, and shoe polish.

USE OF CHEMICAL IRRITANTS

7-13. The use of chemical irritants can be a valuable NL tool for control force leadership to consider during the planning phase. These chemical irritants can drive a threat from an established, enclosed position or deny the rioters access to a certain area without long-lasting effects to those involved. The proper use of chemical irritants may prevent the control force from having to enter a dangerous area; however, improper use can cause injury, death, or property damage.

USE-OF-FORCE GUIDELINES

7-14. The installation commander authorizes the use of chemical irritants after the control force leader determines the type and dosage. The SOP outlines the procedures for securing authorization and provides guidance to help the control force leader make his decisions. The confinement facility commandant should inform the installation commander of the situation, since the installation commander is ultimately responsible. Team members must be aware of how chemical irritants affect personnel and plan accordingly. Inmates not involved must be removed from the area and the chemical cloud's path. The fire department and emergency medical services (EMS) should be on standby during operations involving chemicals.

PEPPER IRRITANT

7-15. The pepper irritant receives its name from varieties of the capsicum plant that is it made from. The active ingredient, Oleoresin Capsicum (OC), is extracted from the plant and micropulverized to make the irritant. It is neither a chemical nor a gas, but an all-natural organic substance. It is effective against individuals who are emotionally disturbed or under the influence of drugs or alcohol. It is effective against domestic and wild animals. OC is often used to divert the threat from making or continuing an assault. OC does not cause permanent damage. Its effects last 30 to 40 minutes after individuals are placed in fresh air.

7-16. OC can have various effects on people. Some effects are—

- Swelling of mucous membranes and the upper respiratory system; however, it does not shut down the system.
- Burning (intense) and discoloration (bright red) of exposed skin.
- Dilating the capillaries.
- Swelling of the eyelids.
- Burning and involuntary closing of the eyes.
- Coughing uncontrollably.
- Gagging.
- Gasping for air.
- Losing strength and coordination (temporarily).

OC has some disadvantages. They are that—

- The canister depressurizes over time if not checked regularly.
- Inmates may gain access to OC and use it against SRT members.
- The canister must be shaken on a regular basis.

DISSEMINATION OF CHEMICAL IRRITANTS

7-17. There are six methods for disseminating chemical irritants.

Pyrotechnic Method

7-18. In the pyrotechnic method of dissemination, the chemical irritant is placed in a canister with an inert material that is ignited when the device is activated. The agent is then carried into the air on the smoke particles of the inert material.

7-19. There are some disadvantages when using a pyrotechnic device indoors. These disadvantages are that the device—

- May cause a fire.
- Has a very slow saturation time (40 to 45 seconds to burn completely).
- Can be thrown back by the threat, even while burning.

Bursting Method

7-20. In the bursting method of dissemination, the irritant and an inert powder are placed in a saw-toothed canister. After a small detonation, the canister splits and expels the irritant in a cloud.

7-21. The bursting method has its advantages and disadvantages. Some advantages are that—

- There is no risk of fire.
- There is no risk of the canister being thrown back.
- The contents disseminate quickly.
- The canister is effective, easy to carry, and easy to control.

Some disadvantages are that—

- There is a slight possibility of fragmentation when detonated.
- The fuse head may separate from the canister.
- Only 50 to 90 percent of the irritant may be disseminated.
- The direction of the cloud formation depends on where the device is detonated.

Aerosol Method

7-22. In the aerosol method of dissemination, the irritant is suspended in an inert liquid that is located in the rear portion of the device. When the projectile penetrates the structure, the rear portion splits open and the irritant is disseminated in a mist. The aerosol method is most appropriate for tactical operations when chemical irritants are used before entry.

7-23. The aerosol method has its advantages and disadvantages. Some advantages are that it—

- Is easily carried.
- Is an excellent deployment system (uses 40-millimeter, 37-millimeter, or 12-gauge ammunition).
- Disperses the contents quickly.
- Is easily controlled and accurate (fin-stabilized).

NOTE: Practice firing is always necessary.

Some disadvantages are the possibility of—

- Injury to individuals if it is fired from close range.
- Damage to the structure if it strikes a weak portion.

Projectiles Method

7-24. Chemical-containing projectiles can be very effective when used correctly. A soldier must know which projectile to use in each unique situation.

7-25. **Thirty-Seven-Millimeter Projectile.** A 37-millimeter projectile can incapacitate anyone in an enclosed area of 4,500 cubic feet. It can penetrate the following:

- A sheet of ¾-inch-thick plywood at 25 yards.
- An automobile windshield at 33 yards.
- A hollow-core door at 58 yards.

7-26. **Forty-Millimeter Projectile.** A 40-millimeter projectile is more effective at greater ranges than a 37-millimeter projectile because of its rifled bore. It can incapacitate anyone in an enclosed area of 4,500 cubic feet. It can penetrate the following:

- A 1-inch-thick plywood at 10 yards.
- An automobile windshield at 55 yards.
- A hollow-core door at 109 yards.

7-27. **Twelve-Gauge Projectile.** A 12-gauge projectile can be loaded into a shotgun with no modifications to the weapon. It can be fired as a single round, or it can be magazine-fed. The 12-gauge projectile can incapacitate anyone in an enclosed area of 1,000 cubic feet. The angle at which the projectile strikes the barrier can affect its penetration capability. It can penetrate the following:

- A sheet of ¾-inch plywood at 33 yards.
- An automobile windshield at 33 yards.
- A hollow-core door at 100 yards.

Fogging Method

7-28. The irritant is suspended in a fogging solution and dispersed with a fogging machine. It is not recommended for tactical operations.

7-29. The fogging method has its advantages and disadvantages. Some advantages are that it—

- Produces enough chemical irritant to cover an enclosed area of 100,000 cubic feet in 26 seconds.
- Disseminates quickly.
- Provides high-volume capabilities.
- Is effective for riot control or other crowd dispersal situations.

Some disadvantages are that it is—

- Noisy.
- Cumbersome.

Multipurpose Grenade Method

7-30. A multipurpose grenade can be hand-thrown or launched from a shotgun with an adapter. It has an extended shelf life of 6 years and an adjustable fuse delay of 2 to 5 seconds. The irritant is located in the cylinder portion of the device and is forced out the bottom of the grenade upon detonation.

WARNING
A multipurpose grenade deployed incorrectly could result in injury. Safety rules should be observed at all times.

RECORD OF EVENTS

7-31. Reporting procedures for serious incidents should be included in the plans. A record of events must be initiated to provide a basis for the preparation and submission of a formal report to higher HQ. As a minimum, include the—

- Time the incident was reported and by whom.
- Time the incident was reported to the facility commander.
- Time the control force was assembled.
- Time the control force entered the facility.
- Weather conditions as they relate to the use of RCAs.
- Number of cadre and inmates injured or killed, including how they were injured or killed and the medical attention given to them.
- Time the operation was completed.
- Time the riot control force restored order.

TRAINING

7-32. Guard units and associated teams must train on a regular basis in the five basic riot control formations. There must be a continuous training program established, to include—

- Interpersonal communication.
- Self-defense.
- Use of force.
- Use of a riot baton.
- Fatal areas of the body to avoid when using the riot baton.
- First aid.
- Emergency plans.
- Intelligence and counterintelligence techniques.
- The use of RCAs and the various methods of dispersing them.
- The use of NLW.
- The use and effects of high-pressure water.
- Riot control formations.

EQUIPMENT

7-33. The recommended basic riot gear used by the confinement riot control force is as follows:

- Riot baton.
- Riot shield.
- Helmet with face shield.
- Groin protector.
- Flak vest.
- Leather gloves.
- Shin protection.

RIOT BATON

7-34. The riot baton is an invaluable weapon in crowd control situations. A riot baton in the hands of a well-trained control force member is the most appropriate weapon, except with extremely violent crowds. The riot baton is not meant to replace NL munitions, but is instead used in situations in which NL munitions are not needed. Losing a riot baton to the crowd does not present a serious threat.

7-35. Use of the riot baton is based on the commander's appraisal of the situation and his choice of force option. Control forces committed with the riot baton must have teams equipped with RCAs, NLW, and lethal coverage. If the confrontation increases in intensity, the commander has these force options available and may rapidly engage the control force.

7-36. A riot baton is a versatile offensive weapon that varies in length from 26 to 42 inches with the optimum length being 36 inches. It is made of hickory and does not shatter or break easily. Both ends of the riot baton are rounded to prevent unnecessary injury. It is approximately 1¼ inches in diameter. Each riot baton has a hole drilled at the grip end with a leather thong threaded through the hole. The thong helps secure the riot baton to the user's hand.

RIOT SHIELD

7-37. The riot shield (see *Figure 7-1*) is an excellent piece of defensive equipment. It is used by control force personnel in the front and on the flanks of the formation.

7-38. The riot shield has its advantages and disadvantages. Some advantages are that—

- OC pepper spray or high-pressure water can be used to back the control force.
- It provides protection from thrown objects.
- It absorbs some of the impact of rioters hurling themselves at the control force.

The disadvantages of the riot shield are that it—

- Severely limits the use of the riot baton.
- May crack if hit with a heavy object.

Figure 7-1. Riot Shields

RIOT CONTROL FORMATIONS

7-39. The five basic formations for civil disturbance operations are the line, wedge, echelon, diamond, and circle. In the correctional environment the basic formations for disturbances are the line, wedge, or echelon with apprehension teams in support. Generally, the diamond and circle formations are not used in the correctional environment. The minimum size force that the commander should consider is an augmented platoon, with a two-to-one advantage over the rioters.

7-40. Correctional facility cadre must be proficient in all riot control formations. These are usually trained at the squad level and above, but performed by a platoon or company-sized unit. Squad line, echelon left, echelon right, and wedge formations form the basis for platoon and company formations. Each squad must be adept in the basic formations before practicing in platoon-sized or larger formations. For more detailed information on formations, see *STPs 19-95B1-SM* and *19-95B24-SM-TG*. In the correctional environment, the support platoon plays a key role by supporting the control formation with apprehension and equipment teams. See *Chapter 6* for an in-depth discussion on civil disturbance formations.

APPREHENSION TEAMS

7-41. Apprehension teams give the control force the ability to extract inmates. Inmates identified as leaders or agitators and inmates that are injured or no longer want to resist can be extracted from the crowd. Apprehension teams can also be used to enter confined areas where the formation cannot operate.

7-42. The apprehension team will consist of eight members and one leader. Each team will have four shield holders, two nonlethal gunners and two apprehension/search members. The squad leader controls the movement of the team and assists the team in exiting the control formation. He maintains communications with the formation and coordinates the team's operations with the formation. The squad leader also ensures that the team uses the minimum amount of force necessary to extract inmates.

7-43. The apprehension team may deploy through the middle of the formation or from around either of the flanks.

7-44. The apprehension team should be deployed within a distance of approximately 10 meters so that the formation can still provide immediate support. If the apprehension team is deployed farther than 10 meters, the formation must advance to close the gap.

7-45. The team will exit the formation in two columns; shield holders in the first and second ranks, nonlethal gunners in the third rank, apprehension/search members in the fourth rank and the leader last. Team members grip the shoulder or equipment of the member in front of them. The shield holders will envelop the inmate that is to be extracted and form a protective barrier. Once the shield holders have enveloped the inmate, the nonlethal gunners will cover their sides with their weapons while the apprehension/search members control and restrain the inmate. When the team has the inmate under control, the squad leader will give the order for the team to withdraw. The team withdraws in reverse order (apprehension/search members with the inmate, followed by nonlethal gunners and shield holders). Team members then move backwards, face the crowd and grip the shoulder or equipment of the member in front of them.

TEAM EQUIPMENT

7-46. The apprehension team uses the following equipment:

- Flex-Cufs.
- Restraint cutters.
- Pepper spray.
- Helmets with face shields.
- Groin protectors.
- Flak vests.
- Elbow pads.
- Leather gloves.
- Shin protectors.
- Knee pads.
- A video camera.

TEAM COMPOSITION

7-47. The apprehension team is composed of five team members. This team consists of four members and a squad leader. Team member one is the team leader.

Squad Leader

7-48. The squad leader assists the team in exiting the control formation and ensures that the team uses the minimum force necessary to subdue inmates. He is also responsible for signaling the PSG if more help is needed.

Team Member One (Team Leader)

7-49. Team member one's responsibility is to make initial contact with the inmate and use pepper spray and apply Flex-Cufs, as needed. He also carries a set of restraint cutters. Once the apprehension team has returned to the relative safety of the control formation, team member one searches the inmate.

7-50. Team member one receives the command to move forward and secure the inmate from the squad leader. This command is usually given by a signal and verbal command. Team member one then notifies the main force commander where his team exits the control force. Once the team is in position to exit the main formation, team member one signals the member of the formation in front of him. This is usually a prearranged signal, such as a tap on the right shoulder. The control force opens and the team exits the formation and makes contact with the assigned inmate. Team members then subdue and control the inmate. If pepper spray is needed, team member five shouts, "Pepper." This gives the other team members time to react to the command and spray the inmate. Team member one then applies Flex-Cufs to the inmate's wrists and ankles. Depending on the mission, the inmate may be left for follow-on security forces or carried behind the control force to be secured by security forces.

7-51. Should the mission require that the inmate be moved, team member one shouts, "Prepare to Move," this allows each member to grab an arm and position the inmate toward the control force formation. When the team is ready to move, team member two shouts, "Ready," alerting team member one that the team is ready for the last command. Team member one then commands, "Move." The team enters the control formation with the inmate. The inmate is searched by team member one and turned over to the security team.

Team Member Two

7-52. Team member two stands directly behind team member five with his hands on team member five's waist. He protects the left side of team member five by moving to team member five's left and helps to secure the inmate after team member one makes initial contact. He repeats team member five's commands and controls the inmate's head after the inmate is on the ground.

Team Member Three

7-53. Team member three stands behind team member two with his hands on team member two's waist. He protects the right side of team member five by moving to team member five's right and helps to secure the inmate after initial contact.

Team Member Four

7-54. Team member four stands behind team member three with his hands on team member three's waist. His task is to help subdue the inmate in any way possible. He usually concentrates on the inmate's body below the knees. He acts as the eyes and ears for the team during control of the inmate.

Team Member Five

7-55. Team member five gives commands to control the inmate.

FORCED CELL MOVE TEAMS

7-56. The intent of a forced cell move is to move an unruly and/or uncooperative inmate from one cell to another. This is a difficult task and must not be taken lightly. A forced cell move must be thoroughly planned, rehearsed, and resourced to ensure the safety of the team members and the inmate.

7-57. A number of reasons may compel the commander to consider a forced cell move. The inmate may refuse to eat, take medication, accept medical attention, get a haircut, keep himself clean, work, or move or rotate cells when ordered to. These are some of the reasons that the commander may order a forced cell move.

TEAM COMPOSITION AND EQUIPMENT

7-58. The FCMT is composed of six corrections cadre and two support cadre from military police and medical sections specially trained in forced cell movement. Primarily, the team consists of the following personnel:

- One officer in charge (OIC) (E7 or above).
- Five corrections cadre.
- One military police investigator.
- One medic.

7-59. The FCMT uses the following standard equipment:

- Helmets with face shields.
- Groin protectors.
- Flak vests.
- Elbow pads.
- Leather gloves.
- Shin protectors.
- Knee pads.
- A pinning shield.
- Restraints (hand or leg irons or Flex-Cufs).
- A video camera.

ACTIONS PRIOR TO TEAM EMPLOYMENT

7-60. The OIC gives an operations order (OPORD) to team member five (the team leader) of the FCMT. As a minimum, the OPORD should answer the following questions about the incident leading up to the move:

- Did the inmate fail to comply with orders?
- Did the inmate assault cadre or other inmates?
- Is the objective of the move to force a shower or haircut?
- Is OC pepper spray authorized for use?
- Has the inmate been moved before?

- What is the condition of the area where the inmate is currently located?
- Has the inmate tested positive for human immunodeficiency virus (HIV)?
- Does the inmate have a weapon or the materials needed to make one?
- Is the area barricaded?
- Has he smeared himself with body fluids or other slippery substances?
- Can you approach the inmate without risk to him and/or others?

7-61. The team is assembled for inspection by the OIC, and the team members are required to brief the OIC on their individual duties and responsibilities. A line formation is the customary formation for this inspection. It should take place out of site of the inmates.

7-62. Each soldier takes one step forward and briefs the OIC as shown in the following example: *"I am (rank and name), team member one. My mission is to pin the inmate using the minimum amount of force necessary. I carry the pinning shield."* He then faces so the camera can record the number printed on the back of his flak jacket and holds up the pinning shield [see *Figure 7-2*]).

Figure 7-2. Recording Team Members' Duty Position

INDIVIDUAL TEAM MEMBER RESPONSIBILITIES

7-63. Each individual team member has his own responsibilities during a forced cell move. Each team member must do his designated job while working as a team.

Team Member One

7-64. Team member one is the eyes and ears of the team. His responsibility is to carry the pinning shield up to the cell door and describe the scene with as much detail as possible (especially including anything that may cause harm to the team). The pinning shield is used to protect the team from body fluids thrown at the them by the inmate.

7-65. Team member one describes the scene in the cell to his teammates by turning his head to the right and speaking in a loud voice so that all team members can hear him. A slap on the right shoulder by team member two alerts him that all team members have heard him. If he receives a slap on the left shoulder from team member two, he repeats the situation report (SITREP) in a loud, clear voice. The decision to enter the inmate's area with or without the pinning shield is team member one's decision, unless the inmate is known to have a weapon. Should he decide to enter without the pinning shield, team member one then enters with both hands held approximately chest high. When team member one is unable to determine if the inmate has a weapon, the pinning shield is used. Charging through the open cell door, team member one drives the inmate to the bunk or ground with the shield. Once pinned, team member one secures the inmate's head and ensures that it is not twisted or turned in relation to its natural form. He may have to use the various pressure points on the inmate's head to help subdue him.

Team Member Two

7-66. Team member two steps forward and briefs the OIC as follows: *"I am (rank and name), team member two. My mission is to secure the inmate's right arm using the minimum amount of force necessary. I carry one set of hand irons."* He then holds the hand irons up for video camera taping (see *Figure 7-3*) and says, *"I also carry one canister of OC pepper spray to be used at the direction of the team OIC."* He then holds up the canister for video camera taping and faces so the camera can record the number on the back of his flak jacket (see *Figure 7-2,* page 7-13.)

Figure 7-3. Recording Team Members' Equipment

7-67. Team member two lines up directly behind team member one. He keeps contact with team member one until he can secure the inmate's strong arm. Team member two taps team member one on his right or left shoulder after being tapped by team member three. Once inside the cell, team member two secures the prisoners strong hand with wrist restraints and announces, "*Hands secure*," to team member five.

Team Member Three

7-68. Team member three steps forward and briefs the OIC as follows: *"I am (rank and name), team member three. My mission is to secure the inmate's left arm using the minimum amount of force necessary."* He then faces so the camera can record the number on the back of his flak jacket (see *Figure 7-4*).

Team member three lines up directly behind team member two. Team member three keeps contact with team member two until he can grasp a portion of the inmate's body. He then concentrates on securing the inmate's weak hand and assisting team member two in applying hand restraints. Team member three taps team member two on his right or left shoulder after being tapped by team member four.

Figure 7-4. Recording Number on Flak Jacket

Team Member Four

7-69. Team member four steps forward and briefs the OIC as follows: "*I am (rank and name), team member four. My mission is to secure the inmate's right leg using the minimum amount of force necessary. I carry one set of leg irons.*" He holds the leg irons up for the camera to record (see *Figure 7-2,* page 7-13). He then faces so the camera can record the number on the back of his flak jacket (see *Figure 7-2*). Team member four lines up directly behind team member three. He taps team member three on his right or left shoulder after being tapped by team member five. Team member four keeps contact with team member three until he can secure a portion of the inmate's body. Team member four concentrates on securing the right leg of the inmate. It is his responsibility to apply the leg irons. Once the legs are secure, he announces, "*Legs secure,*" to team member five.

Team Member Five (Team Leader)

7-70. Team member five steps forward and briefs the OIC as follows: "*I am (rank and name), team member five. My mission is to secure the inmate's left leg using the minimum amount of force necessary. I carry a restraint key (hold up for the camera to view). I am also the team leader for this mission and it is my responsibility to ensure that the team uses the minimum amount of force necessary to subdue the inmate.*" He then faces so the camera can record the number on the back of his flak jacket (see *Figure 7-4*).

7-71. Team member five taps team member four on the right shoulder when he has heard the SITREP from team member one. If he does not hear the report or needs to hear it again, he taps team member four on the left shoulder. Team member five keeps in contact with team member four until he can secure a portion of the inmate's body. He secures the left leg and assists team member four in applying the leg irons. Once he has heard that the hands and legs are secure, team member five signals for the team to conduct an equipment check. The team members check the equipment in place.

NOTE: At no time is the inmate to be released by anyone.

Medic

7-72. The medic briefs the OIC in the following manner: *"I am (rank and name), the medic. My mission is to treat all injuries that may occur during the move."* He follows the instructions given to him by the OIC/NCOIC. His mission is to observe the inmate for signs of injury before the forced cell move and assess the inmate's condition during and after the move.

Military Police Investigator

7-73. The military police investigator (MPI) is responsible for videotaping the entire forced cell move. This includes the OIC/NCOIC briefing the OPORD, each team member briefing his responsibilities to the OIC, and any medical treatment rendered to the inmate.

ACTIONS DURING THE FORCED CELL MOVE

7-74. There are many actions performed during a forced cell move. Each member of the FCMT must complete his task successfully to accomplish the forced cell move.

Officer in Charge or Noncommissioned Officer in Charge

7-75. The OIC, an E7 or above, is responsible for ensuring that the team uses the minimum amount of force necessary to subdue the inmate. He is also responsible for ensuring that all necessary precautions have been taken to ensure the safety of the team and the inmate. The OIC gives the inmate the initial orders to comply and informs him of the consequences of noncompliance. He does this by placing himself at the front of the inmate's cell (see *Figure 7-5*). The orders are not given until it can be enforced by the FCMT. The following prepared statement is read to the inmate by the OIC: *"I am (rank and name), OIC of this FCMT. At this time I am ordering you to lie face down on your bunk (or other designated location) with your head to the rear of the cell. Cross your hands behind your back and cross your feet as the team enters the cell and uses the minimum amount of force necessary to place you in restraints and move you to (give location). Failure on your part to comply with these instructions or to resist the FCMT will result in the use of force (OC pepper spray if applicable). Do you understand these instructions? If so, lie down on the bunk."*

7-76. After reading the instructions to the inmate, the OIC faces the camera and states whether the inmate complied. He then waits 60 seconds and motions for the team to position itself in front of the inmate's cell (see *Figure 7-6*). The OIC directs the opening of the cell when the team is positioned at the entrance to the cell and prepared to enter. He positions himself so he can observe team entry to and departure from the cell. If OC spray is authorized, the OIC tells the team when to spray the inmate with OC spray. This task is done in silence. Communication between team members should be done through hand-and-arm signals, unless noted otherwise.

Figure 7-5. OIC Reading Compliance Statement to Inmate

Figure 7-6. OIC and FCMT

Team Member One

7-77. Once the team is positioned in front of the inmate's cell, team member one starts his report. At a minimum, he must answer the following questions and report them to the team:

- Does the inmate have a weapon, if so, what kind?
- Does the inmate appear dry?
- Is the inmate clothed or wearing extra clothing?
- Has the inmate smeared anything on his body?
- Can you see the inmate's hands?
- Is the cell barricaded or visibly booby-trapped?

- Is the cell dry?
- Are cell furnishings intact?

7-78. Based on the answers to the above questions, team member one must decide whether to enter the cell with or without the shield (see *Figure 7-7*). Team member one's ultimate goal is to pin the inmate on the bunk or floor.

Figure 7-7. FCMT Preparing to Enter a Cell

Team Member Two

7-79. Team member two keeps contact with team member one until he can secure the inmate's strong arm. If he cannot immediately secure the inmate's arm, he secures a portion of the inmate's body and progresses to the inmate's strong arm. Once the inmate's hands are secure, he announces, "*Hands secure,*" to team member five (see *Figure 7-8).*

Team Member Three

7-80. Team member three keeps contact with team member two until he can grasp a portion of the inmate's body. Team member three concentrates on securing the inmate's weak hand and assists team member two in applying Flex-Cufs. Team member three maintains control of the inmate's hands and looks at team member five for further instructions (see *Figure 7-8*).

Team Member Four

7-81. Team member four keeps contact with team member three until he can secure a portion of the inmate's body. Team member four concentrates on securing the inmate's right leg. He applies leg irons and when the inmate's legs are secure, he announces, "*Legs secure,*" (see *Figure 7-8*).

Team Member Five (Team Leader)

7-82. Team member five keeps contact with team member four until he secures a portion of the inmate's body. Team member five concentrates on securing the inmate's left leg and assists with the application of leg irons. When he is informed that the hands and legs are secure, the team searches the inmate for weapons and contraband. Once the search is complete, team member five commands, "Prepare to Lift."

Figure 7-8. FCMT Securing an Inmate

Movement of the Inmate

7-83. Team member one positions his hands to control the inmate's head during the move, turning the inmate's head toward the wall. Team members two and three position themselves on the inmate's right and left sides. They grasp the wrist or forearm of the team member opposite them to form a platform for the inmate to lie on. Team member four positions himself on the inmate's right side even with the inmate's knees. This places him in position to control the inmate's legs. Team member five's position is on the inmate's left side and enables him to control the inmate's lower legs and feet. When in position, team member five ensures that all team members are in place before commanding, "Ready, Lift."

7-84. On the command "Lift," all team members simultaneously stand with the inmate. They should sandwich the inmate between them to facilitate control. Team member five then commands, "Prepare to Turn, Ready—Turn." The inmate is always turned toward the bunk. If a bunk is not available, team member five should specify which direction to turn the inmate. The inmate is then removed from the cell (see *Figure 7-9*, page 7-20).

7-85. Once the inmate is on the tier, team member five commands, "Prepare to Lower, Team—Lower." The team simultaneously lowers the inmate to the floor with the team keeping positive control of the inmate. The medic asks the inmate, *"Do you have any injuries at this time?"* (see *Figure 7-10*, page 7-20). If the inmate has no injuries, the medic looks at the camera and states, *"Inmate (inmate name) appears to have sustained no injuries."* If the inmate has sustained injuries, the medic treats those injuries and completes a sworn statement. If OC was employed, the medic decontaminates the inmate using water and paper towels. The medic instructs the inmate to tilt his head to the side and flush his eyes with water for approximately 5 minutes. The medic then dries the inmate's face.

Figure 7-9. Moving the Inmate

Figure 7-10. Treating Injuries

Placement of the Inmate in a Cell

7-86. Team member five orders an equipment check before returning the inmate to a cell. He then commands, "Prepare to Lift, Team—Lift." Team members position themselves exactly where they were when they removed the inmate from the cell. Next, team members simultaneously lift the inmate and return him to the predetermined cell. They move the inmate into the cell head first with his face toward the bunk. Once inside the cell, team member five commands the team to lower the inmate onto the bunk (see *Figure 7-11,* page 7-20). Team members then remove the inmate's restraints, keeping positive control of the inmate at all times. Team members use a nonverbal signal to notify team member five when the restraints are removed. He signals another equipment check. Team members then account for all equipment in place and prepare to exit the cell. The team exits the cell in the same manner as they entered, but in reverse order.

Figure 7-11. Placing Inmate in Cell

7-87. Team member four moves into position where he can control both of the inmate's legs while team member five moves toward the cell door. When team member four is ready to exit the cell, team member three moves to control the inmate's legs, team member two moves to where he can control both of the inmates arms, and team member one maintains control of the inmate's head. Team member four maintains contact with team member three at all times. When team member three is ready to exit the cell, team member two moves into a position where he can control the inmate's lower body and team member one moves into position to control the inmate's upper body. Using a prearranged, nonverbal signal, the team pulls team members one and two from the cell. The FCMT is prepared to reenter the cell if the inmate attempts to attack team members one and two as they exit the cell. The FCMT remains ready to reenter until the cell door is secured.

Chapter 8

Civil Disturbance Training

Training requires planning and preparation. Unit commanders structure their training program according to the mission essential task list (METL) (see *FM 7-1)* and expected contingencies. These training plans include prior training and the experience and proficiency of their unit. Leaders agree that the operational tempo and expected contingencies for the next 10 to 15 years will likely involve low-intensity military operations other than war (MOOTW). Therefore, civil disturbance training must be included in the commander's annual training program and must cover all aspects of that type of operation. To effectively function as a member of a control force, individual soldiers must be trained in civil disturbance operations. Training must be intensive, realistic, and sustained. Any lack of training (from basic soldier skills to leadership skills) can have an adverse effect to the point of endangering lives, especially the lives of soldiers. Leader involvement in the training process is essential for mission success in crowd control operations.

UNIT TRAINING STRATEGY DEVELOPMENT

8-1. Developing a unit training strategy begins with the mission analysis portion of the MDMP and identifying the GRM. Missions that require soldiers and units to enforce treaties or accords, protect the lives of civilians in uncertain or hostile environments (such as NEOs), or provide large-scale humanitarian assistance demand some sort of graduated-response criteria to maintain order and prevent uncertain environments from becoming hostile.

8-2. When developing training strategies, planners should focus training and evaluation on the echelon level above the target element (for example TF) or the unit identified as needing training.

8-3. Planners must agree on the intent of the GRM. The GRM is for most a training and rehearsal tool. It provides leaders with likely vignettes that can be incorporated into COA analysis, predeployment training, and rehearsals. The GRM becomes a handy reference during situations that require graduated responses.

TRAINING PLAN DEVELOPMENT

8-4. In developing a training plan for civil disturbances, many considerations should be addressed in terms of available resources. Considerations include the number of personnel to train, the equipment needed and its availability, the time available to train to standard, the location of where training will be executed, and the incorporation of SOPs (based on the theater of operations

[CONUS or OCONUS]). A personnel assessment is necessary to determine the training and experience level of soldiers. It is critical to identify and use soldiers with previous real-world experience and leaders that have graduated from the Interservice Nonlethal Individual Weapons Instructor Course (INIWIC) as primary instructors.

8-5. Commanders should structure the unit training program based on the METL, expected contingencies, prior training, and the proficiency of the unit. In developing a training program, use the seven-step approach. For in-depth instructions on training development, see *FMs 7-0* and *7-1*.

8-6. The objective is to train and evaluate units in the performance of civil disturbance procedures. Proficiency in performing civil disturbance procedures is reflected by a unit's ability to quickly and deliberately execute the disturbance control procedures required to prevent loss of life and property or to regain control of a civil disturbance in its assigned area of responsibility.

8-7. Training should focus on, but is not limited to—

- Instructors.
- Individual and collective drills.
- Leaders.
- Equipment.
- Weapons.
- Munitions.
- Live-fire range exercises.
- Qualification and familiarization.
- Soldier training exercises (STXs).
- Culminating exercises.

8-8. There are several successful methods to evaluate trained leader skills and trained soldier skills using scenarios, vignettes, lanes, or STXs. Flexibility is an important key in training plan development due to time restraints.

RANGE SETUP

8-9. The key to setting up a successful range for NL munitions is to ensure that they are in compliance with installation safety regulations. Kinetic energy is the force that pushes the projectiles down range. In most cases, kinetic energy will only carry these NL munitions projectiles 75 meters or less. This gives trainers flexibility as to the locations where a live-fire exercise can be conducted, especially in a field-expedient situation. *Figure 8-1* is an example for conducting a NL live fire range setup.

8-10. As illustrated in *Figure 8-1*, mark off 5-, 10-, 20-, 30-, 40-, and 50-meter distances from the targets with markers. Each of the NL munitions has minimum and maximum effective ranges within the distances of 5 to 50 meters. Impact power and accuracy are greater at close distances and depleted at farther distances. Targets can be anything from the standard E-type to more realistic three-dimensional manikins. Evenly divide your shooters into three groups as shown in *Figure 8-1*. Group one will initially be shield holders, group two will be shooters, and group three will be grenadiers.

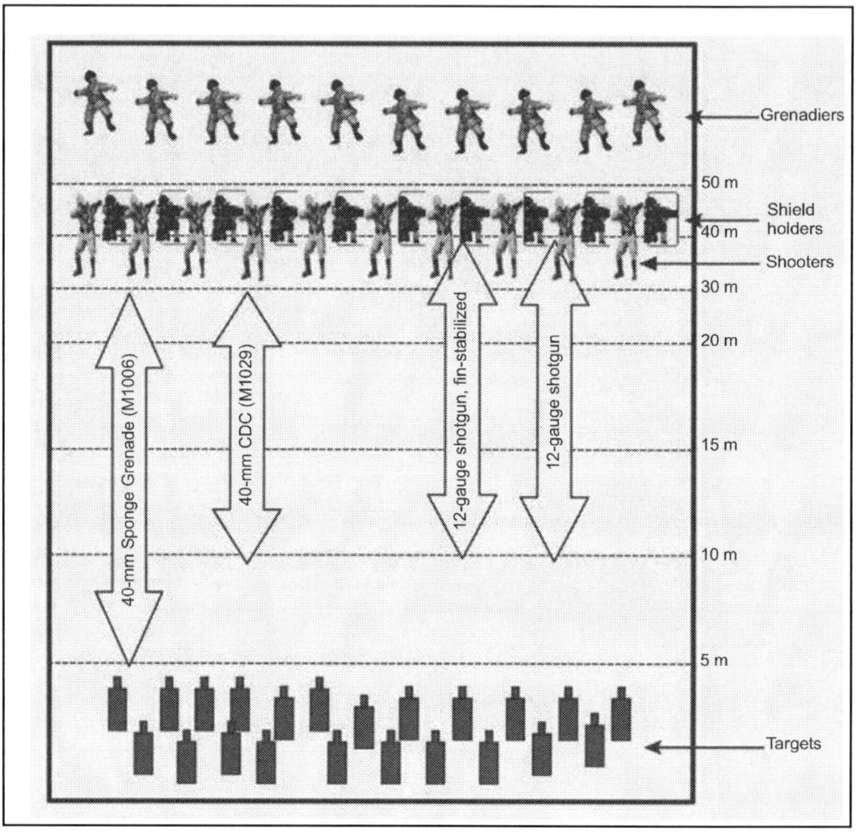

Figure 8-1. NL Munitions Range Setup

8-11. The formation should begin at the farthest effective range firing line (50 meters). After each volley of fire, the formation moves closer to the targets, but stays outside the closest effective range firing line (5 meters). For safety reasons, ensure that the formation stays online throughout each iteration.

8-12. When moving forward to each firing point, it is recommended that the formation always come to a halt before the fire command is given. It is also important to use the tap down drill method (see *Figure 8-1*) rather than having the shooters move forward and fire between the shield holders.

8-13. Range instructors can use verbal commands, such as, "Ready," or "Fire," for each volley. Shooters can be briefed to move forward, tap down the shield holder, and engage targets at the first explosion of distraction grenades thrown by grenadiers. In that instance, the command could simply be "Grenadiers Prepare to Throw, Throw."

8-14. The success of any civil disturbance operation is through effective communication, especially when incorporating the use of NL munitions. For effective training, use verbal commands in conjunction with effective, predetermined hand-and-arm signals during live-fire exercises.

8-15. Each iteration requires two phases of fire. During the first phase, half of the shooters will be firing NL, 40-millimeter rounds using an M203. The others will be firing NL, 12-gauge rounds using a 12-gauge shotgun. In the second phase, shooters switch weapons (safely). It is critical that line safeties are alert during this transition between phases.

8-16. Upon completion of each iteration, clear the weapons and the range. Establish a safe method and return all groups back to the farthest firing line. Once that is accomplished, group one rotates to become group three, group three rotates to become group two, and group two rotates to become group one for the next iteration. Using this method affords a safe and efficient way to get all shooters through the proficiency course. *Figure 8-2* is an example of how to fire each volley based on its characteristics, effectiveness, and capabilities.

Distance	Command	Subject	Actions
10 Meters	12-Gauge Shotgun, Two Point	Shotgun shooters	Load two M1013 rounds.
	M203, Two Point	M203 shooters	Load one M1006 round and have one readily available.
	Ready, Fire	Shotgun shooters	Tap down shields and engage designated targets.
		M203 shooters	Tap down shields, engage designated targets, reload, and reengage.
	Clear Range	Range safeties	Ensure that weapons are clear and safe.
	Safeties, Clear All Weapons	All shooters	Face to right, keep weapons pointed up and downrange.
	Shooters, Right Face, Move	All personnel	Step off shield holder, then shooters, and grenadiers on the command, "Move." Grenadiers reform at 50 meters.

Figure 8-2. Sample Live-Fire Exercise

Distance	Command	Subject	Actions
15 Meters	Shotgun, Two Point	Shotgun shooters	Load two M1013 rounds.
	M203, Two Point	M203 shooters	Load one M1006 round and have another available.
	Grenadiers, Prepare to Throw	Grenadiers	Make distraction grenade ready.
	Throw	Grenadiers	Throw grenade.
		Grenadiers	Throw distraction grenade.
	Ready, Fire	Shot gun shooters	Tap down shields and engage designated targets.
		M203 shooters	Tap down shields, fire at designated targets, reload with M1006 rounds, and engage designated target.
	Clear Range	Range safeties	Ensure that weapons are clear and safe.
	Three Steps Forward	All shooters and grenadiers	Echo preparatory command.
	Move	All shooters and grenadiers	Stay on line and step off three steps, counting each step aloud.
	Clear Range	Range safeties	Ensure that weapons are clear and safe.
	Seven Steps Forward	All shooters and grenadiers	Echo preparatory command.
	Move	All shooters and grenadiers	Stay on line and step off seven steps, counting each step aloud.

Figure 8-2. Sample Live-Fire Exercise (Continued)

Distance	Command	Subject	Actions
20 Meters	Shotgun, Three Point	Shotgun shooters	Load three M1013 rounds.
	M203, One Area, Two Point	M203 shooters	Load one M1026 round and have two M1006 rounds available.
	Grenadiers, Prepare to Throw	Grenadiers	Make distraction grenade ready.
	Throw	Grenadiers	Throw grenade.
		Grenadiers	Throw distraction grenade.
	Ready, Fire	Shotgun shooters	Tap down shields and engage designated targets.
		M203 shooters	Tap down shields, fire area at targets, reload with M1006 rounds, and engage designated target.
	Clear Range	Range safeties	Ensure that weapons are clear and safe.
	Shotgun, Two Point	Shotgun shooters	Load two M1013 rounds.
	M203, Two Point	M203 shooters	Load one M1006 round and have another available.
	Grenadiers, Prepare to Throw	Grenadiers	Make distraction grenade ready.
	Clear Range	Range safeties	Ensure that weapons are clear and safe.
	Three Steps Forward	All shooters and grenadiers	Echo preparatory command.
	Move	All shooters and grenadiers	Stay on line and step off three steps, counting each step aloud.

Figure 8-2. Sample Live-Fire Exercise (Continued)

Distance	Command	Subject	Actions
30 Meters	Shotgun, Two Area, Two Point	Shotgun shooters	Load two M1012 rounds and two M1013 rounds.
	M203, Two Area, Two Point	M203 shooters	Load one round M1026 and have one M1026 and two rounds M1006 available.
	Grenadiers, Prepare to Throw	Grenadiers	Make distraction grenade ready.
	Throw	Grenadiers	Throw grenade.
		Grenadiers	Throw distraction grenade.
	Ready, Fire	Shotgun shooters	Tap down shields, engage targets, and engage designated target with M1013 rounds.
		M203 shooters	Tap down shields, fire area at target, reload with M1006 rounds, and engage designated target.
	Clear Range	Range safeties	Ensure that weapons are clear and safe.
	Seven Steps Forward	All shooters and grenadiers	Echo preparatory command.
	Move	All shooters and grenadiers	Stay on line and step off seven steps, counting each step aloud.

Figure 8-2. Sample Live-Fire Exercise (Continued)

Distance	Command	Subject	Actions
40 Meters	Shotgun, Two Area, One Point	Shotgun shooters	Load two M1012 rounds and one round M1013.
	M203, One Area, One Point	M203 shooters	Load one M1026 round and have one M1006 round available.
	Grenadiers, Prepare to Throw	Grenadiers	Make distraction grenade ready.
	Throw	Grenadiers	Throw grenade.
		Grenadiers	Throw distraction grenade.
	Ready, Fire	Shotgun shooters	Tap down shields and engage targets with M1013 rounds.
		M203 shooters	Tap down shields, fire area at target, reload with M1006 rounds, and engage designated target.
	Clear Range	Range safeties	Ensure that weapons are clear and safe.
	Seven Steps Forward	All shooters and grenadiers	Echo preparatory command.
	Move	All shooters and grenadiers	Stay on line and step off seven steps, counting each step aloud.
50 Meters	Shotgun, One Area	Shotgun shooters	Load one M1012 round.
	M203, One Area, One Point	M203 shooters	Load one M1026 round and have one M1006 round available.
	Grenadiers, Prepare to Throw	Grenadiers	Make distraction grenade ready.
	Throw	Grenadiers	Throw grenade
		Grenadiers	Throw distraction grenade.
	Ready, Fire	Shotgun shooters	Tap down shields, engage targets, and engage designated target with M1013 rounds.
		M203 shooters	Tap down shields, fire area at the target, reload with M1006 rounds, and engage the designated target.
	Clear Range	Range safeties	Ensure that weapons are clear and safe.
	Seven Steps Forward	All shooters and grenadiers	Echo preparatory command.
	Move	All shooters and grenadiers	Stay on line and step off seven steps, counting each step aloud.

Figure 8-2. Sample Live-Fire Exercise (Continued)

NOTE: Shield holders are referred to as group one, shooters are referred to as group two, grenadiers are referred to as group three.

8-17. Currently, there are no qualification standards established in the Army for NL munitions. Area rounds, which contain multiple projectiles designed to affect multiple targets when engaged, only require approximately five rounds. This affords shooters an opportunity to realize the characteristics and capabilities of these rounds.

8-18. Single projectile rounds like the M1013 and the M1006 require a greater degree of proficiency. If situations call for the engagement of targets with these rounds, it is essential that trained shooters be able to hit what they are aiming at. For example, firing 10 rounds from various ranges can provide soldiers with the confidence to be effective shooters. A standard can be established that requires shooters to hit targets with at least 6 out of the 10 rounds used in the example to be considered proficient.

8-19. *Figures 8-3; 8-4,* page 8-11; *8-5,* page 8-21; and *8-6,* page 8-22, give examples of plans to follow if in a situation similar to the following scenario (see *Army Training and Evaluation Program (ARTEP) 19-313-10-MTP*):

A company commander is providing security patrols in support of a joint task force (JTF). Besides providing security patrols for a sector of responsibility, one platoon must always be designated as a quick reaction force (QRF) for civil disturbance operations. For the last 2 weeks the sector has been quiet and the commander has ordered all platoons to conduct civil disturbance training on a rotational basis.

A new order has been sent down to provide a squad-sized element to provide security for the future polling site for an upcoming election. This polling site has yet to open, but intelligence indicates potential problems as the election approaches.

On the day prior to the scheduled election, a crowd of approximately 25 to 30 demonstrators aggressively move toward the polling site, chanting slogans and yelling toward the security forces posted there. While some demonstrators begin throwing whatever they can pick up, the majority are simply yelling, locking arms, and blocking roadways. Hundreds of additional people, not part of the crowd, begin to gather to watch the activities as they unfold.

In an assessment of the demonstrators, it is determined that they intend to stay through the night into the morning of the election with the intention to block access to the polls. It is essential that the crowd is dispersed and that access to the polls is unrestricted in time for the election to begin the next morning.

Evaluation Scheme

The evaluation will consist of five lanes conducted over a 3-day period. Platoons will execute 24-hour operations during the evaluation. Any retraining will be done immediately following the execution of the lane if time permits.

The following tasks will be evaluated:

Lane 1	Cordon and search operations
Lane 2	Civil disturbance QRF
Lane 3	Security and defense of a critical site (polling site)
Lane 4	Night civil disturbance QRF
Lane 5	Night LP/OP (PIO)

Evaluation Schedule

Date and Time	1st Platoon	2d Platoon	3d Platoon	4th Platoon
	Lane 1	Lane 2	Lane 3	Lane 4
	Lane 2	Lane 3	Lane 4	Lane 5
	Lane 3	Lane 4	Lane 5	Lane 1
	Lane 4	Lane 5	Lane 1	Lane 2
	Lane 5	Lane 1	Lane 2	Lane 3

Figure 8-3. Evaluation Scheme

Unit Training Schedule and Firing Tables

Unit Training Schedule 1

Training Day 1

Time	Subject	Location	Uniform	Instructor	Notes
0800	Administrative time	Classroom	U1		
0830	GRM	Classroom	U1		
1000	ROE and use of force	Classroom	U1		
1200	Chow				
1300	MOOTW	Classroom	U1		
1500	Effective communication	Classroom	U1		

Administration:

Uniforms:

U1: BDU

U2: BDU and LBE with harness and canteens

U3: BDU and LBE with harness, canteens, flak jacket, and Kevlar helmet

U4: BDU and LBE with harness, canteens, flak jacket, Kevlar helmet, and assigned weapon

Transportation:

Weapons:

Ammunition:

Required Equipment for Training:

1. Projection screen
2. Video projector with CPU
3. Television
4. VCR

Miscellaneous:

1. All personnel are required to have a notebook and a pencil or pen.
2. Handouts must be prepared and passed out.

Figure 8-4. Sample Two-Week Plan

Training Day 2

Time	Subject	Location	Uniform	Instructor	Notes
0800	NLCS	Classroom	U1		
0900	PPCT and empty hand	Classroom	U1		
1000	PPCT and empty hand (Practical application)	PT field	U2		
1200	Chow				
1300	PPCT and empty hand (Practical application)	PT field	U2		
1700	COB				

Administration:

Uniforms:

U1: BDU

U2: BDU and LBE with harness and canteens

U3: BDU and LBE with harness, canteens, flak jacket, and Kevlar helmet

U4: BDU and LBE with harness, canteens, flak jacket, Kevlar helmet, and assigned weapon

Transportation:

Weapons:

Ammunition:

Required Equipment for Training:

1. Projection screen
2. Video projector with CPU
3. Training strike bags
4. Flex-Cufs

Miscellaneous:

1. All personnel are required to have a notebook and a pencil or pen.
2. Handouts must be prepared and passed out.
3. Medical personnel must be on location.

Figure 8-4. Sample Two-Week Plan (Continued)

Training Day 3					
Time	**Subject**	**Location**	**Uniform**	**Instructor**	**Notes**
0800	PPCT and empty hand (Practical application)	PT field	U2		
1200	Chow				
1300	PPCT and empty hand (Practical application)	PT field	U2		
1700	COB				

Administration:

Uniforms:

U1: BDU

U2: BDU and LBE with harness and canteens

U3: BDU and LBE with harness, canteens, flak jacket, and Kevlar helmet

U4: BDU and LBE with harness, canteens, flak jacket, Kevlar helmet, and assigned weapon

Transportation:

Weapons:

Ammunition:

Required Equipment for Training:

1. Projection screen
2. Video projector with CPU
3. Training strike bags
4. Flex-Cufs

Miscellaneous:

1. All personnel are required to have a notebook and a pencil or pen.
2. Handouts must be prepared and passed out.
3. Medical personnel must be on location.

Training Day 4					
Time	**Subject**	**Location**	**Uniform**	**Instructor**	**Notes**
0800	Riot baton training	Classroom	U1		
0900	Riot baton training (Practical application)	PT field	U2		
1200	Chow				

Figure 8-4. Sample Two-Week Plan (Continued)

1300	Riot baton training (Practical application)	PT field	U2		
1700	COB				

Administration:

Uniforms:

U1: BDU

U2: BDU and LBE with harness and canteens

U3: BDU and LBE with harness, canteens, flak jacket, and Kevlar helmet

U4: BDU and LBE with harness, canteens, flak jacket, Kevlar helmet, and assigned weapon

Transportation:

Weapons:

1. Expandable riot baton with carrier
2. Training riot batons

Ammunition:

Required Equipment for Training:

1. Projection screen
2. Video projector with CPU
3. Training strike bags
4. Flex-Cufs

Miscellaneous:

1. All personnel are required to have a notebook and a pencil or pen.
2. Handouts must be prepared and passed out.
3. Medical personnel must be on location.

Training Day 5

Time	Subject	Location	Uniform	Instructor	Notes
0800	Riot baton training (Practical application)	PT field	U2		
1000	Crowd dynamics	Classroom	U1		
1130	Chow				
1230	Crowd dynamics	Classroom	U1		
1330	CD formations	Classroom	U1		

Figure 8-4. Sample Two-Week Plan (Continued)

1430	RCAs and OC	Classroom	U1		
1700	COB				

Administration:

Uniforms:

U1: BDU

U2: BDU and LBE with harness and canteens

U3: BDU and LBE with harness, canteens, flak jacket, and Kevlar helmet

U4: BDU and LBE with harness, canteens, flak jacket, Kevlar helmet, and assigned weapon

Transportation:

Weapons:

1. Expandable riot baton with carrier
2. Training riot baton

Ammunition:

Required Equipment for Training:

1. Projection screen
2. Video projector with CPU
3. Training strike bags
4. Flex-Cufs

Miscellaneous:

1. All personnel are required to have a notebook and a pencil or pen.
2. Handouts must be prepared and passed out.
3. Medical personnel must be on location.

Training Day 6

Time	Subject	Location	Uniform	Instructor	Notes
0800	RCAs and OC (Practical application)	PT field	U2		
1200	Chow				
1300	RCAs and OC (Practical application)	PT field	U2		
1700	COB				

Figure 8-4. Sample Two-Week Plan (Continued)

Administration:

Uniforms:

U1: BDU

U2: BDU and LBE with harness and canteens

U3: BDU and LBE with harness, canteens, flak jacket, and Kevlar helmet

U4: BDU and LBE with harness, canteens, flak jacket, Kevlar helmet, and assigned weapon

Transportation:

Weapons:

Ammunition:

1. MK inert OC
2. MK 4 live OC disperser
3. M36 live CR disperser

Required Equipment for Training:

1. Five large clean trash cans with liners
2. Two water hoses with running water

Miscellaneous:

Medical personnel must be on location.

Training Day 7

Time	Subject	Location	Uniform	Instructor	Notes
0800	NL munitions employment	Classroom	U1		
1030	Chow				
1115	Safety brief	Range	U4		
1130	NL munitions (live fire)	Range	U4		
TBD	Evening chow				
TBD	Safety brief	Range	U4		
TBD	NL munitions night fire exercise	Range	U4		
TBD	Clean up and COB				

Figure 8-4. Sample Two-Week Plan (Continued)

Administration:

Uniforms:

U1: BDU

U2: BDU and LBE with harness and canteens

U3: BDU and LBE with harness, canteens, flak jacket, and Kevlar helmet

U4: BDU and LBE with harness, canteens, flak jacket, Kevlar helmet, and assigned weapon

Transportation:

Yes

Weapons:

1. M16A2 with M203
2. 12-gauge shotgun

Ammunition:

1. Fin-stabilized M1012
2. M1013 multipellet (rubber)
3. M1006 40-millimeter sponge grenade
4. M1026 40-millimeter multiball (rubber)
5. M84 diversionary grenade

Required Equipment for Training:

1. Fifteen to twenty human form targets
2. Eye and hearing protection
3. Fifteen to twenty riot shields
4. Two MREs per student
5. Video recorder
6. Three to five high-intensity lights
7. All students should be issued face shields, shin guards, and ammunition pouches

Miscellaneous:

1. Medical personnel must be on location.
2. Range safety officers must be available.
3. Primary and secondary means of communication with range control must be available.
4. Operational risk assessments must be conducted.

Figure 8-4. Sample Two-Week Plan (Continued)

Training Day 8					
Time	**Subject**	**Location**	**Uniform**	**Instructor**	**Notes**
0800	CD formations (Practical application)	TA 1	U4		
1200	Chow				
1300	CD formations (Practical application)	TA 1	U4		
1700	COB				

Administration:

Uniforms:

U1: BDU

U2: BDU and LBE with harness and canteens

U3: BDU and LBE with harness, canteens, flak jacket, and Kevlar helmet

U4: BDU and LBE with harness, canteens, flak jacket, Kevlar helmet, and assigned weapon

Transportation:

Yes

Weapons:

1. Training weapons
2. Training riot batons

Ammunition:

Required Equipment for Training:

1. Eye protection
2. Fifteen to twenty riot shields
3. Video recorder
4. Three to five high-intensity lights
5. Face shields, shin guards, and ammunition pouches

Miscellaneous:

1. Medical personnel must be on location.
2. Primary and secondary means of communication must be available.
3. Operational risk assessment must be conducted.
4. Twenty to thirty OPFOR personnel must be available.

Figure 8-4. Sample Two-Week Plan (Continued)

Training Day 9					
Time	Subject	Location	Uniform	Instructor	Notes
0800	Introduction to M5, PVAB, and M315	Classroom 1	U1		
1200	Chow				
1300	M5, PVAB, and M315 (Practical application/LFE)	Range	U4		
1700	COB				

Administration:

Uniforms:

U1: BDU

U2: BDU and LBE with harness and canteens

U3: BDU and LBE with harness, canteens, flak jacket, and Kevlar helmet

U4: BDU and LBE with harness, canteens, flak jacket, Kevlar helmet, and assigned weapon

Transportation:

Yes

Weapons:

Two complete M315s and installation kits

Ammunition:

1. Ten live MCCMs
2. Ten live M315 rounds of each

Required Equipment for Training:

1. Eye protection
2. Ten inert MCCMs
3. Two complete PVAB systems

Miscellaneous:

1. Medical personnel must be on location.
2. Primary and secondary means of communication must be available.
3. Operational risk assessment must be conducted.

Figure 8-4. Sample Two-Week Plan (Continued)

Training Day 10					
Time	**Subject**	**Location**	**Uniform**	**Instructor**	**Notes**
0800	Tactical consideration	MOUT classroom	U1		
0930	Safety brief	MOUT classroom	U4		
1000	Tactical exercise	MOUT Facility	U4		

Administration:

Uniforms:

U1: BDU

U2: BDU and LBE with harness and canteens

U3: BDU and LBE with harness, canteens, flak jacket, and Kevlar helmet

U4: BDU and LBE with harness, canteens, flak jacket, Kevlar helmet, and assigned weapon

Transportation:

Yes

Weapons:

1. Assigned weapons
2. Inert RCA/OC individual dispersers
3. Training riot batons
4. Fifty practice grenades

Ammunition:

Required Equipment for Training:

1. Eye protection
2. Ten inert MCCMs
3. Face shields, shin guards, and ammunition pouches
4. Fifteen to twenty full-length riot shields

Miscellaneous:

1. Medical personnel must be on location.
2. Primary and secondary means of communication must be available.
3. Operational risk assessment must be conducted.
4. A video recorder must be available.

Figure 8-4. Sample Two-Week Plan (Continued)

Sample Unit Training Schedule				
Training Day 1				
Time	**Event**	**Location**	**Instructor**	**Notes**
0730-0830	Force continuum	Classroom		L
0830-0930	Management of aggressive behavior	Classroom		L
0930-1030	ROE	Classroom		L
1030-1130	Crowd control	Classroom		L
1130-1230	Chow			
1230-1700	PPCT and empty hand	PT field		L, PA
Training Day 2				
Time	**Event**	**Location**	**Instructor**	**Notes**
0730-1130	Riot baton training	PT field		L, PA
1130-1230	Chow			
1230-1530	Riot baton training	PT field		PA
1530-1700	CD formations	Classroom		L
Training Day 3				
Time	**Event**	**Location**	**Instructor**	**Notes**
0730-1130	CD formations	PT field		PA
1130-1230	Chow			
1230-1700	NL munitions, NLW	Range		L, PA
Training Day 4				
Time	**Event**	**Location**	**Instructor**	**Notes**
0730-1130	RCAs and OC	Classroom and PT field		L, PA
1130-1230	Chow			
1230-1700	RCAs and OC	PT field		PA

Figure 8-5. Sample One-Week Plan

Training Day 5				
Time	**Event**	**Location**	**Instructor**	**Notes**
0730-1700	Tactical exercise	MOUT facility		OPS
NOTE: Appointed student leaders ensure that troops are fed and rotated for rest.				

Figure 8-5. Sample One-Week Plan (Continued)

Sample Unit Training Schedule				
Training Day 1				
Time	**Event**	**Location**	**Instructor**	**Notes**
	Shield and riot baton techniques	MOUT site		Individual and collective tasks
	NLCS and munitions	Range Classroom		Leader training
	Formation training	MOUT site		Individual and collective tasks
	Chow			
	Formation training	MOUT site		Individual and collective tasks
Training Day 2				
Time	**Event**	**Location**	**Instructor**	**Notes**
0730-1930, 12-hour operations	Live-fire exercise	Range		M203/12-gauge
	QRF exercise	MOUT site		Individual and collective tasks
NOTE: Once weapons and munitions familiarization is complete, shooters join the QRF exercise to be used. For live-fire range setup, see *Figure 8-1*.				

Figure 8-6. Sample Two-Day Plan

Appendix A

Metric Conversion Chart

This appendix complies with current Army directives which state that the metric system will be incorporated into all new publications. *Table A-1* is a metric conversion chart.

Table A-1. Metric Conversion Chart

US Units	Multiplied By	Metric Units
Cubic feet	0.0283	Cubic meters
Feet	0.3048	Meters
Feet per second	0.3048	Meters per second
Inches	0.0254	Meters
Inches	25.4001	Millimeters
Miles per hour	1.6093	Kilometers per hour
Pounds	0.4536	Kilograms
Yards	0.9144	Meters
Metric Units	**Multiplied By**	**US Units**
Cubic meters	35.3146	Cubic feet
Kilograms	2.2046	Pounds
Kilometers per hour	0.6214	Miles per hour
Meters	3.2808	Feet
Meters	39.3701	Inches
Meters	1.0936	Yards
Meters per second	3.2808	Feet per second
Millimeters	0.0394	Inches

Appendix B

Operations and Legal Considerations in the Continental United States

Providing military assistance to state and local governments to assist them in quelling a civil disturbance or riot requires close coordination through a host of state and federal agencies. It requires a thorough briefing of soldiers at all levels on what they can and cannot do with respect to law enforcement. Civil authorities must be briefed on the restrictions placed on federal forces by the *Constitution of the United States* and federal statutes and laws. The national guard (NG), when not in a federal status, operates under the control of the state governor and the adjutant general (AG). The NG has historically been the first military responder during emergencies.

FEDERAL INTERVENTION AND AID

B-1. Under the *Constitution of the United States* and *United States Code*, the President is empowered to direct federal intervention in civil disturbances to—

- Respond to state requests for aid in restoring order.
- Enforce the laws of the United States.
- Protect the civil rights of citizens.
- Protect federal property and functions.

B-2. Under the *Constitution of the United States*, each state is responsible for protecting life and property within its boundaries. State and local governments use their LEA forces to maintain law and order and to quell civil disturbances. However, if civil disturbance support requirements exceed the resources of a state, federal troops may be requested to help restore and maintain law and order.

B-3. The *Constitution of the United States* and federal statutes authorize the President to direct the use of federal armed troops within the 50 states and territories and their political subdivisions. The President is also empowered to federalize the NG of any state to suppress rebellion and enforce federal laws.

B-4. Federal assistance is provided to a state when the state has used all of its resources, including its NG, to quell a disorder and finds the resources insufficient. Usually, active duty federal forces are used to augment the NG of the requesting state. However, the President may choose to federalize the NG of another state and use them alone or with other forces to restore order.

B-5. The President may employ federal troops to ensure the execution of US law when a state opposes or obstructs US law or impedes the course of justice under those laws. The President may employ armed federal troops to suppress insurrection, domestic violence, unlawful assemblies, and conspiracy if such acts deprive the people of their constitutional rights and privileges and the civil authorities of a state cannot or will not provide adequate protection.

B-6. The president is also authorized to use armed federal troops to protect federal property and functions when the need for protection exists and local civil authorities cannot or will not give adequate protection. The right of the United States to protect all federal property and functions regardless of their locations is an accepted principle of our government.

B-7. As a temporary measure, federal military equipment and facilities may be loaned to law enforcement agencies of state and local governmental bodies for use during civil disturbances. These resources may also be loaned to state NG and non-DOD federal agencies. The requesting agencies are expected to provide enough resources of their own to minimize the need for US military resources. In addition, the loan of the resources must not conflict with US military needs.

ROLES AND RESPONSIBILITIES OF VARIOUS AGENCIES

B-8. Several agencies within the federal and state governments play a key role in supporting the request of local authorities for support during a civil disturbance.

SECRETARY OF DEFENSE

B-9. The Secretary of Defense retains approval for federal support to civil authorities involving the use of DOD forces, personnel, and equipment. Approval may also involve DOD support that will result in a planned event with the potential for confrontation with specifically identified individuals and groups or will result in the use of lethal force.

SECRETARY OF THE ARMY

B-10. The Secretary of the Army, as an executive agent for the Secretary of Defense, is the approval authority for federal emergency support in response to natural and man-made disasters (except weapons of mass destruction [WMD]). Military forces respond to directions and guidance from the director of military support (DOMS) (the action agent for the Secretary of the Army).

DIRECTOR OF MILITARY SUPPORT

B-11. The DOMS plans and executes the DOD domestic support mission to civil authorities. He is a general officer appointed by the Secretary of the Army. The DOMS is the DOD primary contact for all federal departments and agencies during DOD involvement in most domestic support operations.

STATE COORDINATING OFFICER

B-12. The governor appoints the state coordinating officer (SCO) to coordinate disaster operations for the state. The SCO is the primary point of contact for the federal coordinating officer (FCO) in facilitating disaster assistance.

DIRECTOR OF THE FEDERAL EMERGENCY MANAGEMENT AGENCY

B-13. The director of the Federal Emergency Management Agency (FEMA), on behalf of the President, appoints the FCO. The FCO coordinates the timely delivery of federal disaster assistance to the affected state and local governments and disaster victims. He works closely with the SCO to determine state requirements and to coordinate national level issues with the catastrophic disaster response group. There must be close and continuous coordination with FEMA to ensure the proper use of military assets in support of civil authorities.

DEFENSE COORDINATING OFFICER

B-14. The defense coordinating officer (DCO) operates under the authority of the President and is the DOD representative designated to coordinate on-the-scene activities with the FCO. FEMA and other federal agencies that request support from DOD go through the DCO for validation and resourcing.

NATIONAL GUARD

B-15. The NG, in state status (see *Title 32, United States Code {USC}),* is the governor's primary military response organization for emergencies and disasters. The NG responds and operates under state laws unless they are federalized, when they operate under the same constraints and laws as federal units.

UNITED STATES ARMY RESERVE

B-16. The US Army Reserve is capable of providing military support. This assistance and support may include the use of equipment and other resources, including units and individuals.

STATE AND LOCAL GOVERNMENT RESPONSIBILITIES

B-17. State and local government officials, operating under the authority granted by state constitutions and the *Constitution of the United States*, are responsible for daily safety and security issues that impact the quality of life of their citizens. State and local officials have primary responsibility for emergency preparedness.

STATE RESPONSIBILITIES

B-18. Many state government agencies are responsible for coming to the aid of state citizens.

Governor

B-19. A state governor is empowered by each state constitution and the *Constitution of the United States* to execute the laws of the state and to command the state NG while it is serving in state status. Governors are also responsible for issuing executive orders or emergency proclamations, such as declaring states of emergency and ensuring that state agencies plan for potential disasters and civil disorder.

B-20. Once a man-made or natural disaster occurs, the governor assesses its extent and determines if local government requests for assistance should be honored. If appropriate, the governor declares a state of emergency, activates the state emergency response plan, and may order the NG into state active duty. The governor assigns the NG its mission through the office of emergency service (OES)/emergency medical dispatching (EMD).

Plans, Operations, and Military Support Officer

B-21. Each state NG has a plans, operations, and military support officer (POMSO) who prepares and coordinates contingency plans for military support to civil authorities (MSCA) during disaster response and recovery operations. The POMSO coordinates training plans and exercises between the state NG and federal, state, and local emergency management agencies. The POMSO serves as the NG point of contact with DOD officials during a declared federal emergency or disaster.

Office of Emergency Services

B-22. Each state has a specific agency and/or office that coordinates emergency preparedness and serves as the governor's coordinating agency during an emergency. The titles of these offices vary from state to state. Some examples are the Division of Emergency Government, Emergency Management Agency, Department of Public Safety, or Office of Emergency Preparedness. This manual refers to this office using the generic term OES/EMD.

B-23. Generally, the OES/EMD is organized as a stand-alone office under the governor or aligned under the AG or the state police, and in some instances, the AG is the OES/EMD director. It operates the state emergency operations center and coordinates with state agencies and federal officials for assistance.

Adjutant General

B-24. The state NG is the governor's primary military response force in an emergency. The AG, through the state area command (STARC) (specifically the POMSO), develops and coordinates emergency response plans for disasters and emergencies. The AG is in command of state NG forces ordered into state active duty.

State Government Agencies

B-25. State government departments and agencies prepare emergency preparedness plans for their areas of specialization in support of the governor's plan. They also participate in emergency preparedness exercises.

LOCAL RESPONSIBILITIES

B-26. Mayors, city managers, local police, fire protection officials, county executives, sheriffs, prosecuting attorneys, and public health officials are some of the people responsible for law enforcement, safety, health, and fire protection on a daily basis. They are responsible for developing appropriate emergency response plans and responding to emergencies within their jurisdictions. Most local jurisdictions have an OES/EMD to plan and coordinate actions in an emergency. In many cases, local jurisdictions have mutual aid agreements with other jurisdictions that allow for firefighting, medical, and police assistance across jurisdictional boundaries. Once local officials determine that an emergency is beyond the scope of their resources or ability to respond, the senior local official is responsible for requesting additional assistance from the next agency in the assistance request process.

LEGAL CONSIDERATIONS AND CONSTRAINTS

B-27. The *Constitution of the United States*, laws, regulations, policies, and other legal issues limit the use of federal military personnel in domestic support operations. Any Army involvement in civil disturbance operations involves many legal issues requiring comprehensive legal reviews.

B-28. Under the *Constitution of the United States*, Congress has the authority to raise and support an army, provide and maintain a navy, and make rules for governing and regulating the land and naval forces. The *Constitution of the United States* places the military under civilian control and designates the President as commander-in-chief. Statutes provide for civilian leadership in the office of the secretary of defense, service secretaries, and various other civil authorities. Under the *Constitution of the United States*, the civilian government must preserve public order and carry out governmental operations within its territorial limits. The *Constitution of the United States* allows the use of the federal military to execute or enforce the law when necessary to protect federal or civilian property and functions. However, significant restrictions exist on employing federal military forces within the United States.

B-29. The unique capabilities of the military enable it to support federal, state, or local civilian agencies. In most circumstances, the DOD is one of many federal agencies reacting to a domestic emergency or crisis, playing a subordinate, supporting role to a lead civilian agency.

LAW

B-30. In the United States, civilian agencies provide for the needs of citizens. Civilian, federal, state, and local government and law enforcement agencies execute US laws. Laws governing the use of the military in domestic operations are complex, subtle, and constantly changing. Commanders involved in civil disturbance operations should staff plans, policies, programs, exercises, funding, operations, constraints, and limitations with their local SJA to ensure conformity with legal requirements. They should scrutinize each request for aid (whether it is for equipment, individuals, or training) to ensure that it conforms to statutory requirements. The following references are a snapshot of the law on military support to civilian authorities:

- *Section 331, Chapter 15, Part 1, Subtitle A, Title 10 USC.*
- *Section 332, Chapter 15, Part 1, Subtitle A, Title 10 USC.*
- *Section 333, Chapter 15, Part 1, Subtitle A, Title 10 USC.*
- *Section 334, Chapter 15, Part 1, Subtitle A, Title 10 USC.*
- *Section 1385, Chapter 67, Part 1, Title 18 USC.*
- *DOD Directive 3025.12.*

- *DOD Directive 3025.15*
- *DOD Directive 5525.5*
- *DOD Civil Disturbance Plan 55-2*
- *NGR 500-1/ANGI 10-8101*
- *FM 3-19.15*

INFORMATION OPERATIONS

B-31. Information superiority helps forces anticipate problems and requirements. It allows commanders to control situations earlier and with less force, creating the conditions necessary to achieve the end state. Public affairs, PSYOP, and civil military operations are activities that will allow the commander to control situations earlier and with less force.

B-32. Information is available from a multitude of sources, the primary being open sources, law enforcement, and the military. A diversity of sources is the best approach because it prevents biased behavior.

Open sources include—

- Libraries.
- Newspapers and news periodicals.
- Radio and television.

Law enforcement sources include—

- Local law enforcement agencies.
- National law enforcement agencies.

Military sources include—

- DOD intelligence community (most restrictive source).
- Local MI field offices.

B-33. The DOD intelligence community operates under limitations imposed by regulations and executive orders. Attempts to skirt these restrictions may violate regulations or federal statutes.

RESTRICTIONS

B-34. Strict limits are placed on collecting information related to a civil disturbance in order to protect the civil rights of people and organizations not affiliated with the DOD. Civil disturbance plans and materials must not include lists of groups or people not affiliated with the DOD. Information on civilians and civilian organizations can be collected only with specific authorization from the Secretary or the Under Secretary of the Army. Conditions for collecting information include the existence of threats against Army personnel, functions, or property (see *ARs 380-10* and *380-13).* Civil disturbance information (available in public documents) or open-source information may be collected. However, specific rules regarding its storage must be followed. Commanders must coordinate with SJA, MI, and USACIDC before collecting any such information.

B-35. Civilian law enforcement agencies may limit access to law enforcement intelligence. In such circumstances, the military may request that agencies release the unclassified version.

B-36. If the Department of Justice (DOJ) determines that federal intervention into a civil disturbance is likely, then information relating to the disturbance is provided to the Army Assistant Chief of Staff for Intelligence. The information is analyzed and then passed through channels to the TF commander for planning purposes.

B-37. Military intelligence collection efforts, except liaison, may begin only when DA orders. The orders will come through the CSA's personal liaison officer and the TF commander. Covert operations to gather information on non-DOD individuals and groups must be approved on an operation-by-operation basis by the Undersecretary of the Army.

B-38. Once DA approves the collection effort, MI elements establish and maintain liaison with the appropriate local, state, and federal authorities. Then they collect information pertaining to the incident and general situation. Their estimate of the situation is passed to the TF commander and DA.

POSSE COMITATUS

B-39. Generally, federal military forces may not give law enforcement assistance to civil authorities without conflicting with the *Posse Comitatus Act*. However, constitutional and statutory exceptions to this prohibition do exist. The recent emphasis on drug interdiction has led to an increase in those exceptions.

B-40. The *Judiciary Act of 1789* allowed US marshals to call upon the military as a posse comitatus. This continued until after the Civil War, when the federal government used the Army to execute reconstruction era policies. The southern states regarded the use of the military for this purpose as abusive and repressive, and in 1878, President Rutherford B. Hayes signed the original bill ending the practice.

B-41. The *Posse Comitatus Act* prescribes criminal penalties for the use of the US Army or Air Force to execute the laws of or to perform civilian law enforcement functions within the United States. DOD policy extends this prohibition to the US Navy and Marine Corps. Prohibiting the military from executing the laws means that military personnel may not participate directly in—

- An arrest; a search and seizure; a stop and frisk; or an interdiction of vessels, aircraft, or vehicles.
- A surveillance or pursuit.
- A civilian legal case or any other civilian law enforcement activity as informants, undercover agents, or investigators.

B-42. The *Posse Comitatus Act* does not apply to the following:

- Members of the NG when in state service.
- Members of a Reserve Component when not on active duty or active duty for training.
- DOD civilians, unless under the direct command and control of an active duty officer.

- A soldier when off duty and acting only in a private capacity.
- Soldiers taking action for the primary purpose of furthering a military or foreign affairs function of the United States; for example, enforcing military justice, maintaining law and order on military installations, and protecting classified materials.

CONSTITUTIONAL EXCEPTIONS TO THE POSSE COMITATUS ACT

B-43. Under its inherent authority, the US government is responsible for preserving public order and carrying out governmental operations within its territorial limits, by force if necessary. Under the *Constitution of the United States*, exceptions allow the use of military force to execute or enforce the law. Some of the exceptions are to—

- Protect civilian property and functions.
- Ensure preservation of public order and carry out government operations, by force if necessary.
- Protect civil rights where local authorities do not or cannot protect them.
- Protect federal property and functions.
- Quell civil disturbances and labor strife that rises to the level of civil disorder.

B-44. The President may order the armed forces to support state civil authorities suffering from an insurrection or civil disturbance. He must act personally by first issuing a proclamation calling on insurgents to disperse and retire peacefully within a limited time. This occurred when federal forces were called in to counter the Los Angeles riots in 1992.

Protect Civilian Property and Functions

B-45. A sudden and unexpected civil disturbance, disaster, or calamity may seriously endanger life and property and disrupt normal governmental functions to such an extent that local authorities cannot control the situation. At such times, the federal government may use military force to prevent loss of life or wanton destruction of property and to restore government functions and public order. This exception has rarely been used.

Protect Federal Property and Functions

B-46. The federal government may use military force to protect federal property and federal government functions when local authorities cannot or decline to provide adequate protection.

ARMED FORCES

B-47. Basic guidance for the interaction of the military, the Reserve Component, and civilian LEAs is given in *10 USC Chapter 18*. Guidelines on reimbursement and restrictions, direct participation in law enforcement activities, the use of information collected during military operations, and the use of military equipment and facilities are some of the topics covered. Title 10 prohibits the military from directly participating in arrests, searches, seizures, or other similar activity unless authorized by law (such as arrests on military property). The *Fiscal Year 1989 Act* and subsequent *National Defense Authorization Acts* have authorized the DOD to provide more support to LEAs in the counterdrug effort.

NATIONAL GUARD

B-48. For a description of how the secretary of defense may provide funds to state governments (including the District of Columbia, Commonwealth of Puerto Rico, and US territories) for civil disturbance operations performed by the National Guard when not in federal service, see 32 *USC*. The *Posse Comitatus Act* does not apply to the NG when not in federal service. Each state specifies how NG may be employed.

SUPPORT TO CIVIL LAW ENFORCEMENT

B-49. It is DOD policy to cooperate with civilian law enforcement officials to the greatest practical extent. However, cooperation must consistently meet the needs of national security and military preparedness, the historic tradition of limiting direct military involvement in civilian LEAs, and the requirements of applicable law. US military forces are never placed under the command of civilian law enforcement officers or nonfederalized NG. DOD policy concerning the provision of military support to LEAs, including personnel and equipment, are contained in *DOD Directive 5525.5*.

B-50. The Army assists civilian law enforcement by providing personnel, equipment, training, and expert advice within the limits of applicable law. Army NG units in state status (see 32 *USC*) provide the primary source of military assistance. They may be able to provide assistance to civil authorities in instances when federal units are precluded due to the restrictions of the *Posse Comitatus Act*.

ROLES OF THE NATIONAL GUARD AND FEDERAL FORCES

B-51. The preservation of law and order in the civilian community is the responsibility of state and local governments and law enforcement authorities. When a civil disturbance in a civilian community turns to widespread rioting that includes arson, looting, and acts of violence, the civil authorities may decide that they do not have the resources to quell the riot. They may then turn to the NG and federal forces to support the civil authorities in restoring law and order. Civil disturbances in any form are prejudicial to public law and order.

NATIONAL GUARD FORCES

B-52. The NG (as a state organization) responds to the governor according to state law for civil disturbance operations. NG regulations direct planning and training for the civil disturbance mission. During most civil disturbance situations, the NG will be the first military responder and will usually remain in state active duty status (according to 32 *USC*) throughout the operation. In extreme circumstances, the NG can be brought on federal service for civil disturbance operations when ordered to under the appropriate federal statute by the President. During the 1992 Los Angeles civil disturbance, the President federalized the California NG. Once the NG has been deployed to assist the civil authorities with a civil disturbance, it is too far into the deployment cycle to start determining what tasks the NG can do as a state or as a federally controlled force. It is the commander's responsibility not to accept missions that the unit is not adequately trained to do. The following are examples of appropriate missions for the NG:

- Manning traffic control points.
- Providing building security.
- Providing area security and patrols.
- Providing security at custody facilities.
- Providing security and escort for emergency personnel and equipment.
- Protecting sensitive sites.
- Transporting law enforcement personnel.
- Providing show of force.
- Dispersing crowds.
- Employing riot control munitions.
- Providing very important person (VIP) protection and escort.
- Providing quick reaction and reserve force.

B-53. Examples of inappropriate taskings:

- Negotiating hostage situations.
- Barricading suspects.
- Participating in criminal investigations.
- Acting as federal Army forces.

B-54. The commitment of federal troops to deal with domestic civil disturbances must be viewed as a drastic measure of last resort. Their role should never be greater than what is absolutely necessary under the circumstances. Commanders should take every measure to avoid the perception of an invading force. A JTF designated to respond to a civil disturbance should project the image of a restrained and well-disciplined force whose sole purpose is to help restore law and order with minimal harm to the people and property and with due respect for all law-abiding citizens.

B-55. The role of federal Army forces is to assist civil authorities in restoring law and order when the magnitude of the disturbance exceeds the capabilities of local and state LEA, including the NG. Under the provisions of the *Constitution of the United States* and selected federal statutes, the President may order the employment of the federal armed forces to aid local and state civil authorities to protect the Constitutional rights of citizens. Federal military forces may also protect federal facilities and installations in any state, territory, or possession.

B-56. The *DA Civil Disturbance Plan 55-2* gives guidance for Army forces directed to quell civil disturbances. In addition, *AR 500-50* outlines responsibilities, prescribes policies, and provides guidance for the use of Army resources in response to civil disturbances. It outlines statutory restrictions and designates staff and command responsibilities for planning and executing civil disturbance operations.

B-57. The JTF commander exercises control of all federal military forces (including NG in federal status) committed to assist civil authorities. Federal military forces remain under the military chain of command during civil disturbance operations. Federal forces will not be placed under the command of either civil officials or NG commanders in a nonfederal status. Civil authorities retain control of their state and local law enforcement agencies. The JTF commander establishes liaison with the senior representative of the attorney general (SCRAG) and other appropriate federal, state, and local civil authorities.

B-58. When a JTF is being assembled it must deploy with a provost marshal (PM) who will be the POC for situation assessment, operational information, and security matters. He also serves as a liaison with civilian and law enforcement agencies.

B-59. Federal military forces must be tailored to the specific civil disturbance situation. Sufficient combat support and combat service support units will be required to sustain the force throughout the deployment. Coordination with civil officials may allow the force to draw on resources available from state and local agencies. Close and continuous coordination between federal military forces and LEAs will provide the commander with the detailed and timely information required to employ and protect the force effectively.

B-60. In supporting *DA Civil Disturbance Plan 55-2,* intelligence personnel may conduct close and continuous liaison with LEAs and military police to ensure that their units receive the information needed to allow the commander to adequately protect the force. The JTF commander should staff intelligence support missions with his senior intelligence officer and legal counsel before approving the mission.

B-61. Federal military forces must be employed in tasks or missions appropriate to their organization and training; they must not be employed in ways that violate the legal restrictions in effect. Military forces may be used to disperse unlawful assemblies and to patrol disturbed areas to prevent unlawful acts. They may be used to assist in the distribution of essential goods and the maintenance of essential services. Forces may also establish traffic control points, cordon off areas, release smoke and obscurants, and serve as security or quick reaction forces. Certain types of missions (such as gathering intelligence on civilians) are always inappropriate for military forces during civil disturbance operations.

B-62. Requests for specific military missions are typically passed through a single state or federal law enforcement coordinating officer, as approved by the SCRAG. Validated requests are transmitted to the JTF commander and his HQ for staffing and coordination. Approved missions are assigned through the military chain of command to the appropriate element or unit for execution. Units and soldiers will not accept taskings or missions directly from law enforcement or civilian officials, except in a direct support relationship as approved and ordered through the military chain of command.

B-63. It is important that a military liaison be provided to each LEA HQ generating requests for support. This liaison can assist LEA officials in determining the types and quantities of military support to be requested. JTF HQ can facilitate this mission assignment process by providing LEAs with a detailed listing of the types of missions military forces may conduct.

B-64. The AO for a deployed unit should coincide with the jurisdiction or subdivision boundaries of the LEA it supports. This arrangement facilitates liaison and coordination between law enforcement and military chains of command.

LEAD AGENCY CONCEPT AND ROLE OF MILITARY

B-65. The DOJ is the lead federal agency for civil disturbance operations. The Attorney General's on-scene representative is the SCRAG. The DOD has designated DA as their executive agent for military assistance for civil disturbance (MACDIS) (see *FM 27-100.*)

JOINT TASK FORCE

B-66. Military support to civil authorities in disasters and domestic emergencies is a DOD (not a service component) responsibility. The supported commander may decide that the severity and scope of a disaster require a joint response. In civil assistance, as in operational level warfighting, the commander-in-chief (CINC) uses the different and complementary capabilities of each service to accomplish the mission. Such use requires knowledge of both the capability and the availability of all service component assets, to include their agencies and installation.

B-67. The CINC may establish a JTF to provide comprehensive military support. Establishing a JTF may provide the best DOD response to civil emergencies. A JTF is established to execute a specific mission limited in scope and duration. The JTF objective in civil emergencies is to deploy forces to the area rapidly to assist immediately in saving lives and safeguarding property and to continue providing assistance required by the federal response plan (FRP) and the particular situation.

B-68. The JTF is configured for each specific mission. In some civil emergencies, the JTF may require a greater proportion of combat service support type units and capabilities than in typical warfighting deployments. The JTF must be able to provide emergency assistance across all lines of support.

JOINT TASK FORCE CONTROL ELEMENT

B-69. Responsibility for controlling the civil disturbance resides with the civil authorities. The JTF is organized to support and lend assistance to them in their effort to return civil order to the community.

B-70. The control element consists of the command group and the crisis management team (CMT). The command group is made up of city, county, state, and military command personnel. They set policy and issue directives, coordinate the activities of civil and military support agencies, supervise the CMT, and coordinate with outside agencies. The CMT is made up of representatives of civilian and military staff sections. The CMT advises the command group and coordinates operations and support for the action element of the TF. The control elements are located with the emergency operations center (EOC) to facilitate information processing, resource management, and operational control. The JTF command element should colocate with the civilian command element at the EOC. This will expedite the exchange of information. If an EOC has not been set up, the commander may establish one while making provisions for including civil authorities to ensure a unity of effort.

B-71. Not all CMT members are located at the EOC. Some key people may use liaisons to represent them at the EOC. Key personnel can then research and discuss ideas freely, away from the confusion associated with crisis management. The liaison can transmit guidance and answers to the EOC. Some agencies may not be needed in the EOC at first, but they must be prepared for inclusion. The decision to include an agency in the EOC is based on the likelihood that they may have to take an action or a support role on the importance of the agency to the mission.

Appendix C
Urban Terrain Analysis

Of all the environments in which the Army may need to conduct operations, the urban environment confronts commanders with a combination of difficulties rarely found in other environments. The distinct characteristics of the environment result from the combination of man-made features and supporting infrastructure superimposed on the existing natural terrain and the density of civilians in close proximity to combat forces. The human dimension it represents is potentially the most important and perplexing for commanders and their staffs to understand and evaluate. Commanders at all levels must make extraordinary efforts to assess and understand their particular urban environment to plan, prepare for, and execute effective civil disturbance operations.

SIGNIFICANT FACTORS

C-1. Although urban areas throughout the world possess similar characteristics, no two are alike. The sprawl of Los Angeles, for example, bears little physical resemblance to New Delhi. The most significant factors affecting the uniqueness and complexity of each area are its societal characteristics. While the terrain can be extremely complex, information about this aspect, its potential effects on operations, and how it changes over time may be determined with some degree of certainty. However, the human dimension is more difficult to understand and assess, particularly its effect on military operations. The urban environment is not neutral—it helps or hinders. The commander who can best understand and exploit the effects of the urban environment has the best chance of success.

C-2. Whether a metropolitan area or a village, each urban environment has an identifiable system of urban components that constantly change and interact. This system of components consists of the terrain, the society, and the infrastructure that provides the critical link between the two (see *Figure C-1*, page C-2).

C-3. These systems are not separate and distinct categories but instead are overlapping and interdependent. A thorough analysis of these systems contributes to the commander's situational understanding and allows him to develop COAs that apply appropriate resources to control a crowd and prevent a major civil disturbance. This understanding allows commanders to engage decisive points critical to maintaining peace or restoring normalcy to the urban environment.

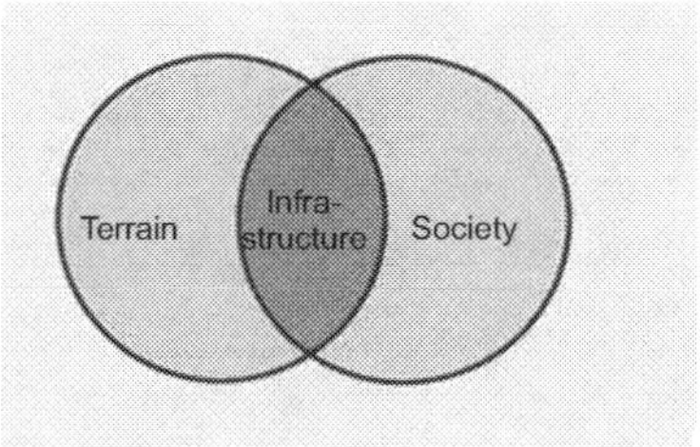

Figure C-1. Keys to Understanding the Urban Environment

C-4. The social aspects and the integrating infrastructure are the most difficult for commanders to understand and assess. These issues require greater dependence on LEAs and local governmental agencies for their information, knowledge, and expertise.

URBAN TERRAIN TYPES

C-5. The physical terrain is the most basic and easily recognizable aspect of the urban area; however, truly understanding it requires comprehending its multidimensional nature, urban and street patterns, urban form and function, and size. The infinite ways in which these factors (see *paragraph C-15)* can intertwine make it impossible to describe a typical urban area. However, these factors provide a framework for understanding the complex terrain within an urban environment. Although the terrain consists of both natural and man-made features, the latter dominates the physical characteristics of the urban environment and has significant effects on military systems and soldiers, tactics, and operations.

URBAN PATTERNS

C-6. The combinations of urban areas, lines of communications (LOCs), and natural terrain result in the identification of four urban patterns (see *Figure C-2*) that will influence civil disturbance operations. An integral part of each urban pattern is the hub or the central urban area around which outlying urban areas radiate. In SOSO, the hub is important, as it is often the economic, political, or cultural center of the surrounding area.

Satellite Pattern

C-7. This pattern consists of a central hub surrounded by smaller, dependent urban areas. LOCs tend to converge on the hub. The natural terrain throughout this pattern is relatively homogenous. Outlying or satellite urban areas often support the principal urban area at the hub.

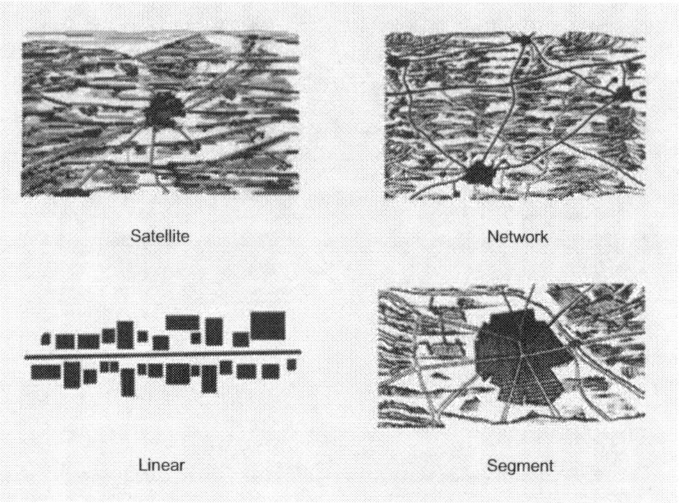

Figure C-2. Urban Patterns

Network Pattern

C-8. This pattern represents the interlocking of the primary hubs of subordinate satellite patterns. Its elements are more self-sufficient and less supportive of each other, although a dominant hub may exist. Major LOCs within a network are more extensive than in a satellite pattern and take a rectangular rather than a convergent form. The natural terrain within a network may vary more than in a single satellite array.

Linear Pattern

C-9. This pattern is a potential subelement of the previous two patterns. It may form one ray of the satellite pattern or be found along connecting links between the hubs of a network. Most frequently, however, this pattern results from the string of minor urban areas aligning a confined natural terrain corridor such as an elongated valley, a body of water, or a man-made communications route.

Segment Pattern

C-10. When dominant natural terrain (such as a river or man-made features [canals, major highways, or railways] divides an urban area, it creates a segmented pattern). This pattern makes it easier to assign areas of responsibility to units. However, if operations are in support of the city LEA, then units should be assigned to a police precinct to avoid crossing several precinct lines, as would happen when assigning units by dominant terrain features.

STREET PATTERNS

C-11. Patterns within the urban area result from the arrangement of the streets, roads, highways, and other thoroughfares. They are a result of the influences of natural terrain, the original designer's personal prejudices, and the changing needs of the inhabitants. The commander must understand the street patterns in the area in which his control force will operate. This will offer force protection, preventing the control formation from being outmaneuvered, flanked, or surrounded by a hostile crowd. Urban areas can display any of the three basic patterns or any combination of the three (see *Figure C-3).*

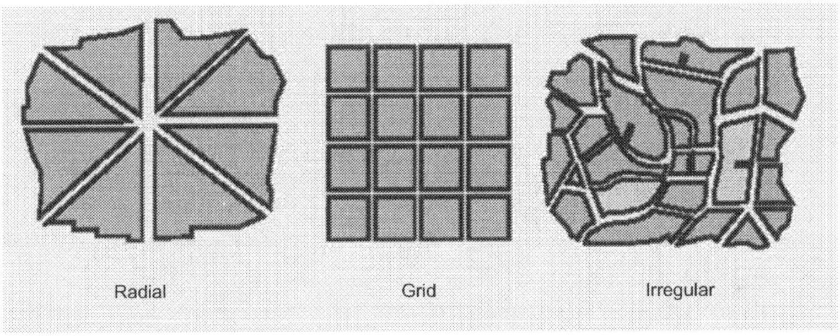

Figure C-3. Basic Internal Street Patterns

Radial

C-12. Societies of highly concentrated religious or secular power often construct urban areas with a radial design with all primary thoroughfares radiating out from the center of power. Terrain permitting, these streets may extend outward in a complete circle or may form a semicircle or arc when a focal point abuts a natural barrier such as a coastline or mountain. To increase mobility and traffic flow, societies often add concentric loops or rings to larger radial patterns.

Grid

C-13. The most adaptable and universal form for urban areas is the grid pattern (lines of streets at right angles to one another forming blocks similar to the pattern of a chessboard). A grid pattern can fill in and eventually take over an original radial pattern. Most urban areas, regardless of the original intent, plan, or vision, emerge from successive plans overlaid on one another.

Areas are well planned to fit with previous plans and others a haphazard response to explosive urban growth. The results may mix different patterns, blend patterns in symmetric combinations, or have no discernable geometric pattern (the irregular pattern).

Irregular

C-14. Irregular patterns can be specifically designed for aesthetic reasons (as in many suburban housing developments). They are sometimes used to conform to marked terrain relief (see *FMs 3-06* and *3-06.11*).

URBAN FORM AND FUNCTION

C-15. Throughout the world, urban areas have similar form and function. In form, urban areas contain like characteristics, readily divisible into distinct sections or areas. Functionally, they tend to be the center of population, finance, politics, transportation, industry, and culture.

- **Core.** The core is the heart of the urban area, the downtown, or central business district. Relatively small and compact, it contains a large percentage of the urban area's shops, offices, and public institutions.
- **Core Periphery.** As the name implies, the core periphery forms the outer edges of the core of the urban area. Generally, it has undergone less change than the core, resulting in buildings of uniform height (generally two to three stories in towns and five to ten stories in larger urban areas).
- **Industrial Area.** Industrial areas generally develop on the outskirts of the urban areas, where commercial transportation is easiest (along airfields and major sea, river, rail, and highway routes). The open/ dispersed pattern of the buildings provides sufficient space for large cargoes, trucks, and materials-handling equipment.
- **High-Rise Area.** High-rise areas consist of multistoried apartments, commercial offices, and businesses separated by large open areas such as parking lots, parks, and individual one-story buildings. The automobile, mass transit systems, and improved road networks encourage these areas to grow and function further from the urban core.
- **Residential Area.** Residential areas can be found dispersed throughout the urban area; however, large suburban areas (or sprawl) normally form on the outskirts. The combined population of surrounding suburban areas often outnumbers that of the urban area. Proper and specific suburbs tend toward homogeneity based on ethnicity, religion, economics, or some other social aspect.
- **Commercial Ribbon Area.** Commercial ribbon areas are rows of stores, shops, and restaurants built along both sides of major streets that run through urban areas. These same types of areas often develop along the roads that connect one urban area to another (strip areas).

URBAN SOCIETY

C-16. Although complex, understanding the urban terrain is relatively straightforward in comparison to comprehending the multifaceted nature of urban society. As urban areas increase in size, they generally lose homogeneity; therefore, commanders must understand and account for the characteristics of a diverse population whose beliefs may vary. The behavior of civilian populations within an urban area is dynamic and poses a special challenge to commanders.

C-17. The center of gravity during an urban operation may be the civilian inhabitants themselves. The side that enjoys the support of the population has many advantages. To gain and retain this support, commanders must first understand the complex nature and character of the urban society. Second, they must understand and accept that every military action (or inaction) may influence the relationship between the urban population and Army forces, and by extension, mission success.

C-18. The identification of groups helps commanders focus on specific segments of the urban society to determine their beliefs and needs and how those beliefs and needs motivate them to future action (or inaction). Groups may be categorized based on race, religion, national origin, tribe, clan, class, party affiliation, education level, union memberships, occupation, age, or any other significant social grouping.

C-19. Commanders should consider political, economic, and historical factors in this analysis. These factors usually affect all groups to some extent and often provide the basis for many of their beliefs, needs (actual or perceived), and subsequent behavior. In most cases, training and discipline, grounded in cultural understanding and sensitivity, will help mitigate many of the potential adverse effects resulting from military and civilian interaction.

URBAN INFRASTRUCTURE

C-20. Urban infrastructures are those systems that support the inhabitants, their economy, and their government. They provide the link between the physical terrain and the urban society. Hundreds of systems may exist. Each system has a critical role in the smooth functioning of the urban area. All systems fit into the following five broad categories: communications and information, transportation and distribution, energy, commerce, and human services.

Appendix D

Practical Application

It is always best to derive lessons learned from practical operations, in the field. These are operations where a measure of success was acquired, along with important lessons learned. The following reflects an actual TF application of the doctrine highlighted in the contents of this manual.

OVERVIEW

D-1. The purpose of this manual is to provide TF units and elements with the latest guidance on providing crowd control and preventing civil disturbances.

D-2. Local civilians often gather in towns or along strategic locations throughout an AO to demonstrate grievances or other causes (for example, protesting the apprehension of a fellow citizen, policies of US forces, and policies of government services). Although civilians have the right to assemble, crowd gatherings are predominantly ethnically based, motivated, and influenced. Local civilians can apply for a permit to legally conduct a rally 72 hours in advance as long as their assembly is peaceful and unobtrusive to government services, freedom of movement, or the rights of others. On occasion, intelligence collection can indicate a possibility for violence. In these instances or during spontaneous crowd formations, the TF must provide forces in support of a safe and secure environment or to assist civilian police in preventing unlawful acts. If unlawful acts are committed, detain violators for adjudication under the legal system.

PROCEDURES

D-3. Take precautions to avoid becoming a catalyst to the civil disturbance. During lawful assemblies, the TF must make every effort to treat HN individuals with respect. This means that the TF must engage in peaceful gatherings using interpersonal communication skills. This ensures that demonstrators understand that TF personnel respect their right to lawfully assemble during peaceful demonstrations. When possible, TF engagement of a peaceful assembly should be unobtrusive.

D-4. Do not display civil disturbance equipment or weapons, overtly prepare defenses against a crowd, or aggressively violate crowd space, unless presented with information, intelligence, or negative indicators of pending violence.

NEGATIVE INDICATORS

D-5. Any response to negative indicators should be sequential, measured, and focused on deescalating rather than escalating the crowd.

NEGATIVE INDICATORS OF A POTENTIAL CIVIL DISTURBANCE

D-6. Predictive intelligence is the best indicator of a potential civil disturbance and may provide insight on internal planning, external influences, environmental conditions, or catalyst events that should influence TF planning and response.

D-7. Other indicators of a potential civil disturbance are people beginning to assemble during a scheduled or spontaneous rally. When people begin to assemble, leaders should pay particular attention to crowd dynamics. As the crowd demeanor or dynamics change from peaceful assembly to unruly behavior or unlawful acts, negative indicators should provide an advance warning and prompt a measured response. Negative indicators of a potential change in crowd dynamics and unlawful activities should prompt immediate countermeasures (see *Figure D-1*).

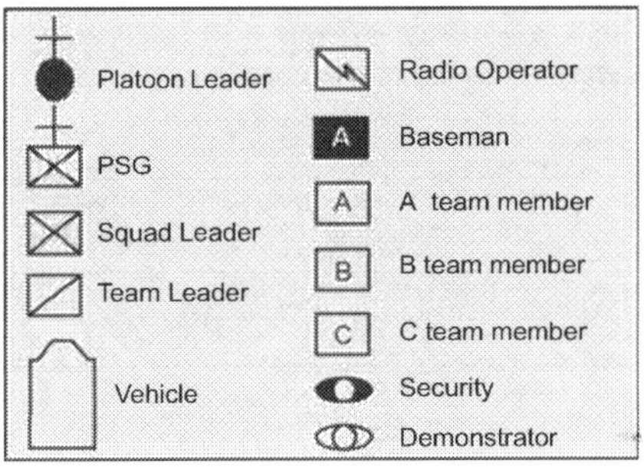

Figure D-1. Symbols for a Formation Element

Orientation

D-8. Crowd dynamics change from people milling around and talking among themselves in isolated groups to a more collective focus concentrated on a single objective (for example, TF personnel, specific agitators or magnets, a government figure, or other activity).

Massing

D-9. The crowd begins to mass or tighten into a large contiguous body from a loose formation to a static congregation located in a more concentrated area. Unchecked, massing can provide anonymity and a collective feeling of invulnerability. This change in group dynamics could provide ideal conditions for a civil disturbance (see *Figure D-2*).

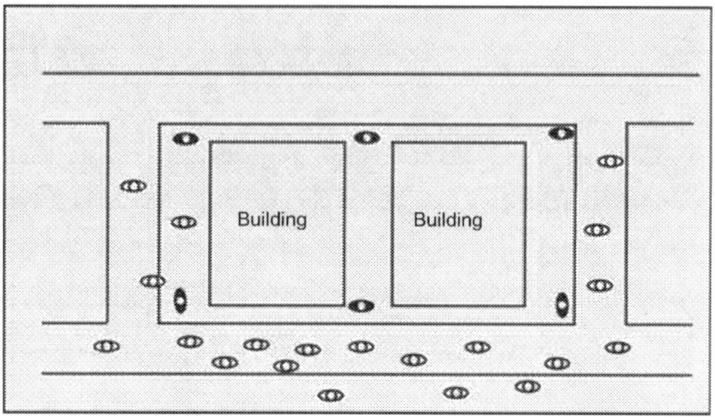

Figure D-2. Crowd Massing

Chanting

D-10. People may begin chanting slogans or yelling ethnically charged rhetoric or obscenities. Because chanting can indicate an increase in crowd intensity, interpreters should be positioned in strategic locations to interpret crowd rhetoric.

Signaling

D-11. Signaling occurs when members of the crowd position themselves to communicate and synchronize crowd actions tied to a plan with specific objectives. Signaling may indicate the intent to escalate the message through the use of violent or unlawful acts.

Blocking

D-12. The crowd may attempt to block public thoroughfares (for example, entrances and exits to public buildings, MSRs, community streets, or other traffic areas that may degrade freedom of movement or similar actions). This escalates the level of attention given to the crowd's message or prevents freedom of movement to opposing persons or groups.

Arming

D-13. Any signs that the crowd is arming itself with objects (for example, bottles, rocks, and sticks) is a clear indication of impending danger to the TF and other civilians. In fact, during a recent civil disturbance, some members of the crowd were seen handing out bottles to others in the crowd. Arming demonstrates intent toward violence is unlawful.

Outside Influence

D-14. The presence of people from outside the community may indicate more sophisticated planning and resources than would otherwise be expected from local community members. Outside people may feel insulated from responsibility for local actions and, therefore, provide a dangerous catalyst for a violent civil disturbance.

Agitators and Instigators

D-15. Agitators and instigators are common elements associated with civil disturbances. They often provide a catalyst to incite the crowd toward violence. Whether spewing rhetoric or committing unlawful acts, their continued presence will fuel potential strife.

Absence of Children

D-16. Recent civil disturbances have shown that even when children were initially present at a rally, they were absent before the civil disturbance erupted with violence. While the presence of children is not a guarantee of a peaceful assembly, the removal of children from a rally may provide early warning for a pending escalation of hostilities or a full-blown riot.

PREVENTION OF CROWD ESCALATION

D-17. If TF personnel believe, based on intelligence or negative indicators, that a crowd is planning to, or may spontaneously, riot they should immediately engage crowd participants using the preventive TTP measures.

D-18. The force executing these preventive TTP can include a number of civil disturbance teams and elements capable of executing the tasks shown in *Figures D-2,* page D-3; *D-3; D-4,* page D-6; and *D-5,* page D-8. These figures demonstrate a full range of crowd control teams and elements, to include camera teams with interpreters, checkpoints, QRF response, riot control formation, NLW teams, snatch and detain teams, blocking teams, and detainee evacuation, as required. *Figures D-2, D-3, D-4,* and *D-5,* illustrate the evolution of a crowd formation and the recommended positioning of teams and elements to engage people as they congregate. Included with each security team is a designated camera or video recorder.

D-19. Depending on the situation and crowd dynamics, it may not be necessary to establish all crowd control teams and elements, but they should be included in planning. Regardless of task organization, teams and elements must be prepared to execute the crowd control and civil disturbance TTP. The order to execute the following TTP should be tailored based on local conditions.

Early Contact

D-20. Engaging people as a crowd forms is essential to prevention. Whether based on intelligence or negative indicators, people should be engaged early, before the crowd masses as shown in *Figure D-2,* rather than waiting until after the crowd masses as shown in *Figure D-3.* Early contact will assist TFs in personalizing their interaction and ensuring that people understand their presence is known, documented, and can be linked to their actions.

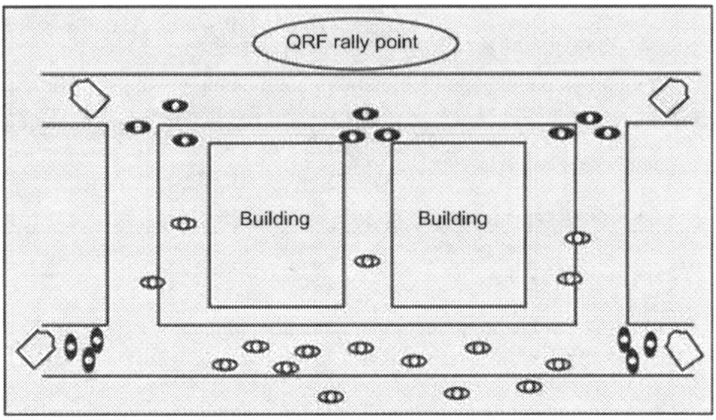

Figure D-3. Early Contact and Unobtrusive Crowd Control Measures

D-21. Cameras are one of the best tools for preventing civil disturbances. The use of cameras to document crowd formations, particularly lens-size groups (small enough to fit in a single photo), will personalize contact and eliminate feelings of anonymity among the crowd. People whose identity has been documented will think twice before committing unlawful acts. Even without documenting names, photo analysis can identify people associated with the civil disturbance, especially those who unlawfully participated as crowd instigators, agitators, or magnets or may commit unlawful acts.

D-22. As people begin to gather in and along streets, camera teams should be deployed as unobtrusively as possible to engage people before they congregate in mass. As depicted in *Figure D-3*, camera teams should consist of two soldiers with an interpreter and be employed in a widely dispersed pattern. Such small, dispersed camera teams are less likely to intrude on lawful rallies or create invasive friction that might become a catalyst to a civil disturbance.

D-23. As the crowd begins to demonstrate negative indicators for a possible civil disturbance (*Figure D-4*, page D-6), camera teams should become more invasive by actively engaging groups to take their pictures. Camera teams should focus cameras on instigators, agitators, or magnets (personnel that garner respect or compliance from the crowd). If a person is purposely avoiding the camera, even if his activities are lawful, the team should take special measures to obtain his picture. Each time a camera team takes a picture of a person or group of people, they must immediately engage them using the warning outlined in *paragraph D-26*.

D-24. Once the crowd has massed and displays negative indicators or begins to riot, camera teams should immediately move behind the riot formation line (see *Figure D-4*). It is essential, however, that camera teams continue to document the event, focusing on people who are committing illegal acts such as arming themselves, throwing objects, and assaulting TF personnel.

D-25. Immediately following a crowd control or civil disturbance operation, photos must be consolidated, processed, and submitted to the TF analysis and control element (ACE) for analysis. A near real-time analysis will provide immediate intelligence for ongoing operations to prevent subsequent civil disturbances or to detain violators from the previous operation.

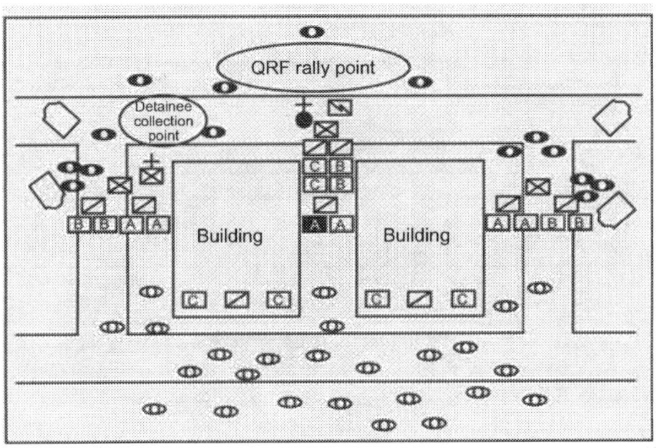

Figure D-4. More Obtrusive Engagement as the Crowd Demonstrates Negative Indicators

Crowd Warning

D-26. Once the team has taken a picture of a person or group of people, they should warn the crowd that as individuals they would be held accountable for their actions. Using an interpreter, the camera team leader will make solid eye contact with the person or group and state the following warning: *"We have your picture. You will be held accountable for your actions. If you commit an unlawful act, you will be detained, if not today, tomorrow. Blocking roads or access to public buildings is unlawful."*

Limit Crowd Size

D-27. Persons seeking to conduct a rally must apply for a permit through civilian authorities at least 72 hours prior to the meeting. If a rally has not been lawfully scheduled through civilian authorities, or if it has, but is demonstrating negative indicators and is turning into a civil disturbance, TF units and elements should attempt to limit the crowd or separate gatherings under 10 people (see *Figure D-3*, page D-5). Walking patrols can assist camera teams in breaking up large groups. Walking patrols should operate in conjunction with camera teams. This will ensure that camera teams can accurately document and provide the aforementioned warning to prevent further escalation of crowd size or behavior.

Clear Routes

D-28. Routes should be considered key terrain and must be controlled by the TF to retain the initiative. Note how a QRF divides in order to clear side streets between protected buildings in *Figure D-4.* Blocking routes or public thoroughfares is not only illegal, but severely restrictive to civil disturbance operations. Routes must remain open to allow movement of TFs and elements, whether it is for camera teams during early contact or for the QRF to respond during an actual civil disturbance. Because blocking routes is illegal, camera teams must document responsible persons for immediate or subsequent detention and legal processing.

Isolate the Area

D-29. As depicted in *Figure D-3,* page D-5, checkpoints should be established to isolate the area. While checkpoints can prevent people from entering the isolated area, they should never prevent people from exiting the area. When trapped, people may escalate the level of violence out of fear, anger, or desperation. When placing checkpoints, leaders should find a location that is both close enough to limit the number of persons being isolated, but far enough away that they will not draw a crowd from people within the isolated area. Checkpoints are essential to preventing others from joining the already gathered crowds. These checkpoints must provide storage room for processing vehicles in both directions; but when limited by space or other considerations, it should simply block traffic trying to enter the isolated area. Improve force protection measures in and around checkpoints as necessary. Strategically placed checkpoints can provide numerous countermeasures to prevent and control a civil disturbance or to identify and detain persons who may have committed unlawful acts during the riot. Use checkpoints to prevent potential participants from joining the crowd or civil disturbance. Checkpoints should also prevent any activity that may commingle ethnic groups. The checkpoint must process all people departing the area by documenting personal and vehicle information with the pictures of all occupants in front of or adjacent to the vehicle.

Establish Lethal Overwatch

D-30. The mentality of "taking the high ground" is just as important in civil disturbance operations as in any other type of operation. Although QRF assets on the ground have both NL and lethal capabilities, it is critical to ensure their safety by deploying marksman/observer teams as shown with the C-team designates on top of the buildings in *Figures D-4* and *D-5,* page D-8. Detain persons committing unlawful acts. The detention and legal processing of persons who commit unlawful acts, against TFs or local laws sends a clear message that all people will be held accountable for their actions. Immediately detain persons who are committing unlawful acts and are isolated or easily accessible, unless the detain operation would create a catalyst event or endanger the apprehension team. If it is not feasible to immediately detain them, a picture with relevant information should be forwarded for analysis and inclusion on the detain list for their detention at a later date and time. Remember, as stated in the warning, they will be detained, if not today, tomorrow. Detainees should be quickly relocated to the forward processing site for detainee evacuation (see *Figure D-5*). Detainee evacuation sites should be behind the control formations to prevent any interference from the local population. Sites should be positioned at two separate locations, such as at the end of a town or located along an MSR. All people who are detained based on

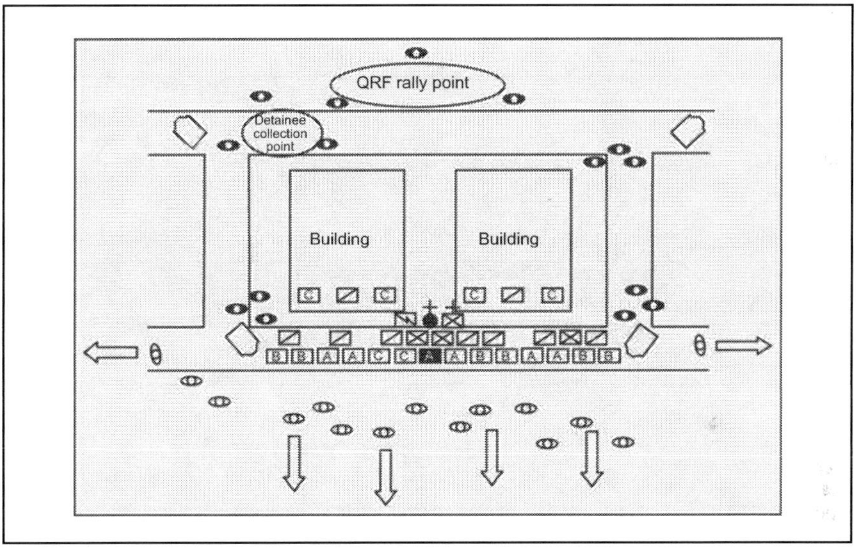

Figure D-5. Full Engagement as Crowd Unlawfully Demonstrates

solid evidence of a crime (for example, pictures of them armed with dangerous objects, blocking a road, assaulting TF members) will be turned over to local authorities for legal processing.

Detain List

D-31. When executing operations based on the detain list, do not show people the list or separate pictures from the list.

Quick Reaction Force Position

D-32. The QRF should be positioned with easy access to decisive points to interdict formations or gathering of crowds, support checkpoints, or provide sustained operations in support of early contact teams. Notice in *Figure D-1*, page D-2, how the QRF is positioned out of direct sight of the crowd until employment becomes necessary. QRF personnel are readily organized, equipped, and easily moved into position to move between buildings and through obstacles (see *Figure D-4,* page D-6).

D-33. At this point, if the crowd ignores verbal orders, NL munitions from appropriate standoff distances could be used to influence and motivate the crowd to comply. Use of NL munitions could prevent the necessity for QRF personnel to come into direct physical contact with the crowd. Early positioning of the QRF could become a catalyst event that may instigate a civil disturbance.

D-34. Assembling in mass (see *Figure D-5*), directly in front of the protected buildings, illustrates to the crowd that there is strength in numbers and the QRF is determined to disperse the crowd from the area. As a reminder, remove the flanking checkpoints that may block the dispersal of the crowd.

Glossary

A2C2	Army airspace command and control
AA	assembly area
AFL-CIO	American Federation of Labor-Congress of Industrial Organizations
AG	Adjutant General
AO	area of operations
ARTEP	Army training and evaluation program
AS	area security
ASV	armored security vehicle
ATTN	attention
BDU	battle dress uniform
CA	civil affairs
CD	combat development
CDC	crowd dispersal cartridge
CFR	Code of Federal Regulations
CINC	commander-in-chief
CMT	crisis management team
COA	course of action
COB	close of business
CONUS	continental United States
COTS	commercial off-the-shelf
CP	candle power
CPU	central processing unit
DA	Department of the Army
DCO	defense coordinating officer
DD	Department of Defense
DM	designated marksmen
DOD	Department of Defense
DOJ	Department of Justice
DOMS	director of military support
EMD	emergency medical dispatching
EMS	emergency medical services
EOC	emergency operations center
FCMT	forced cell move team
FCO	federal coordinating officer
FEMA	Federal Emergency Management Agency
FM	field manual

FRP	federal response plan
FSE	fire support element
FSO	fire support officer
G2	Assistant Chief of Staff, G3 (Intelligence)
G3	Assistant Chief of Staff, G3 (Operations and Plans)
G5	Assistant Chief of Staff, G5 (Civil Affairs)
G6	Assistant Chief of Staff, G6 (Information Management)
GOTS	government off-the-shelf
GRM	graduated response matrix
HIV	human immunodeficiency virus
HMMWV	high-mobility multipurpose wheeled vehicle
HN	host nation
HQ	headquarters
I/R	internment and resettlement
INIWIC	Interservice Nonlethal Individual Weapons Instructor Course
IO	information operations
IPB	intelligence preparation of the battlefield
JMC	joint military commission
JTF	joint task force
L&O	law and order
LBE	load-bearing equipment
LBV	load-bearing vest
LEA	law enforcement agency
LOC	lines of communication
LOS	line of sight
LP	listening post
MACDIS	military assistance for civil disturbance
MANSCEN	Maneuver Support Center
MCCM-GE	modified crowd control munition-ground emplacement
MCM	Manual for Courts-Martial
MDMP	military decision-making process
METL	mission essential task list
METT-TC	mission, enemy, terrain, troops, time available, and civilian considerations
MI	military intelligence
mm	millimeter
MO	Missouri
MOOTW	military operations other than war
MOUT	military operations on urbanized terrain
MPI	military police investigator
MRE	meal, ready to eat

MS	mobility support
MSCA	military support to civil authorities
MTP	mission training plan
MWD	military working dog
NCO	noncommissioned officer
NEO	noncombatant evacuation operations
NG	National Guard
NL	nonlethal
NLCS	nonlethal capabilities set
NLW	nonlethal weapons
OC	Oleoresin Capsicum
OCONUS	outside continental United States
OES	office of emergency services
OIC	officer in charge
OP	observation post
OPFOR	opposing forces
OPORD	operations order
PA	public address
PAO	Public Affairs Office
PIO	police intelligence operations
PM	provost marshal
PLT	platoon
PMO	Provost Marshal Office
POMSO	plans, operations, and military support officer
PPCT	pressure point control tactic
PPE	personal protective equipment
PSG	platoon sergeant
PSYOP	psychological operations
PT	physical training
PVAB	portable vehicle arrest barrier
QRF	quick reaction force
RCA	riot control agent
ROE	rules of engagement
RUF	rule for the use of force
S2	Intelligence Officer (US Army)
SCO	state coordinating officer
SCRAG	senior representative of the attorney general
SITREP	situation report
SJA	Staff Judge Advocate
SOFA	status of forces agreement

SOMA	status of mission agreement
SOP	standing operating procedure
SOSO	stability operations and support operations
SQD	squad
STARC	state area command
STP	soldier training publication
STX	soldier training exercise
TA	Training Area
TF	task force
TLP	troop leading procedure
TM	team
TPT	tactical psychological operations team
TRADOC	United States Army Training and Doctrine Command
TSOP	tactical standing operating procedure
TTP	tactics, techniques, and procedures
UCMJ	Uniform Code of Military Justice
UN	United Nations
UNGA	United Nations General Assembly
UNHCR	United Nations High Commissioner for Refugees
UNSC	United Nations Security Council
US	United States
USAMPS	United States Army Military Police School
VCR	video cassette recorder
VIP	very important person
WMD	weapons of mass destruction
WTO	World Trade Organization

Bibliography

AR 380-10. *Foreign Disclosure and Contacts With Foreign Representatives.* 6 June 2003.

AR 380-13. *Acquisition and Storage of Information Concerning Nonaffiliated Persons and Organizations.* 30 September 1974.

Code of Federal Regulations.

Constitution of the United States.

DA Form 2028. *Recommended Changes to Publications and Blank Forms.*

DA Form 4137. *Evidence/Property Document.*

DD Form 2708. *Receipt For Inmate or Detained Person.*

DOD Civil Disturbance Plan 55-2. *Operation Garden Plot.* 15 February 1991.

DOD Directive 3025.12. *Military Assistance for Civil Disturbances.* 4 February 1994.

DOD Directive 3025.15. *Military Assistance to Civil Authorities.* 18 February 1997.

DOD Directive 5525.5. *DOD Cooperation With Civilian Law Enforcement Officials.* 15 January 1986.

Executive Order 12333. *United States Intelligence Activities.* 4 December 1981.

Fiscal Year 1989 Act.

FM 3-06. *Urban Operations.* 1 June 2003.

FM 3-06.11. *Combined Arms Operations in Urban Terrain.* 28 February 2002.

FM 19.10. *Military Police Law and Order Operations.* 30 September 1987.

FM 3-19.4. *Military Police Leaders' Handbook.* 4 March 2002.

FM 3-19.40. *Military Policy Internment/Resettlement Operations.* 1 August 2001.

FM 3-21.5. *Drills and Ceremonies.* 7 July 2003.

FM 3-22.40. *Multiservice Tactics, Techniques, and Procedures (MTTP) For Tactical Employment of Nonlethal Weapons (NLW).* 15 January 2003.

FM 7-0. *Training the Force.* 22 October 2002.

FM 7-1. *Battle Focused Training.* 15 September 2003.

FM 27-100. *Legal Support to Operations.* 1 March 2000.

Judiciary Act of 1789.

MCM.

National Defense Authorization Acts.

NGR 500-1/ANGI 10-8101. *Military Support to Civil Authorities.* 1 February 1996.

Posse Comitatus Act. 1878.

STP 19-95B1-SM. *Soldier's Manual for MOS 95B Military Police Skill Level 1.* 14 January 2003.

STP 19-95B24-SM-TG. *Soldier's Manual and Trainer's Guide for MOS 95B, Military Police, Skill Levels 2/3/4.* 2 December 2002.

STP 19-95C1-SM. *MOS 95C, Corrections Specialist, Skill Level 1 Soldier's Manual.* 30 October 2003.

STP 19-95C24-SM-TG. *MOS 95C, Corrections Specialist, Skill Levels 2/3/4, Soldier's Manual and Trainer's Guide.* 30 October 2003.

Uniform Code of Military Justice.

USC, Title 32. *National Guard.*

USC, Title 10, Chapter 18. *Military Support for Civilian Law Enforcement Agencies.*

USC, Title 10, Subtitle A, Part I, Chapter 15, Section 331. *Federal Aid for State Governments.*

USC, Title 10, Subtitle A, Part I, Chapter 15, Section 332. *Use of Militia and Armed Forces to Enforce Federal Authority.*

USC, Title 10, Subtitle A, Part I, Chapter 15, Section 333. *Interference With State and Federal Law.*

USC, Title 10, Subtitle A, Part I, Chapter 15, Section 334. *Proclamation to Disperse.*

USC, Title 18, Part I, Chapter 67, Section 1385. *Use of Army and Air Force as a Posse Comitatus.*

Index

Made in the USA
Monee, IL
05 November 2026

67811406R00240